New Concrete Materials: Performance Analysis and Research

New Concrete Materials: Performance Analysis and Research

Editor

Zhe Xiong

MDPI

Basel • Beijing • Wuhan • Barcelona • Belgrade • Novi Sad • Cluj • Manchester

Editor
Zhe Xiong
School of Civil and
Transportation Engineering
Guangdong University of
Technology
Guangzhou
China

Editorial Office
MDPI AG
Grosspeteranlage 5
4052 Basel, Switzerland

This is a reprint of articles from the Special Issue published online in the open access journal *Buildings* (ISSN 2075-5309) (available at: https://www.mdpi.com/journal/buildings/special_issues/3I866IYAH7).

For citation purposes, cite each article independently as indicated on the article page online and as indicated below:

Lastname, A.A.; Lastname, B.B. Article Title. *Journal Name* **Year**, *Volume Number*, Page Range.

ISBN 978-3-7258-2018-4 (Hbk)
ISBN 978-3-7258-2017-7 (PDF)
doi.org/10.3390/books978-3-7258-2017-7

Contents

Preface

With the development of civil engineering, the requirements for the performance of concrete are also increasing. In order to meet the requirements of use, researchers continue to develop new concrete materials with higher strength and better durability. Reasonable use of new concrete materials can greatly improve the quality of civil engineering projects. The recycling of solid waste plays a very important role in reducing costs and promoting environmental protection. Before applying new concrete materials to practical engineering, it is necessary to explore the basic mechanical properties, durability, and bonding performance with reinforcement of the new concrete materials. By combining macroscopic mechanical tests with microscopic analysis (e.g., scanning electron microscopy, X-ray diffraction, etc.), we can gain a deeper understanding of the mechanical properties and failure mechanisms of new concrete materials. This Special Issue aims to encourage scientists and researchers to publish their experimental and theoretical findings or solutions for new concrete materials. This Special Isue contains 12 papers, and I would like to thank the researchers who have contributed to this Special Issue as well as the reviewers, editors, and administrators who participated in its production.

Zhe Xiong
Editor

buildings

MDPI

Article

Experimental Study on Performance and Mechanism of High-Strength Artificial Blocks Based on Dredged Silt

Qingsong Zheng [1], Piaoyang Cai [2], Yourong Zeng [1], Yun Wang [1], Ming Feng [2] and Chaohua Jiang [2,*]

1 Fujian Communications Planning and Design Institute Co., Ltd., No. 43 Jiaotong Road, Taijiang District, Fuzhou 350004, China; qingsongzheng90@gmail.com (Q.Z.); shizhihao886@gmail.com (Y.Z.); wangyun9889@gmail.com (Y.W.)
2 College of Port, Coastal and Offshore Engineering, Hohai University, No. 1 Xikang Road, Nanjing 210098, China; 231303020060@hhu.edu.cn (P.C.); 211603010014@hhu.edu.cn (M.F.)
* Correspondence: chaohuajiang@hhu.edu.cn

Abstract: This paper investigates the preparation and properties of high-strength artificial blocks made from dredged silt with a clay content of 52.0%. A comparative analysis of the mechanical properties of dredged silt blocks produced using semi-dry pressing and vibration molding methods was conducted. The study examined the effects of using fly ash (FA) and ground granulated blast-furnace slag (GGBS) as substitutes for cement on the compressive strength, splitting tensile strength, and dry shrinkage of the blocks. Additionally, the microstructure of the dredged silt blocks was analyzed using scanning electron microscopy (SEM), mercury intrusion porosimetry (MIP), and thermogravimetric analysis. The results show that specimens prepared using the pressing method exhibit better mechanical performance with compressive and splitting tensile strength reaching 64.8 MPa and 5.6 MPa at 28 d, respectively, which increased by 111.07% and 143.48% compared to specimens prepared through vibration molding. The addition of FA and GGBS reduces the early strength of the block to a certain extent but without a significant adverse effect on later strength. GGBS demonstrates faster hydration and a better filling effect. The addition of GGBS or FA refines the pore structure and reduces the diameter of pores in the paste, which is beneficial for improving the dry shrinkage performance of the block. At 120 d, the dry shrinkage of blocks containing 50% FA and GGBS shows a reduction of 29.7% and 27.1%, respectively, compared to blocks made with cement. The properties of the silt blocks can be notably enhanced through mechanical force, particle gradation, and hydration action. The preparation of artificial blocks such as road bricks and ballast blocks using dredged soil as the main raw material has been applied in projects such as the Yangtze River waterway regulation in China and Skikda Port in Algeria.

Keywords: dredged silt; mechanical properties; drying shrinkage; microstructure

Citation: Zheng, Q.; Cai, P.; Zeng, Y.; Wang, Y.; Feng, M.; Jiang, C. Experimental Study on Performance and Mechanism of High-Strength Artificial Blocks Based on Dredged Silt. *Buildings* **2024**, *14*, 2407. https://doi.org/10.3390/buildings14082407

Academic Editor: Abdelhafid Khelidj

Received: 16 June 2024
Revised: 21 July 2024
Accepted: 29 July 2024
Published: 4 August 2024

1. Introduction

In recent years, the resource utilization of waste materials has emerged as a critical issue [1,2]. By crushing and processing waste concrete, mortar, bricks, tiles, and similar materials into recycled aggregates, these can be used to completely or partially replace natural aggregates in concrete production. This recycled aggregate concrete can then be used to create artificial blocks, such as protective bricks, and can potentially be employed in load-bearing structures like beams, deep beams, and corbels [3,4]. Alternatively, lightweight aggregates made from solid waste such as fly ash can be used to prepare lightweight blocks or walls. The incorporation of solid waste into building materials not only consumes a large amount of industrial waste but also reduces the environmental impact, conserves land, lowers project costs, alleviates the shortage and depletion of natural sand and gravel resources, and contributes positively to ecological protection.

In addition, the rapid development of water transportation has led to extensive construction and maintenance of ports and waterways, resulting in significant volumes of dredged soil. China has become a major dredging country worldwide, with an annual production of over 1 billion cubic meters of dredging soil. Traditionally, much of this dredged soil has been discarded in deep channel waters, leading to inefficient utilization. While some dredged soil is used for land reclamation, the methods employed are often simplistic and extensive, causing a waste of soil resources and secondary pollution to the environment [5]. Therefore, the rational, effective, and environmentally friendly utilization of dredged soil resources to play a positive role has become a new problem that urgently needs to be solved in the water transportation industry. The resource utilization of dredged soil has attracted widespread attention [6].

The use of dredged soil in construction is primarily divided into two main approaches. The first involves the use of enhanced dredged soil in the restoration and reclamation of abandoned mines and industrial sites through landfill techniques. The second approach utilizes dredged soil as an aggregate in the manufacture of concrete blocks that possess sufficient strength for use in road construction and municipal engineering projects. Through field planting experiments, Dong et al. [7] confirmed the practical effectiveness of utilizing dredged sediments for waste disposal through mine reclamation. Their results revealed that dredged sediments provided ample organic matter and nitrogen, resulting in enhanced plant growth and increased density of photosynthetic carbon fixation. Furthermore, Chu et al. [8] explored the combined use of dredged river sediment, iron ore tailings, and calcium carbide slag as fill materials to tackle ground subsidence issues. They found that an optimal mix consisting of a 7:3 mass ratio of dredged river to iron ore tailings along with a cement content of 16.7% yielded the highest flowability. Li et al. [9] conducted an extensive study on the processing of dredged sediments, which included dehydration, detoxification, and calcination to create ceramic particles. These particles were then used for backfilling and capping in dredged areas, leading to significant enhancements in bed stability and a reduction in sediment erosion by more than 70%. Additionally, the study demonstrated the effective inhibition of phosphorus release from contaminated sediments, with an inhibition efficiency of up to 80%. In another study, Yu et al. [10] proposed a method for recycling dredged sludge by incorporating ground steel slag into the slurry. The sludge was then subjected to continuous processing through foam drying and carbonization stabilization; the maximum unconfined compressive strength of the modified soil reached 2.01 MPa, which was 8.5 times higher than the strength before carbonization stabilization. Hussain et al. [11] explored the potential of utilizing harbor and river sediments in the experimental production of soil-cured bricks. The study utilized sediment from Dunkirk harbor in France and replicated the brick-making process on sediments from the Usamazinta River, and they compared the mechanical properties of bricks made from the two types of sediments, including tensile strength, toughness, and fiber distribution. The results indicate that the addition of natural fibers can transform earth blocks into ductile materials, enhancing the toughness of the bricks. Moreover, the tensile strength meets the minimum recommended tensile strength of 0.25 MPa. Meanwhile, Yang et al. [12] harnessed dredged sediments as lightweight aggregates in the production of non-sintered permeable bricks. Their experiments revealed optimal performance conditions, specifically with a water-to-cement ratio of 0.34, forming pressure of 3 MPa, and porosity level of 15%. Based on the above, it is evident that the resource utilization of dredged soil primarily revolves around methods such as stabilization, solidification, and sintering to improve its properties.

Numerous studies have been conducted on the mechanical properties and solidification potential of dredged soil when used as fine aggregate in concrete, with various researchers utilizing cementitious materials to enhance its capabilities. Ponnada et al. [13] experimented with a mixture of dredged sand and quarry dust as a partial replacement for river sand in concrete, finding that a 50% replacement dosage yielded a compressive strength of 38.7 MPa. Concurrently, Bhairappanavar et al. [14] crafted bricks from dredged materials, assessing their compressive strength, water absorption, and freeze–thaw durability. Results showed compressive strengths ranging from 10.3 to 17.2 MPa, water absorption between 13% and 18%, and minimal weight loss of 0.1% to 0.4% after 50 freeze–thaw cycles. Xu et al. [15] designed concrete mixtures with different dredged sand contents and subjected them to various numbers of freeze–thaw cycles to study the frost resistance and fracture behavior of dredged sand concrete. The study revealed that an augmented dredged sand content led to an improved pore structure, thereby enhancing the fracture behavior and frost resistance of the concrete. The study revealed that an augmented dredged sand content led to an improved pore structure, thereby enhancing the fracture behavior and frost resistance of the concrete. Vinothkumar et al. [16] found that dredged marine sand performs well as a substitute for fine aggregate in concrete, improving its physical, mechanical, and mineralogical properties. The presence of calcite and calcium enhances the strength of the concrete, while silica acts as a filler material. In another approach, Huang et al. [17] prepared high-performance concrete by substituting 25% of ordinary sand with dredged sand, which improved the concrete's density, reduced its porosity, and enhanced the pore distribution uniformity and interface structure. Meanwhile, Lee et al. [18] studied the influence of different mix proportions on the shrinkage behavior of lightweight dredged soil concrete at early and long-term ages. The shrinkage strain at 28 d was found to be 39% to 55% of the shrinkage strain at 210 d, indicating that the internal curing effect of lightweight aggregates delayed the onset of the acceleration period.

Despite the existing research, there remains a noticeable gap in the study of high-strength blocks using high-clay-content dredged soil as the primary material. The assurance of quality and reliability necessitates extensive experimentation coupled with effective control measures. While chemical, physical, and mechanical properties of dredged sediment blocks have been studied, the mechanisms behind the mechanical properties and microstructural characteristics, such as phase composition and pore structure, have not been thoroughly revealed. Research on different molding methods (such as compaction molding and vibration molding) and evaluations related to drying shrinkage are still relatively limited. Further research is needed to deepen our understanding of concrete preparation using dredged soil, and it will provide important scientific foundations for addressing related issues.

Dredged silt is a major type of dredged soil, characterized by small pores, poor structure, and weak permeability. Currently, there is no specialized research on the preparation of blocks using silt as a substitute for traditional sand and gravel materials. This study aims to address this gap by utilizing industrial waste materials, such as silicon aluminum compounds (e.g., fly ash) and mineral powder, to partially replace cement. Dredged silt is used as the primary raw material instead of ordinary sand and gravel to prepare artificial blocks. The mechanical properties of the silt blocks under ordinary vibration molding and semi-dry pressing molding are compared and analyzed. The study evaluates the compressive strength, splitting tensile strength, and dry shrinkage performance of the dredged silt blocks with varying cement replacement levels of fly ash and mineral powder (0%, 20%, 30%, 40%, and 50%). Additionally, micro-mechanism research such as SEM, MIP and TG-DSC is carried out, which can provide an effective way for the high-value-added utilization of dredging silt blocks and lay a foundation for the subsequent promotion and application of dredging silt blocks.

2. Experimental Programs

2.1. Materials

2.1.1. Dredged Silt

According to JTS 181-5-2012 [19], silt mainly refers to soil with a particle size greater than 0.075 mm, with a particle content of less than 50% by total mass and a plasticity index less than 10. The dredged silt (Figure 1) is characterized by small pores, low permeability, and the presence of capillary water. The dredged silt used in this article is from the northern part of the Beibu Gulf in China, with a particle content greater than 0.075 mm of 48.0%, which corresponds to a clay content of 52.0% (Figure 1). Its plasticity index is 9.5. It can be observed that the dredged silt in Beibu Gulf contains a wide variety of mineral components (Figure 2), including, in order of their content from high to low: Kato stone (43.41%), calcite

(27.92%), hematite (12.85%), goethite (8.04%), hard water alumina (3.88%), calcite (1.82%), gibbste (1.12%), and quartz (0.95%). The chemical composition analysis of the dredged silt was carried out according to the "Methods of Chemical Analysis of Clay" GB/T 16399-2021 [20], and the specimens were all passed through a square-hole sieve with a pore size of 0.088 mm and dried for 2 h (to a constant amount) in a 105 °C ± 5°C drying oven for chemical composition testing. The test results indicate that the dredged silt in Beibu Gulf mainly contains chemical components such as SiO_2, Al_2O_3, Fe_2O_3, CaO, MgO, K_2O, Na_2O, and others (see Table 1). The mass fraction of oxide components exceeds 99%, indicating a relatively low presence of other substances and organic matter in the silt sample. The silt sample contains alkaline substances K_2O and Na_2O, suggesting that the soil sample is slightly alkaline in nature.

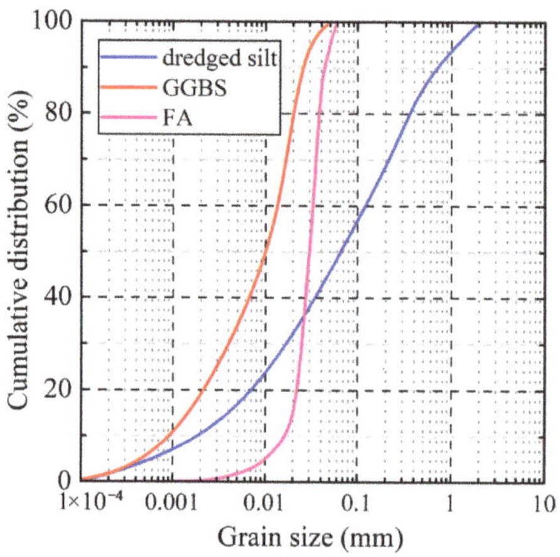

Figure 1. Size distributions of the studied dredged silt and SCMs (FA and GGBS).

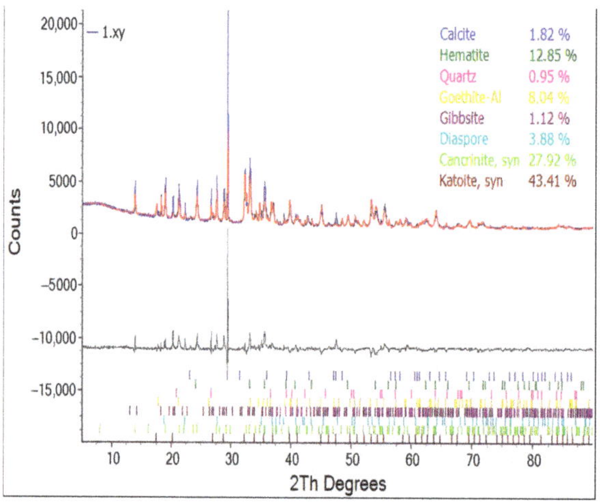

Figure 2. XRD diffraction image of dredged silt in Beibu Gulf.

Table 1. The chemical compositions and physical properties of OPC, GGBS and FA.

	Dredged Silt	OPC	GGBS	FA
LOI	1.20	2.92	1.25	0.75
SO_3	0.08	2.55	0.50	0.27
SiO_2	60.14	21.2	51.89	33.71
Fe_2O_3	5.78	3.30	6.59	0.50
Al_2O_3	16.23	4.65	26.98	15.23
CaO	8.12	62.59	7.91	38.16
MgO	3.45	1.11	1.44	8.76
K_2O	3.12	0.41	1.17	0.49
Na_2O	1.15	0.06	0.71	0.21
TiO_2	0.73	0.21	1.04	0.73
Density (g/cm^3)	2.65	3.1	2.8	2.3
Specific surface (m^2/kg)	200	350	425	375
Water requirement (%)	-	-	92	95
Activity index	-	-	93	80

2.1.2. Ordinary Portland Cement (OPC)

Conch brand OPC42.5 Grade was utilized for this study. Table 1 provides the chemical compositions and physical properties of the OPC, which are in accordance with JTG 3420-2020 [21].

2.1.3. Supplementary Cementitious Materials (SCMs)

In this study, FA and GGBS were, respectively, used as supplementary cementitious materials in a certain proportion to replace ordinary cement. FA used in the study conformed to the grade II level specified with a specific surface area of 375 m^2/kg. The GGBS, sourced from Nanjing Meibao Company, was S95 grade with a specific surface area of 425 m^2/kg. Notably, all GGBS particles are smaller than 0.05 mm, which is significantly finer than the particle size of dredged silt. Table 1 presents the characteristics of both GGBS and FA. All particle sizes of GGBS are less than 0.05 mm, which is much smaller than the particle size of dredging silt. Table 1 presents the characteristics of GGBS and FA.

2.2. *Mix Design and Preparation of the Specimens*

The dredged silt specimens were formed using two different methods: semi-dry pressing and vibration molding. In the pressing process, pressure was applied by employing a universal testing machine, gradually increasing it at a rate of 600 N/s until reaching 10 MPa. The pressure was then maintained for a duration of 2 min, followed by an increase to 15 MPa at a rate of 400 N/s, and another 2-min maintenance of pressure. Laboratory pressing molds and methods are shown in Figure 3. In the case of vibration molding, the identical mixture proportions were applied, supplemented by the inclusion of a 1% superplasticizer, proportionate to the mass of the cementitious material. To meet the design requirement of achieving a compressive strength exceeding 30 MPa after 28 d, with a fixed mass of dredged silt and OPC, the ratio of cementitious material to dredged silt was established at 1:2.5. Various replacement percentages of FA or GGBS were employed instead of OPC, namely 0%, 20%, 30%, 40%, and 50%. The compressive and splitting tensile strength of the specimens were assessed at 3, 28, 90 and 180 d. Each set of strength results represents the average value obtained from three samples.

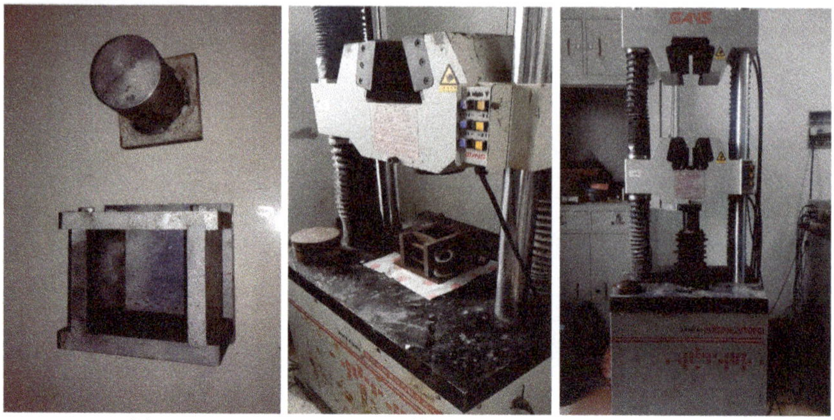

(a) (b) (c)

Figure 3. Laboratory pressing tools and method. (**a**) Pressing grinding tools. (**b**) The laboratory presses on a universal testing machine. (**c**) Test piece compression process.

2.3. *Testing Methods*

2.3.1. Compressive Strength

According to SL/T 352-2020 [22], the specimen was a 100 mm cube. Three cubes were prepared for each mix proportion, and the average compressive strength was obtained after curing at 3, 28 and 90 d, respectively. The compressive strength was then tested on a SHT4305 universal testing machine, and the load increased at a rate of 0.3–0.5 MPa/s until the samples failed.

2.3.2. Splitting Tensile Strength

The splitting tensile strength of the 100 mm cube was tested on a SHT4305 microcomputer-controlled (Mechanical Testing & Simulation; shanghai; China) electro-hydraulic servo universal testing machine in accordance with SL/T 352-2020 at 3, 28 and 90 d. A square steel pad with a length of 100 mm and 5 mm × 5 mm sections was used between the upper and lower platens and the test specimens. Each test result is the average value of the three samples.

2.3.3. Dry Shrinkage

The shrinkage test of dredged silt blocks was carried out according to the SL/T 352-2020. Each test result is the average value of three 40 mm × 40 mm × 160 mm prismatic specimens at 1, 3, 7, 14, 21, 28, 42, 56, 90 and 120 d, respectively, in order to comprehensively detect the drying shrinkage of dredged silt blocks in the early, middle, and later stages of hydration. The initial lengths of the specimens were measured after curing in the laboratory for 24 h. After demolding, the samples were transported to a chamber at 20–25 °C and a relative humidity of 50–55% until the testing age.

2.3.4. Mercury Intrusion Porosimetry (MIP)

The performance of the block is closely related to its pore structure, including factors such as porosity, pore size distribution, and connectivity. Mercury intrusion porosimetry (MIP) is a widely used technique for characterizing the pore size distribution of cement-based materials. By introducing mercury into the material and measuring the pressure changes within the pores, the size, shape, and distribution of the pores can be inferred. Pore structure analysis was carried out using a Poremaster GT-60 MIP (Quantachrome Instruments; Shanghai; China). The sample size was spherical particles with a diameter of 5 mm and a mercury pressure range of 0–200 MPa. The samples underwent a controlled drying process at temperatures not exceeding 60 °C to remove moisture or other liquids from the pores, ensuring the accuracy of the test results.

2.3.5. Scanning Electron Microscopy (SEM) Analysis

SEM analysis was performed using the cross-section and polished section, respectively, using JMS-5600LV (Nanjing Suozheng Automation Instrument Co., Ltd; Nanjing; China). Polished surfaces are typically more suitable for observing interfaces such as aggregate–paste interfaces, while direct cross-sections are better for observing product morphology. The cross-section samples were prepared by soaking the test blocks, which were cured to a certain age in alcohol for 3 days, then placing them in a vacuum drying oven to dry. After drying, the test blocks were knocked open and a suitable sized fragment was taken from the center as the cross-sectional sample. The polished section samples were prepared by cutting the test block that had already reached the curing age into parallel and flat sections on the upper and lower surfaces. Then, the samples surfaces were polished with sandpaper, gently brushed with a fine brush under tap water, and finally dried.

2.3.6. Thermogravimetric Analysis- Differential Scanning Calorimetry (TG-DSC)

Thermal analysis of the sample was carried out using an STA PT1600TG-DSC/DTA synchronous thermal analyzer produced in Germany (Linseis; Ingolstadt, Germany), with a temperature range of −150~1750 °C, a sample range of 0~25 mg, and a heating rate of 0.1~100 °C/min.

3. Results and Discussion

3.1. Influence of Different Molding Methods on Mechanical Properties of Dredged Silt Blocks

Blocks were produced using semi-dry pressing and vibration molding techniques, with a binder composition of 70% cement and 30% ground granulated blast-furnace slag (GGBS). The experimental results for the relevant mechanical properties are depicted in Figure 4. Compared to the specimens prepared using vibration molding, the specimens prepared using the pressing method exhibits significant improvements in compressive strength and splitting tensile strength. The compressive strength and splitting tensile strength at 28 d reached 64.8 MPa and 5.6 MPa, respectively. These values denote a substantial improvement, registering remarkable increases of 110.5% and 142.8% compared to specimens prepared using vibration molding with identical mix proportions.

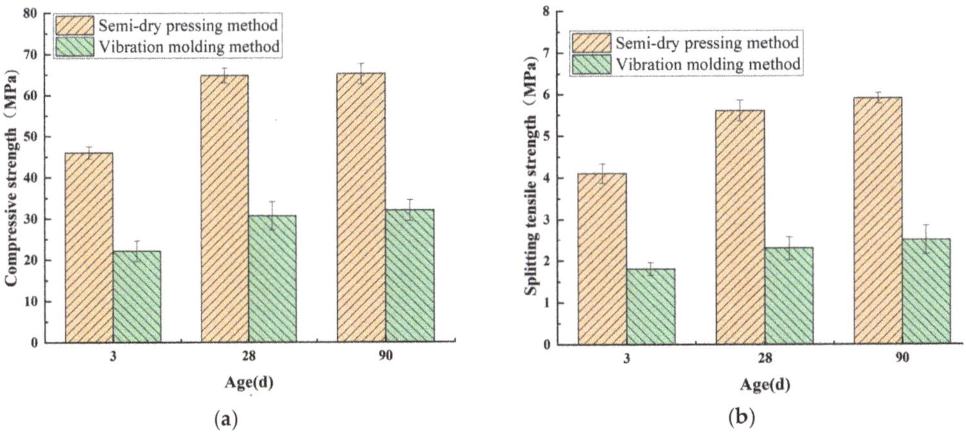

Figure 4. The influence of compression method and vibration molding method on compressive strength (**a**) and splitting tensile strength (**b**).

The preparation of ordinary concrete typically involves vibration molding, a process in which the material is liquefied under high-frequency vibration, allowing it to densely fill the mold under its own weight and external forces. This method effectively expels air between particles, resulting in a material with uniform density and reliable strength. Compression molding is the process of using mechanical force to squeeze out the air wrapped in the mixture, reducing the number and volume of pores, expelling excess water, and eliminating some micro cracks caused by chemical shrinkage of the slurry, thereby improving the compactness and strength of the block. Similarly, semi-dry compression molding utilizes mechanical force to extrude the air enclosed in the mixture, reduce the number and volume of pores, and expel excess moisture, thereby enhancing the density and strength of the block. Due to the high compressibility and difficulty in compacting the dredged mud, compression molding can significantly improve the strength of the block, especially its early strength. Additionally, it allows for immediate demolding, which accelerates the turnover rate of the grinding tool. During the process of pressing molding, applying a significant amount of pressure can result in tighter contact and denser arrangement of the dredged silt particles, increasing both the frictional and cohesive forces between them. Moreover, OPC and GGBS, serving as binders, facilitate the bonding of these particles, further enhancing the block's integrity. For the dredged silt used in this study, the presence of sand particles within the silt provides a skeletal support during block preparation. Compared to the dredging soil mainly composed of clay particles, it is easier to play a role in particle grading such as particle matching and embedding, thereby improving the strength and durability of the blocks. Consequently, the strength of dredged silt blocks can be significantly improved through the combined effects of mechanical force, particle size distribution, and hydration.

The block pressing production process mainly includes weighing, dry mixing, wet mixing, compression molding, demolding, curing and other processes. The prepared block can be demolded immediately, simplifying the process and accelerating mold turnover. The specific pressing production process of dredging soil blocks is shown in Figure 5.

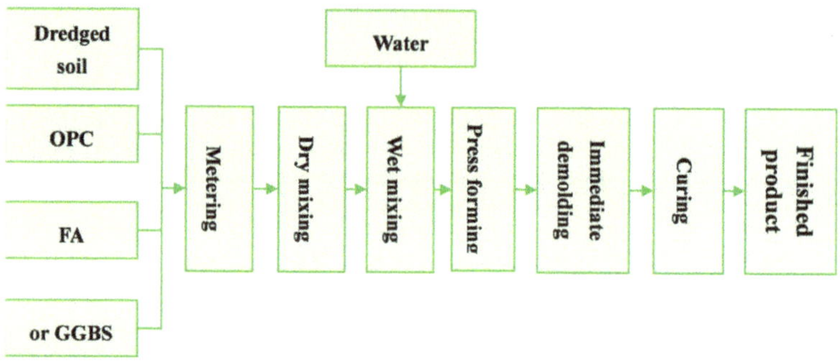

Figure 5. Dredged soil block manufacturing process diagram.

3.2. *Influence of Different Proportions of Fly Ash and Ground Granulated Blast Furnace Slag, Respectively, on Mechanical Properties*

The compressive strength and splitting tensile strength of the dredged silt blocks were systematically examined using the pressing method, as depicted in Figures 6 and 7, incorporating partial replacements of OPC with FA and GGBS. Both FA and GBSS addition can lead to a decrease in the early strength of the block. GBSS reduces the strength of the block at 3 d, and after 28 d, the strength can approach that of OPC. Compared with pure cement test blocks, the addition of 50% GBSS reduces the compressive strength by 37.97% and 8.57% at 3 and 28 d, respectively. The graphical representation of the observed trends indicates a consistent increase in compressive strength across various mixtures as the curing period extends, without indication of a strength reverse shrinkage phenomenon during this progressive period of curing.

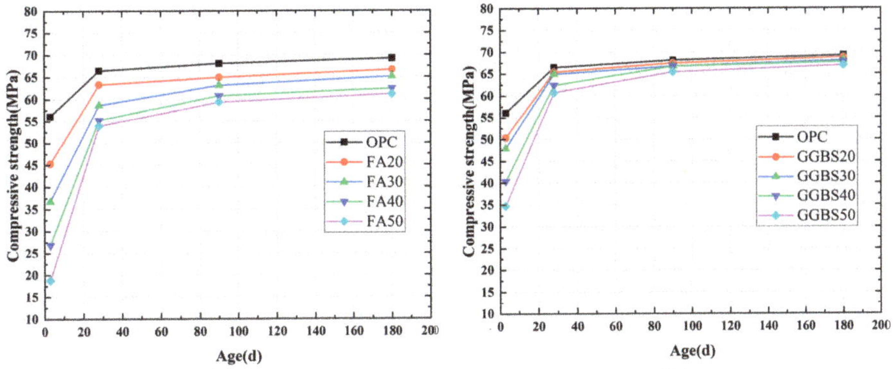

(a) (b)

Figure 6. Influence of different FA (**a**) and GGBS (**b**) contents on compressive strength of dredged silt blocks.

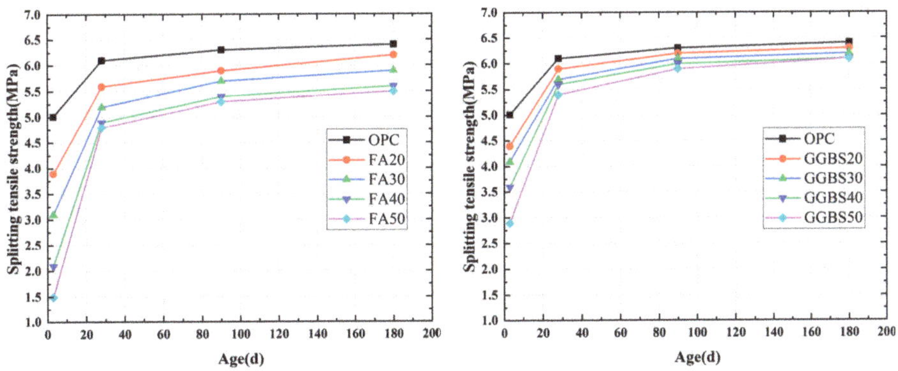

(a) (b)

Figure 7. Influence of different FA (**a**) and GGBS (**b**) contents on splitting tensile strength of dredged silt blocks.

The compressive strengths of the blocks without SCMs at 3, 28, and 90 d were 51.5 MPa, 62.2 MPa, and 63.7 MPa, respectively. These data indicate that the majority of strength development for blocks with OPC occurs before the 28-day mark, with only modest gains in strength thereafter. In contrast, FA exhibits lower early reactivity, which leads to a significant reduction in the initial compressive strength. As the FA content increases, there is a noticeable decline in strength, particularly at early stages. For instance, blocks with 50% FA content displayed the lowest early compressive strength at 3 d, recording only 19.1 MPa. This reduction is attributed to FA's slower hydration reaction, which produces fewer hydration products in the initial stages, resulting in weaker bonding. This is because FA produces fewer hydration products in the early stages, resulting in less compact bonding. While the strength development curve of OPC tends to plateau, the inclusion of FA still yields a significant strength increase after the 28-day mark. FA necessitates an extended curing period to fully manifest its reactivity, accounting for the observed lower early strength. However, as hydration reactions proceed, FA generates more hydration products, filling the slurry structure and improving block strength. The early weakness of FA limits its application in OPC, but its strength continues to increase with a significant growth rate, making it suitable for situations with high demands for later-stage strength. Similarly to FA, GGBS reduces early strength (at 3 d), albeit to a lesser extent compared to an equivalent

FA content. At 28 d, the strength of the silt blocks containing GGBS is slightly lower than that of the silt blocks comprising only OPC. There is rapid strength development from 3 d to 28 d, indicating the quick hydration reaction of the GGBS. GGBS exhibits high reactivity, and its hydration degree is high at 28 d, with strength close to OPC when used at 50% content. While GGBS decreases strength at 3 d, it gradually approaches the strength of OPC after 28 d, whereas FA reduces strength at all ages. Compared to FA, GGBS contains not only SiO_2 and Al_2O_3 but also a higher amount of CaO, belonging to the calcium–silicon–aluminum system.

By incorporating FA and GGBS as fine admixtures in the system, a portion of OPC is replaced, leading to the addition of mineral SCMs that fill the small voids between OPC and its hydration products. This results in enhanced compactness of the paste. Furthermore, the introduction of mineral admixtures alters the particle size distribution of the cementitious material system, effectively eliminating existing large-pore voids and promoting a more uniform distribution of pores [23]. This not only increases the quantity of formed gel but also ensures a more even distribution of hydration products throughout the entire interface transition zone.

3.3. *Influence of Different Proportions of Fly Ash and Ground Granulated Blast-Furnace Slag, Respectively, on Dry Shrinkage Performance*

The small particle size of dredged silt inherently exacerbates issues related to shrinkage, leading to enhanced water adsorption on the surface of these fine particles. This leads to poor compaction and ineffective reduction in capillary pore water evaporation, resulting in significant drying shrinkage [18]. Drying shrinkage, a crucial parameter influencing concrete durability, plays a pivotal role in determining the longevity of concrete structures. A smaller drying shrinkage rate corresponds to a reduced degree of volume reduction and diminished formation of microcracks within the concrete matrix. This limited development of microcracks serves to decrease the permeability and permeation parameters of the concrete, ultimately yielding durable concrete materials with enhanced longevity [24]. Consequently, by reducing the drying shrinkage rate, not only can the overall quality of concrete be enhanced, but its resistance to external factors can also be strengthened, ultimately achieving long-term reliable performance.

Figure 8 shows the drying shrinkage values of dredged silt blocks with different proportions of SCMs. Research results showed that the drying shrinkage curves of dredged silt blocks containing FA and GGBS were essentially the same within the first 10 d, with minimal differences. This indicates that the addition of GGBS and FA can reduce the drying shrinkage of the silt blocks in the later stages. As time progresses, the drying shrinkage of the dredged silt blocks with FA begins to decelerate after 20 d, while the blocks without FA continue to shrink until 40 d before slowing down. The drying shrinkage of the blocks containing FA is lower than that of the blocks without FA. Specifically, at 120 d, the blocks incorporating 20%, 30%, 40%, and 50% FA demonstrated reductions in drying shrinkage values of 12.1%, 16.2%, 27.1%, and 29.7%, respectively. This indicates that a higher FA content exhibits a more pronounced inhibiting effect on drying shrinkage. The primary reason for this behavior is that FA has a slower rate and degree of participation in hydration reactions, thereby slowing down the overall hydration process within the system. This reduction in the rate of water consumption by hydration reactions, combined with the gradual hydration process of FA, acts as a skeletal structure, limiting the shrinkage of the silt blocks. In the early stages, the high elastic modulus of FA particles plays a role in restraining shrinkage. With prolonged time and increased FA content, the effect of restricting drying shrinkage becomes more significant.

The blocks containing 20%, 30%, 40%, and 50% GGBS showed reductions in drying shrinkage values of 8.1%, 14.8%, 21.6%, and 27.1%, respectively, at 120 d. This trend demonstrates that a higher GGBS content enhances the effect on inhibiting drying shrinkage. The main reason is that the addition of GGBS reduces the evaporation of water from the silt blocks under drying conditions. The active components in GGBS react with the hydration products of OPC, forming gel-like substances such as hydrated calcium silicate gel. These gel-like substances have cementitious properties and can fill the microscopic pores within the concrete, reducing the connectivity of the pores. As a result, the loss of moisture and the occurrence of drying shrinkage are reduced.

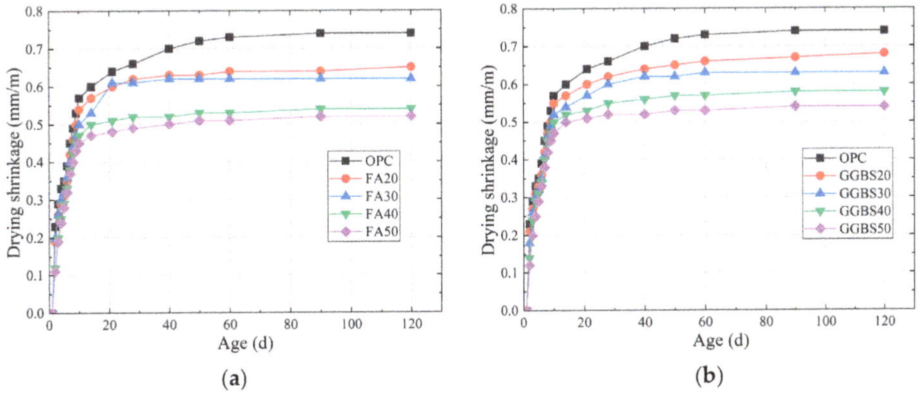

Figure 8. Effect of FA (**a**) and GGBS (**b**) content on dry shrinkage of dredged silt blocks.

Furthermore, the addition of GGBS refines the pore structure within the concrete mix, making it more challenging for water to migrate under drying conditions. This is supported by Saluja's research [25]; introducing extremely fine GGBS as a substitute for OPC in ordinary concrete containing crushed gravel aggregates has been found to generate a denser concrete matrix. This type of concrete exhibits finer capillary pores, resulting in a reduced availability of free water and consequently creating higher capillary pore water pressure, leading to increased shrinkage strain. Similarly, Shariq et al. [26] also found that as the substitution level of OPC with GGBS increased, the shrinkage strain also increased. The reason for the difference between this article and their research findings is attributed to the utilization of fine dredged clay particles, significant filling of GGBS in the pores, and a notable enhancement in the particle gradation effect of the system.

3.4. Thermogravimetric Analysis

The chemical composition, pozzolanic reactivity, and particle dimensions of FA and GBSS are dissimilar, culminating in distinct hydration byproducts. The secondary pozzolanic reactions of FA and GGBS deplete the calcium hydroxide (CH) present within the paste, concurrently modifying the quantity and constitution of the resultant hydration byproducts. Figure 9 presents the differential scanning calorimetry (DSC) and thermal gravity analysis (TG) curves of the hydration products of binary cementitious materials containing 50% FA and GGBS at 28 d. The DSC curve shows the main endothermic peaks of different samples occurring at approximately 90 °C, 430 °C, and 670 °C, which correspond to the decomposition processes of CH and calcium carbonate ($CaCO_3$). The temperature ranges of these peaks are approximately 40 °C to 220 °C, 420 °C to 470 °C, and 620 °C to 700 °C, respectively. These ranges are indicative of the evaporation of water in varying states within the cementitious matrix, the decomposition of CH, and the disintegration of $CaCO_3$, respectively.

In the paste, water exists in two states: physically bound water and chemically bound water. These two states cannot be easily distinguished during the evaporation process. To facilitate research, a definition is provided: water that is removed above 105 °C is referred to as non-evaporable water, while water removed below 105 °C is referred to as evaporable water.

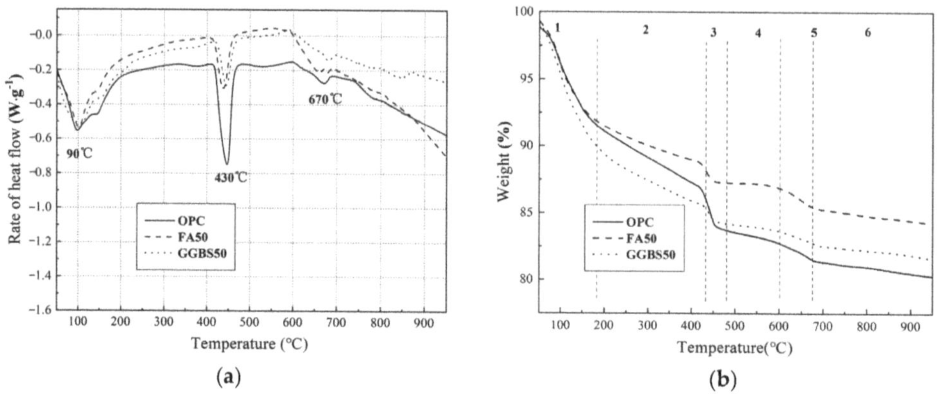

Figure 9. (**a**) DSC and (**b**) TG curves of the slurry blended with FA and GGBS at 28 d.

The TG curve can be divided into six stages based on different weight loss mechanisms. Stage 1 (40 °C to 105 °C) corresponds to the loss of evaporable water, which refers to the moisture absorbed from the air prior to testing and subsequently dried out from the specimen. This weight loss is primarily influenced by the specific surface area of the binary cementitious paste samples.

Stage 2 (105 °C to 420 °C) corresponds to the loss of non-evaporable water, primarily attributed to the dehydration of the hydrates in the paste. The amount of non-evaporable water lost in the paste of OPC and 50% GGBS is similar and significantly higher than that in the paste containing 50% FA. This indicates that the OPC and 50% GGBS paste have a significantly higher amount of hydrated gel products compared to the paste with FA. Furthermore, FA exhibits the slowest hydration rate, with only a small proportion of FA particles undergoing hydration and minimal gel product formation at the 28-day mark. This analysis provides insights into the principal rationale for the pronounced reduction in block strength during the early phases following the substitution of cement with FA.

The thermogravimetric decomposition of pure calcium hydroxide typically occurs within the temperature range of 500 °C to 800 °C. However, when calcium hydroxide is mixed with other hydrated products and substances, its decomposition temperature may be lower than 500 °C. Consequently, Stage 3 (420 °C to 470 °C) primarily involves the weight loss due to the decomposition of CH in the paste. Compared to OPC, the pozzolanic activity of FA significantly consumes the CH content present in the paste. As the temperature continues to rise beyond 500 °C, the residual CH within the paste undergoes further decomposition, resulting in continuous mass loss throughout this stage.

The weight loss caused by water loss in Stage 4 (470 °C to 620 °C) and Stage 6 (700 °C to 950 °C) is much lower than that in Stage 2. The differences among the three types of pastes are not significant, indicating that the dehydration mechanisms of these pastes are similar at high temperatures.

Stage 5 (620 °C to 700 °C) represents the weight loss due to the decomposition of calcium carbonate ($CaCO_3$) in the paste. Compared to OPC, the weight loss caused by the thermal decomposition of $CaCO_3$ is slightly reduced in the pastes with solely FA or solely GGBS. This reduction is likely related to the decrease in CH content and the improved resistance to carbonation in the paste matrix.

The findings delineated in Sections 3.2 and 3.3 of this manuscript demonstrate that the utilization of FA and GGBS as substitutes for cement reduces the initial strength of silty soil blocks, without negatively impacting their ultimate strength. In comparison to FA, the incorporation of GBSS significantly augments the strength of the blocks. Moreover, the concurrent addition of both GBSS and FA ameliorates the desiccation shrinkage of the blocks. Analysis via TG-DSC above unveils variations in the hydration byproducts within a binary cementitious system inclusive of FA and GBSS, facilitating a comprehensive understanding of the principal causations and mechanisms underpinning these outcomes.

3.5. Pore Structure Analysis

Based on the results shown in Figure 10, the binary cementitious material system with 50% FA and GGBS undergoes changes in pore size distribution after 28 d of hydration. It can be observed that the addition of different GGBS and FA leads to a reduction in pore size and an increase in the number of small pores, resulting in a finer pore structure of the paste. Compared to FA, GGBS demonstrates a better effect in refining the pore size and shows superior pore refinement.

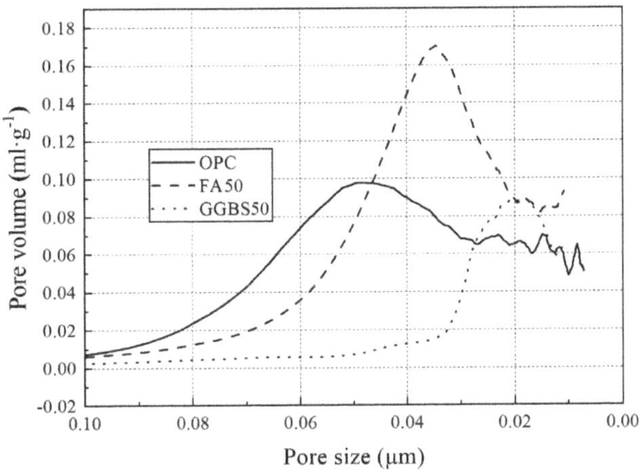

Figure 10. Pore size distribution of OPC blended with SCMs.

The specific surface area of OPC given in Table 1 is 350 m^2/kg, while the specific surface area of FA is 375 m^2/kg; the difference between the two is not significant. Furthermore, as depicted in Figure 11, the majority of FA particles fall within the size range of 0.01 mm to 0.1 mm, indicating that the particle size distribution of FA is of the same magnitude as OPC cement and incapable of filling the smaller interstitial voids. Additionally, due to the low reactivity of FA, it cannot produce a substantial amount of cementitious material to fill the gaps. Moreover, FA contains a considerable proportion of inert (non-reactive) crystalline phases, and the slow hydration reaction leads to the sedimentation of FA particles and inert crystalline materials within the existing pore structure, potentially increasing the heterogeneity of the pore structure [27,28]. In contrast, GGBS has a faster hydration rate and undergoes a more complete secondary pozzolanic reaction by 28 d, producing a significant amount of gel. GGBS refines the critical pore throats, which are the smallest continuous channels within the pore structure. The GGBS also alters the chemical composition and growth pattern of C-S-H, which in turn affects the pore structure. These gels can fill more large pores. The size and connectivity of pores play crucial roles in determining the drying shrinkage of the specimens. Larger pore diameters facilitate quicker moisture loss from the interior of the specimens, which in turn increases drying shrinkage. Conversely, a denser pore structure with limited connectivity restricts the movement of water molecules, potentially beneficial in reducing moisture loss, but can also increase capillary stress within the material. An overly dense pore structure increases capillary stress within the system and may lead to the development of internal cracks, thereby reducing the strength of the material.

Figure 11. Dredging silt.

3.6. Scanning Electron Microscopy Analysis

Figure 12 presents the SEM images of the hydration of a binary cementitious material system at 28 d. In Figure 12a, the microstructure of the OPC paste at 28 d reveals the presence of abundant CH crystals. However, the C-S-H gel content is insufficient to fill all the pores, resulting in numerous unhydrated micrometer-scale voids. After adequate hydration reaction, the paste exhibits increased compactness, with hydration products nearly filling all the micrometer-scale voids and no apparent defects. Figure 12b shows the hydration of a paste containing 50% FA, which appears porous and loosely packed. The image reveals the presence of numerous unhydrated FA particles on the surface, along with well-crystallized CH. The interfaces between C-S-H gel, unhydrated FA, and CH particles are clearly defined. At 28 d of hydration, the crystalline CH phase does not show a significant reduction, and the pozzolanic effect of FA is still limited. There is poor bonding between FA particles and the paste, with a large number of exposed, smooth-surfaced FA spheres and semispherical pits resulting from FA detachment. The hydration products exhibit complete crystallinity, and there are large hexagonal plate-shaped CH particles that are stacked and distributed in layers. It can be observed that the hydration of FA at 28 d is minimal, leading to a decrease in early strength of FA-containing pastes and limiting the effectiveness of FA. However, the hydration degree increases significantly in the later stages, resulting in significant strength development. The morphology effect, pozzolanic activity, and micro-aggregate effect of FA all have an impact on the composition, structure, and properties of cement pastes, particularly favoring the development of later-stage strength. Enhancing the early strength of high-volume FA cement pastes is crucial for promoting the widespread application of FA.

Figure 12c,d are the polished surface and cross-section SEM images, respectively, of the hydration of a binary GGBS–cement system at 28 d. From the polished section SEM image in Figure 12c, it can be seen that there are still some large GGBS particles (above 10 μm) that have not participated in the reaction. Although the content of finer GGBS particles (below 10 μm) in the paste is relatively small, these particles exhibit higher reactivity. Unhydrated GGBS particles show good bonding with the paste interface. Hydration products such as calcium silicate hydrate (C-S-H) gels fill the large pores, reducing defects in the paste and improving the bonding between different phase interfaces. From the cross-section SEM image in Figure 12d, it is evident that the GGBS–cement paste exhibits good compactness, and intact crystalline calcium hydroxide (CH) is difficult to identify within the paste. This indicates that a secondary pozzolanic reaction occurred, with GGBS and CH reacting to form C-S-H. The physical filling and pozzolanic effect of GGBS optimize and compact the microstructure of the paste.

Figure 12. SEM patterns of cement and binary cementitious system blended with SCMs. (**a**) OPC 28 d; (**b**) FA50 28 d; (**c**) GGBS50 28 d polished surface; (**d**) GGBS50 28 d fracture surface.

4. Conclusions

(1) The molding method plays a pivotal role in the mechanical properties of dredged silt blocks. Compared to conventional vibratory molding, the compaction method significantly improves the compressive and splitting tensile strengths of the specimens. For instance, blocks made from silt with 70% cement and 30% GGBS as cementitious materials achieved compressive and splitting tensile strengths of 64.8 MPa and 5.6 MPa, respectively, after 28 days of curing. These represent improvements of 111.07% and 143.48%, respectively, compared to blocks prepared with the same mix ratio using vibratory molding. The presence of sand particles within the dredged silt provides a skeletal support and better particle matching, which contribute to improving properties of block. The properties of the silt blocks can be significantly enhanced through mechanical force, particle gradation, and hydration action.

(2) The addition of FA and GGBS reduces the early strength of the block to a certain extent but without a significant adverse effect on later strength. At 28 d, the compressive strength of blocks containing 50% FA and GGBS reached 54.0 MPa and 60.8 MPa, respectively. Both GGBS and FA exert notable impacts on reducing drying shrinkage. The dry shrinkage values of dredged soil blocks with 20%, 30%, 40%, and 50% FA and GBSS added at 120 d. were reduced by 12.1%, 16.2%, 27.1%, 29.7%, and 8.1%, 14.8%, 21.6%, and 27.1%, respectively, compared to the dredged blocks without FA. The optimal mix ratio that meets the requirements of blocks such as protective tiles and pavement tiles is that the mass ratio of binder: dredged silt = 1:2.5, OPC: GGBS or FA = 5:5, water-to-binder and silt ratio of 0.15 and semi-dry compression molding with a forming pressure of 15 MPa.

(3) TG-DSC analysis reveals that the quantity of gel formed in OPC and 50% GGBS paste is significantly higher than that in the FA paste, with a notable loss of non-evaporable

water. GGBS or FA reacting with CH to form C-S-H significantly consumes the CH present in the paste. MIP analysis indicated that GGBS exhibits superior pore structure refinement compared to FA, which restricts water migration, contributing to reduced drying shrinkage. FA slows down hydration reactions and provides a framework to limit block shrinkage. SEM analysis shows that GGBS–cement paste exhibits good compactness, while FA paste appears porous, loosely packed, and has some unhydrated particles on the surface. Physical filling and the volcanic ash effect of FA and GGBS can optimize and compact the microstructure of paste. Hydration products such as C-S-H gels fill the large pores to reduce defects in the paste and improve the bonding between different phase interfaces.

(4) Utilizing discarded dredged soil as a substitute for conventional sand and gravel materials, and employing large quantities of industrial by-products such as GGBS and FA to replace conventional high-energy-consuming cement, high-strength artificial blocks can be fabricated through compaction molding to replace ordinary concrete. These blocks can be used for the production of revetment bricks, pavement bricks, ballast blocks, etc., and applied in proximity to channel regulation projects, roads, or municipal engineering. The preparation of artificial blocks such as road bricks and ballast blocks using dredged soil as the main raw material has been applied in projects such as the Yangtze River waterway regulation in China and Skikda Port in Algeria. Future tests can be conducted to evaluate the water erosion resistance, chloride ion erosion resistance, and the durability performance against freezing and thawing, carbonation, and permeation of the dredged soil blocks. These tests will lay the groundwork for their comprehensive promotion and application in water-related areas, as well as in cold and coastal regions.

Author Contributions: Conceptualization, Q.Z.; funding acquisition, Q.Z.; data curation, Q.Z. and M.F.; visualization, Q.Z; methodology, P.C.; resources, P.C.; formal analysis, M.F.; validation, Y.Z.; investigation, Y.Z.; writing—original draft, Y.W.; writing—review and editing C.J.; project administration, C.J. All authors have read and agreed to the published version of the manuscript.

Funding: This research work was funded by Fujian Provincial Science and Technology Plan Project (2023Y0041) and the Fujian Provincial Transportation Technology Project (2022Y040).

Data Availability Statement: Available from the authors.

Conflicts of Interest: Authors Qingsong Zheng, Yourong Zeng and Yun Wang were employed by the company Fujian Communications Planning and Design Institute Co., Ltd. The remaining authors declare that the research was conducted in the absence of any commercial or financial relationships that could be construed as a potential conflict of interest. The authors declare no conflict of interest.

Abbreviations

Definition of all abbreviations and notations.

abbreviations or notations	full words
FA	fly ash
GGBS	ground granulate blast-furnace slag
SCMs	supplementary cementitious materials
OPC	ordinary Portland cement
MIP	mercury intrusion porosimetry
SEM	scanning electron microscopy
TG-DSC	thermogravimetric analysis–differential scanning calorimetry

References

1. Bian, Z.; Bai, Y.; Douglas, W.S.; Maher, A.; Liu, X. Multi-year planning for optimal navigation channel dredging and dredged material management. *Transp. Res. Part E Logist. Transp. Rev.* **2022**, *159*, 102618. [CrossRef]
2. El-Seidy, E.; Chougan, M.; Sambucci, M.; Al-Kheetan, M.J.; Biblioteca, I.; Valente, M.; Ghaffar, S.H. Lightweight alkali-activated materials and ordinary Portland cement composites using recycled polyvinyl chloride and waste glass aggregates to fully replace natural sand. *Constr. Build. Mater.* **2023**, *368*, 130399. [CrossRef]

3. Shakir, Q.M.; Alghazali, A.F. Hybrid curved precast deep beams composed partially from concrete made with recycled concrete aggregate. In *E3S Web of Conferences*; EDP Sciences: Najaf, Iraq, 2023; Volume 427, p. 02025.
4. Shakir, Q.M. New model of eco-friendly hybrid deep beams with wastes of crushed concrete. *J. Teknol.* **2023**, *85*, 145–154.
5. Mostafa, A.; Farzad, H. Prediction of mechanical and durability characteristics of concrete including slag and recycled aggregate concrete with artificial neural networks (ANNs). *Constr. Build. Mater.* **2022**, *325*, 126839.
6. Wang, Y.; Wang, G.; Wan, Y.; Yu, X.; Zhao, J.; Shao, J. Recycling of dredged river silt reinforced by an eco-friendly technology as microbial induced calcium carbonate precipitation (MICP). *Soils Found.* **2022**, *62*, 101216. [CrossRef]
7. Dong, Y.; Xu, F.; Liang, X.; Huang, J.; Yan, J.; Wang, H.; Hou, Y. Beneficial use of dredged sediments as a resource for mine reclamation: A case study of Lake Dianchi's management in China. *Waste Manag.* **2023**, *167*, 81–91. [CrossRef]
8. Chu, C.; Deng, Y.; Zhou, A.; Feng, Q.; Ye, H.; Zha, F. Backfilling performance of mixtures of dredged river sediment and iron tailing slag stabilized by calcium carbide slag in mine goaf. *Constr. Build. Mater.* **2018**, *189*, 849–856. [CrossRef]
9. Li, X.; Huang, L.; Reible, D.; Zeng, X.; Liu, S.; Fu, J.; Wang, K.; Fang, H. Inhibition of sediment erosion and phosphorus release by remediation strategy of contaminated sediment backfilling. *Water Res.* **2023**, *239*, 120055. [CrossRef] [PubMed]
10. Yu, C.; Cui, C.; Zhao, J.; Zheng, J.A. novel approach to utilizing dredged materials at the laboratory scale. *Constr. Build. Mater.* **2021**, *313*, 125568. [CrossRef]
11. Hussain, M.; Levacher, D.; Leblanc, N.; Zmamou, H.; Djeran-Maigre, I.; Razakamanantsoa, A.; Saouti, L. Reuse of harbour and river dredged sediments in adobe bricks. *Clean. Mater.* **2022**, *3*, 100046. [CrossRef]
12. Yang, M.; Ju, C.; Xue, K.; Peng, Y.; Han, H.; Wan, Q.; Peng, X.; Wu, Y. Environmental-friendly non-sintered permeable bricks: Preparation from wrap-shell lightweight aggregates of dredged sediments and its performance. *Constr. Build. Mater.* **2021**, *273*, 121751. [CrossRef]
13. Ponnada, S.; Sankar Cheela, V.R.; Gopala Raju, S. Investigation on mechanical properties of composite concrete containing untreated sea sand and quarry dust for 100% replacement of fine aggregate. *Mater. Today Proc.* **2020**, *32*, 989–996. [CrossRef]
14. Bhairappanavar, S.; Liu, R.; Shakoor, A. Eco-friendly dredged material-cement bricks. *Constr. Build. Mater.* **2021**, *271*, 121524. [CrossRef]
15. Xu, H.; Bu, J.; Chen, X.; Chen, Q.; Xu, B. Fracture behavior of dredged sand concrete under freeze–thaw cycles. *Constr. Build. Mater.* **2023**, *366*, 130192. [CrossRef]
16. Vinothkumar, K.; Vijayakumar, G. Study on the physical, mechanical and mineralogical properties of dredged marine sand as a partial replacement for fine aggregate in concrete. *Mater. Today Proc.* **2023**. [CrossRef]
17. Huang, W.; Chen, X.; Luan, J.; Ning, Y.; Ji, T.; Shi, Z. Deep insight into mechanical behaviour and microstructure mechanism of dredged sand in the lower reaches of the Yangtze River and manufactured sand concrete. *J. Build. Eng.* **2023**, *68*, 106105. [CrossRef]
18. Lee, K.-H.; Yang, K.-H.; Yoon, H.-S. Shrinkage strains of lightweight aggregate concrete using expanded bottom ash and dredged soil granules. *Constr. Build. Mater.* **2018**, *188*, 934–945. [CrossRef]
19. *JTS 181-5-2012*; Design Code for Dredging and Reclamation Works. China Communication Press: Beijing, China, 2012.
20. *GB/T 16399-2021*; Method for Chemical Analysis of Clay. Standards Press of China: Beijing, China, 2021.
21. *JTG 3420-2020*; Testing Methods of Cement and Concrete for Highway Engineering. Ministry of Transport of the People's Republic of China: Beijing, China, 2020.
22. *SL/T 352-2020*; Test Code for Hydraulic Concrete. China Water & Power Press: Beijing, China; China Communications Press: Beijing, China, 2020.
23. Chaohua, J.; Xiaobin, Z.; Guilan, T.; Da, C. Experimental Study on the Performance and Microstructure of Cementitious Materials Made with Dune Sand. *Adv. Mater. Sci. Eng.* **2016**, *10*, 2158706.
24. Ponraj, S.L.; Aruna, G.; Krishna Rao, A.; Srinivas Kadrekar, K. Studies on drying shrinkage and water permeability of fine fly ash high performance concrete. *Mater. Today Proc.* **2021**, *46*, 930–933.
25. Saluja, S.; Kaur, K.; Goyal, S.; Bhattacharjee, B. Assessing the effect of GGBS content and aggregate characteristics on drying shrinkage of roller compacted concrete. *Constr. Build. Mater.* **2019**, *201*, 72–80. [CrossRef]
26. Shariq, M.; Prasad, J.; Abbas, H. Creep and drying shrinkage of concrete containing GGBFS. *Cem. Concr. Compos.* **2016**, *68*, 35–45. [CrossRef]
27. Linderoth, O.; Johansson, P.; Wadsö, L. Development of pore structure, moisture sorption and transport properties in fly ash blended cement-based materials. *Constr. Build. Mater.* **2020**, *261*, 120007. [CrossRef]
28. Berodier, E.; Scrivener, K. Evolution of pore structure in blended systems. *Cem. Concr. Res.* **2015**, *73*, 25–35. [CrossRef]

buildings

MDPI

Article

Reduction in the Constrained Shrinkage and Crack Risk of Different-Strength-Level Concretes with Saturated Ceramsite

Pengpeng Nan [1,*], Bin Zuo [1], Song Han [2], Tao Huang [1] and Erpo Xiao [3]

1 China State Construction Engineering Corporation, No. 5 Anding Road, Chaoyang District, Beijing 100029, China
2 School of Civil Engineering, Beijing Jiaotong University, No. 3 Shangyuancun, Haidian District, Beijing 100044, China
3 China Construction Second Engineering Bureau Ltd., No. 6 Haiying Road, Fengtai District, Beijing 100070, China
* Correspondence: nanpengchd@outlook.com

Abstract: To reduce the shrinkage and cracking risk of concrete, internal curing technology was applied to modify concretes with different strength levels (C30 and C60) by incorporating saturated ceramsite. Three kinds of tests were carried out to study the effects of the incorporation of saturated ceramsite on the compressive strength, hydration temperature rise, and shrinkage behavior of the concretes, respectively. It was found that the internal hydration temperature rise of the concrete could be delayed and reduced due to the internal curing effect of ceramsite. Moreover, both the early-stage and long-term constrained shrinkage behavior of the internal cured concretes were monitored with embedded strain sensors and compared with free shrinkage behavior. For the C30 concrete system, with the incorporation of saturated ceramsite, its constrained shrinkage at 96 h was reduced greatly from 181 $\mu\varepsilon$ to 36 $\mu\varepsilon$. Furthermore, the surface-attached-sensor method was also used to study the shrinkage behavior of the concrete beams and it was found that the sensor location affected the measured shrinkage values greatly. The model beams of both the C30 and C60 concrete systems shrunk significantly in the first month, and the highest cracking risk occurred in this period as well. The internal curing effect of saturated ceramite could significantly reduce the constrained shrinkage of the concrete beam, evidently diminishing the cracking risk. More importantly, compared to the ordinary concrete C30 system, it was revealed that such an internal curing effect was more effective in promoting the performance of the higher-strength concrete (C60). With this effect, the cracking risk of C60 concrete was reduced from 0.69 to 0.37 at 300 days and changed little from then on.

Keywords: internal curing; hydration temperature rise; autogenous shrinkage; cracking risk; saturated ceramsite; concrete

Citation: Nan, P.; Zuo, B.; Han, S.; Huang, T.; Xiao, E. Reduction in the Constrained Shrinkage and Crack Risk of Different-Strength-Level Concretes with Saturated Ceramsite. *Buildings* **2024**, *14*, 1526. https://doi.org/10.3390/buildings14061526

Academic Editor: Jan Fořt

Received: 10 April 2024
Revised: 14 May 2024
Accepted: 19 May 2024
Published: 24 May 2024

1. Introduction

It is well known that the quality of concrete is difficult to control. Due to various practical problems, including the accumulation of air bubbles during construction, the resulting concrete is prone to shrinkage and cracking. The shrinkage of concrete could lead to many problems, such as volume stability, steel corrosion, structural leakage and damage, and even catastrophic collapse; therefore, it has a great influence on the building's structural health and service life [1]. For instance, the bridge guardrail is one of the important structural elements of a bridge and its cracking could be observed occasionally.

As the maintenance conditions are often limited at the construction site, an effective way to overcome the easy-cracking problem of guardrail concrete has been developed from the point of view of materials. It was suggested that internal curing techniques were efficient to reduce the self-shrinkage of concrete and diminish the concrete cracking risk that originated from insufficient curing [2]. During the long-term service life of concrete, the pre-incorporated internal curing component could provide internal water to continuously cure the hardened concrete [1,3,4]. Such internal curing materials can be categorized into two types, i.e., organic and inorganic materials. The organic materials are usually highly absorbent polymers or resins [5–9], such as super-absorbing polymers (SAPs). In the concrete, SAPs could release the pre-absorbed water for internal curing of concrete; however, this likely left more defective pores and led to the reduced strength of concrete generally [1]. The inorganic materials are mainly lightweight aggregates (LWA) [4,10–14]. Among these, ceramsite is one of the most important LWA for concretes [13–22], as it has its own advantages over those SAPs. Firstly, as a typical inorganic ceramic material, it has more compatibility with the inorganic cementitious matrix materials of concrete. Secondly, it can replace part of the coarse aggregates and thus reduce the structural deadweight without bringing evidently negative effects on the mechanical properties of concretes [23]. As revealed recently [9], the compressive strength of concretes incorporated with ceramsite was decreased much less than those incorporated with the SAP. Thirdly, the thermal conductivity coefficient of the ceramsite concrete is usually small and thus has a good thermal insulation performance [24]. It was found that the moisture content of LWAs had a great effect on the concrete shrinkage [11], and the concrete incorporated with saturated LWAs displayed evidently reduced self-shrinkage [12]. Ceramsite could be used singly or combined with other fillers to improve the performance of concretes [20–22], such as polymer fibers and rubber particles. Despite lots of research on the self-shrinkage of ceramsite concrete, more work needs to be carried out on this important internal curing technology, such as on the influence of the location of sensors on monitoring the long-term shrinkage behavior and the relative predicted cracking risk of the internally cured concrete system with different strength levels.

In this study, saturated ceramsite was used for internally curing the concrete with different strength levels. The effect of the internal curing effect on the mechanical properties of the resulting concretes and their hydration temperature rise will be studied. Moreover, both the early-stage and long-term shrinkage behavior of the internally cured concrete will be monitored and analyzed by model beam tests, as well as the different behavior between the constrained shrinkage and free shrinkage. Based on these analyses, the incorporation of saturated ceramsite on the cracking risk of concrete will be revealed. In the meanwhile, the internal curing effect of saturated ceramsite on the performance of the concrete with different strength levels will also be clarified.

2. Materials and Methods

2.1. *Raw Materials*

P.O 42.5 ordinary silicate cement was produced by Bozhou Conch Cement Company (Bozhou, China), and its main physical parameters are shown in Table 1. The mineral admixtures included the grade II fly ash produced by Huainan Changhua Electric Industrial Corporation (Huainan, China); S95 grade mineral powder with an alkali content of 0.4%, loss on burning of 0.06%, water demand of 99%, and an activity index of 84% for 7 days and 107% for 28 days, respectively. The detailed properties of fly ash and mineral powder are shown in Table 2. Two concretes with different compressive strength levels, i.e., C30 (compressive strength of ~30 MPa) and C60 (compressive strength of ~60 MPa), were designed and utilized to study the internal curing effect of saturated ceramsite on their various properties.

Table 1. The physical properties of cement.

Raw Material	Fineness (cm^2/g)	Time of Initial Condensation (min)	Time of Final Condensation (min)	Standardized Consistency (%)	Heat Loss (%)
C30	3400	160	220	27	0.5
C60	3790	122	160	29	0.61

Table 2. The physical properties of fly ash and slag.

Raw Material	Density (g/cm^3)	Specific Surface Area (cm^2/g)	Cl^- (%)	SO_3 (%)
fly ash	2.1	350	0.005	0.9
mineral powder	2.9	461	0.006	1

The fine aggregate was medium-coarse sand with a fineness modulus of 2.54, which was produced in Ganzhou, China. And the coarse aggregate was a mixture of the gravels with sizes of 5–16 mm and 16–31.5 mm. Their detailed properties are listed in Table 3.

Table 3. The physical properties of aggregate.

Raw Material	Apparent Density (g/cm^3)	Bulk Density (g/cm^3)	Clay Content (%)
Fine aggregate	2.62	1.62	0.1
Coarse aggregate	2.68	1.5	0.8

For the preparation of C60 concrete, silica fume was added and it had an average particle size of 0.31 μm, a density of 2.2 g/cm^3, and a specific surface area of 143,100 cm^2/g. The water-reducing agent was a polycarboxylic acid product with a water reduction rate of 29% and solid content of 25% and it was provided by Sika (Jiangsu, China) Building Materials Co.

For the internal curing materials, ceramsite clay particles with a size of less than 10 mm were used. They had a stacking density of 2.8 g/cm^3 and barrel compression strength of 2.1 MPa with a saturated water absorption rate of 22.54%. The mixing water was just tap water.

2.2. Proportioning

Two types of concrete systems with different strength levels, i.e., C30 and C60, were chosen and utilized in this work. Different contents (0, 10%, 20%, and 30%) of saturated ceramsite were used to replace the coarse aggregate in the C30 concrete formulation, and the resulting samples were labeled as C30-0, C30-1, C30-2, and C30-3, respectively. A similar operation was also applied for the C60 concrete formulation. The detailed mix ratios of these two concrete component systems are given in Table 4.

Table 4. The mix ratio of C30 and C60 concrete component system.

Sample	Cement	Aggregate	Gravel	Fly Ash	Mineral Powder	Silicon Powder	Water	Water Reducing Agent	Saturated Ceramsite
C30-0	216	1117	809	108	36	0	140	9	0
C30-1	216	1005	809	108	36	0	140	9	34
C30-2	216	894	809	108	36	0	140	9	67
C30-3	216	782	809	108	36	0	140	9	101
C60-0	312	998	722	104	52	52	130	12.5	0
C60-1	312	898	722	104	52	52	130	12.5	30
C60-2	312	898	722	104	52	52	130	12.5	60
C60-3	312	789	722	104	52	52	130	12.5	90

Note: C30 concrete: total cementitious amount 360 kg/m^3, gravel content 42%, mineral admixture 40%, fly ash:mineral powder = 3:1; C60 concrete: total cementitious amount 520 kg/m^3, gravel content 42%, mineral admixture 40%, fly ash:silicon powder:mineral powder = 2:1:1.

2.3. Test Methods

To simulate the constraint conditions of the actual concrete structure on the guardrail site, model concrete beams were fabricated for testing. Early shrinkage tests were carried out according to the code for "ordinary concrete long-term performance and durability test method standard" (GB/T50082-2009 [25]). Referring to GB/T50081-2002 [26] the code for

"Standard for Test Methods of Mechanical Properties of Ordinary Concrete", the compressive strength tests were performed and the cubic specimens of 100 × 100 × 100 mm were manufactured and tested on a WAW-600D-type electrohydraulic servo system (Beijng Zhongsheng Huaye Technology, Co. LTD., Beijing, China). According to the model beam test used by Bazant P. et al. [27,28], rectangular beams with a design span of 2000 mm and dimensions of 2000 mm × 400 mm × 200 mm were fabricated and the beam ends were secured with anchor bolts and steel plates, as shown in Figure 1.

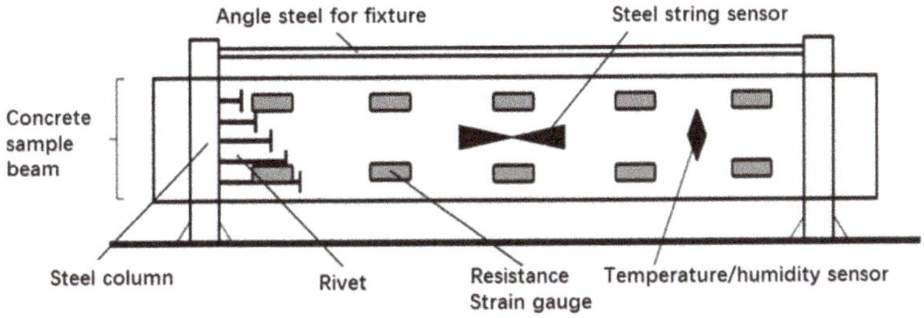

Figure 1. Schematic diagram of the configuration of the concrete sample beam for shrinkage test.

The 215HAT-type high-precision steel string sensors(Kingmach Measurement & Monitoring Technology Co., Ltd., Changsha, China), resistance strain gauges, temperature and humidity sensors (Beijing Power Bausheng Technology Development Co., Ltd., Beijing, China), and the corresponding data acquisition devices (Figure 2) were utilized to monitor the temperature change and the shrinkage behavior of the concrete beams with time. The high-precision steel string sensor is capable of responding to a small load (<1 N); therefore, it works well in monitoring the shrinkage of concrete. All the collected data were analyzed with Origin software (version 8.0). For the internal temperature test of the concrete sample, an SCTH 2002 (Sonbest Company of Shanghai, Shanghai, China) data acquisition device was used to collect the data.

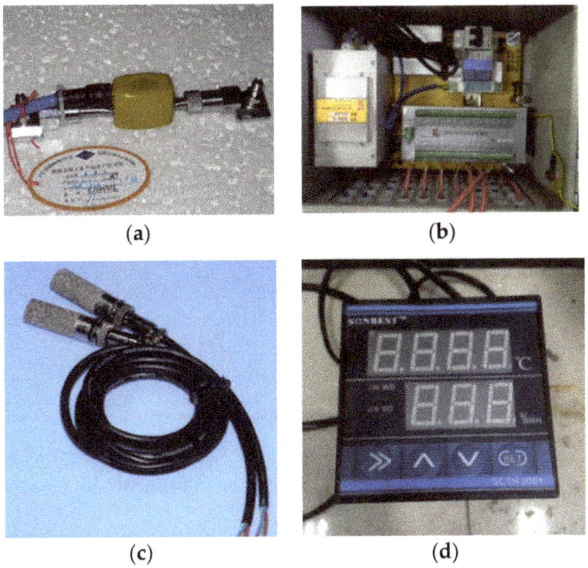

Figure 2. (**a**) Strain gauges and (**b**) the data acquisition device; (**c**) temperature/humidity sensors, and (**d**) their data acquisition device.

As shown in Figure 3, the temperature sensors were embedded at the end of the girder prior to the concrete placement to avoid the influence of early cracking of concrete on them. The data were recorded hourly within 48 h after the concrete placement, and daily thereafter. Shrinkage measurements were achieved through either the sensor embedded in the concrete or the ones patched externally on the beam surface, as shown in Figure 3. The strain gauges were placed before the concrete placement and fixed at the lower reinforcement area of the full girder to record the early shrinkage of the concrete in time. However, the external patched ones should be mounted immediately after 24 h once the concrete beam was demolded. For the sensor location on the side and bottom surface of the beam, a group of strain gauges was distributed every 500 mm, and each group of strain gauges included one at the upper edge and another at the lower edge. Additionally, a set of strain gauges was mounted in the middle of the beam. Long-term data were measured up to 365 days.

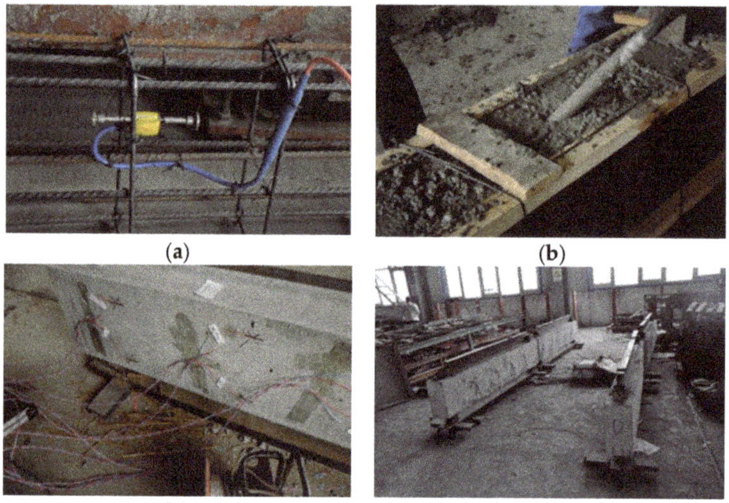

Figure 3. (**a**) Embedding sensors; (**b**) the concrete placement; (**c**) strain gauges applied after demolding; and (**d**) test of model beams.

3. Results and Discussion

3.1. Compressive Strength of the Concrete Specimens

The effect of incorporating different contents of saturated ceramsite on the compressive strength of both normal and high-strength concrete systems was studied and at least three specimens were tested for every sample. For the ordinary low-strength concrete system (C30), the compressive strength of the internally cured concrete at 7 days, 28 days, and 90 days all decreased gradually with the increasing content of ceramsite, as can be seen in Table 5. For the high-strength concrete system (C60), the same phenomenon occurred as well. However, its decrease was much less than that of its C30 counterpart. For instance, by incorporating 30% ceramsite, the compressive strength of the resulting C60-3 at 28 days was reduced by 6.15 MPa from the reference sample C60-0 (82.62 MPa), which was even less than half of that for the reduction in the C30 concrete system (12.75 MPa). This indicates the better applicability of the saturated ceramsite in the high-strength concrete system. It is noted that introducing 20% content of saturated ceramsite has a smaller negative effect on the mechanical properties of concrete. Taking the strength at 28 days as the example, the compressive strength value for C30-2 was 8.5 MPa less than that of C30-0, implying a reduction of 19.5%, while C60-2 was just 4.1 MPa lower than C60-0 and its reduction was only 4.9%. Therefore, the content of saturated ceramsite in the concrete was set to be 20% in the following research.

Table 5. Compressive strength of the different concrete samples.

Sample	Compressive Strength at 7 Days (MPa)	Compressive Strength at 28 Days (MPa)	Compressive Strength at 90 Days (MPa)
C30-0	31.11 ± 1.08	42.22 ± 0.46	46.27 ± 1.61
C30-1	28.13 ± 2.16	37.97 ± 1.71	42.77 ± 2.47
C30-2	25.11 ± 1.88	33.72 ± 1.42	39.27 ± 1.63
C30-3	22.12 ± 1.35	29.47 ± 3.15	35.77 ± 1.87
C60-0	69.57 ± 3.53	82.62 ± 2.14	87.63 ± 3.04
C60-1	68.37 ± 5.18	80.57 ± 3.32	84.83 ± 4.67
C60-2	67.17 ± 2.81	78.52 ± 2.97	82.03 ± 3.52
C60-3	65.97 ± 4.07	76.47 ± 3.33	79.23 ± 4.73

3.2. Hydration Temperature Rise Inside the Model Concrete Beams

The hydration temperature inside the model concrete beam was monitored in situ with the embedded temperature sensor over a 96 h period, and the corresponding curve of the temperature vs the time is shown in Figure 4.

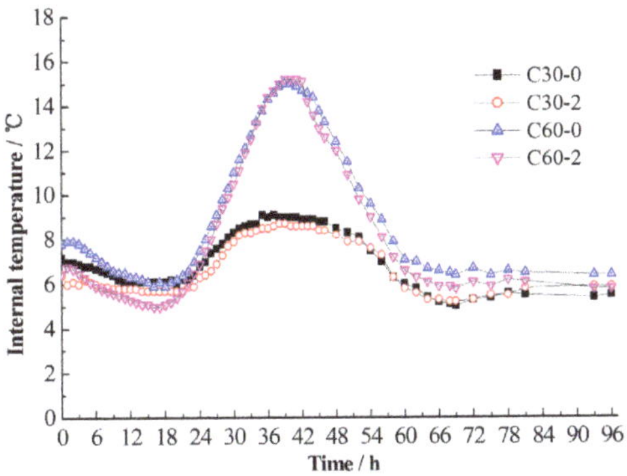

Figure 4. The dependence of the internal hydration temperature of the model concrete beam on the time.

As shown in Figure 4, the hydration temperature rise of concrete was not significant within the first 18 h after the concrete placement. After 18 h, the internal temperature of all the concrete samples began to increase. Throughout the entire test time, the internal temperature of the C30 concrete group increased maximally by ~6 °C. In comparison, the maximum temperature rise of the C60 concrete group increased by about 11 °C, which was significantly higher than that of the C30 concrete system. This implies that the larger thermal stress accumulated in the high-strength concretes leads to their probable larger autogenous shrinkage and higher cracking risk, which will be studied in the following research. Moreover, it was found that with the incorporation of saturated ceramsite in the concrete, the rapid rise in the internal temperature of the specimens was somewhat delayed by about 1–2 h for both concrete systems. The hydration temperature curves of their concrete samples were always below those of the ones without saturated ceramsite, despite that the differences were not so significant. This suggested that the internal curing effect of the saturated ceramsite was able to delay and reduce the internal hydration temperature rise during the hydration and curing of concrete. As suggested recently, the delayed and reduced hydration temperature rise was in favor of reducing the cracking risk of concretes [29]. Theoretically, the internal temperature rise will lead to the uneven curing of mass concrete and the accumulation of excessive thermal stress, which likely results in its autogenous shrinkage and cracking. The internal curing effect of the saturated ceramsite can effectively reduce the heat release rate of cement, regulate its curing and hydration process, delay and reduce its internal hydration temperature rise, and, therefore, reduce the shrinkage of the concrete beams.

3.3. Early Shrinkage Behavior of Concrete Beams

Early shrinkage values of concrete were measured continuously using internally embedded steel chord sensors in their model beams. The test results of the 4 samples within 96 h were shown in Figure 5, as well as those of the free shrinkage counterparts.

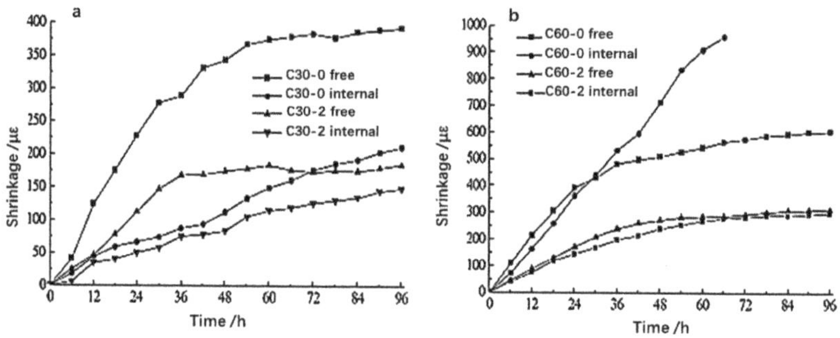

Figure 5. The internal shrinkage and free shrinkage of the model beam of concretes: (**a**) C30 series; (**b**) C60 series.

As can be seen from Figure 5, the internal shrinkage behavior of the concrete model beam was basically consistent with that of free shrinkage. All the shrinkage of the concrete increased generally with time. From 0 to 96 h, the internal concrete shrinkage values of the C30-0 and C30-2 model beams were 211 με and 148 με, respectively, while the corresponding free shrinkage values of the concrete material were 392 με and 184 με, respectively (Figure 5a), confirming that the early internal shrinkage of concrete was constrained and much less than the corresponding free shrinkage. This could be attributed to the presence of the steel columns and rivets at the ends of the beams, as well as the reinforcement rebars. It is well known that such constraints may lead to internal stresses within the concrete materials and even cause significant adverse consequences to the building structure. At 96 h, the constrained shrinkage was calculated as 392 με − 211 με = 181 με. After the incorporation of saturated ceramsite into the concrete, the constrained shrinkage of the resulting C30-2 was calculated as 184 με − 148 με = 36 με. These demonstrated that the internal curing effect of the ceramsite can significantly reduce the confined shrinkage and the internal stresses triggered by early shrinkage of concrete, and, therefore, diminish the early cracking risk. As shown in Figure 5b, similar shrinkage behavior was also found for the C60 concrete system, indicating that the internal curing technique also works well for high-strength concrete. Interestingly, the internal shrinkage curve of C60-2 (C60-2 internal) is almost overlapped with that of the free shrinkage curve of C60-2 (C60-2 free), confirming an even greater internal curing effect for the high-strength concrete compared to the ordinary C-30 concrete system.

3.4. Long-Term Shrinkage of Concrete Beams

The long-term internal shrinkage behavior of concrete beams was monitored with the internally embedded steel string sensors, and the data were recorded once a month over a period of 12 months. As shown in Figure 6, both the internal and free shrinkage of all four concrete beams increased abruptly within the first month and then changed little during the following time.

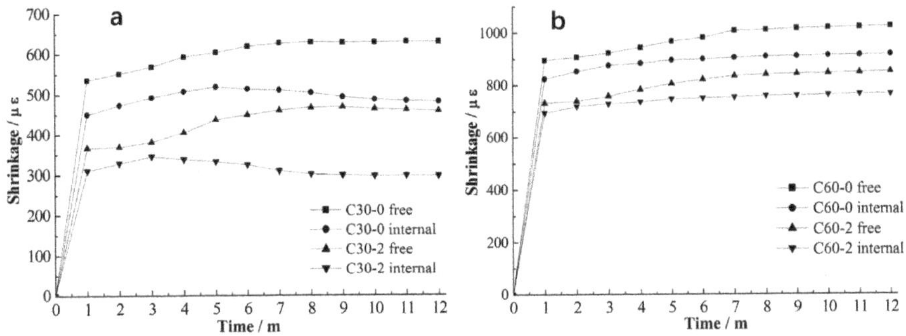

Figure 6. Comparison of internal shrinkage and free shrinkage of concrete beams within 12 months: (**a**) C30 series; (**b**) C60 series.

As shown in Figure 6a, the C30 concrete series showed shrinkage values of 300–600 με in the first month, while their maximal shrinkage change in the following 11 months was only about 100 με. For the C60 concrete series, they had shrinkage values of 700–900 με in the first month (Figure 6b), with a similar maximal shrinkage change to that of C30 concrete in the following time. This means that the shrinkage of the concrete model beams could be almost completed in the early stage of curing and their highest cracking risk should also occur in this time.

For all the concrete beams, it can be seen that all the internal shrinkage curves were significantly lower than those for the free shrinkage values. Specifically, the internal shrinkage value was found to be even decreased in some concrete groups (for instance, C30-0 and C30-2), which could be caused by the restraints of the model beams and the creep relaxation of the concrete materials due to the extended stresses. As the concrete inside the model beam is restrained by the steel reinforcement, their contraction is also restrained, which leads to the accumulation of internal stress. Although such internal stress may not seriously affect the deformation of concrete beams in the early stage, the creep relaxation of concrete occurs under the extended action of internal stress and results in shrinkage reduction. Moreover, it is noted that the shrinkage values of both C30 and C60 concrete incorporated with saturated ceramsite are always lower than those of the reference counterpart during the long-term shrinkage tests, indicating that the internal curing effect of the saturated ceramsite can significantly reduce the long-term shrinkage of concrete in model beams as well.

3.5. Long-Term Shrinkage Behavior of Concrete Beams Monitored by Surface-Attached Strain Sensors

Besides the long-term shrinkage measurements of concrete beams by the internally embedded strain sensors, this shrinkage behavior could also be monitored with the surface-attached strain gauges, which should be more convenient from the point of view of practical applications on-site. The strain gauges were attached at three characteristic locations: (1) two strain gauges were located at the center of the beam; (2) four strain gauges were located 1/4 length from the end of the beam; and (3) four strain gauges at the girder end of the beam. The average data obtained from all the sensors at the same location were taken as the final strain value. As can be seen in Figure 7, for certain concrete beams, the maximum and the minimum shrinkage occurred at the center and the ends of the beam, respectively, while the shrinkage values at 1/4 length to the beam ends were always the medium ones. This could be related to the different restrain at different beam locations. As the model beam is restrained by fixed supports at the beam ends, early contraction of the concrete materials causes the tensile stresses within the beam and significant volume deformation as well. Since the support constraints at the beam ends are realized by steel columns and rivets, the shrinkage of concrete is more constrained at the beam ends; therefore, the corresponding shrinkage values are lower. In contrast, due to the fewer constraints in the middle of the beam, this part bears most of the early contraction of the entire beam, which results in its larger shrinkage values. Also, for this reason, the early cracks in the model concrete beams usually occur in the middle.

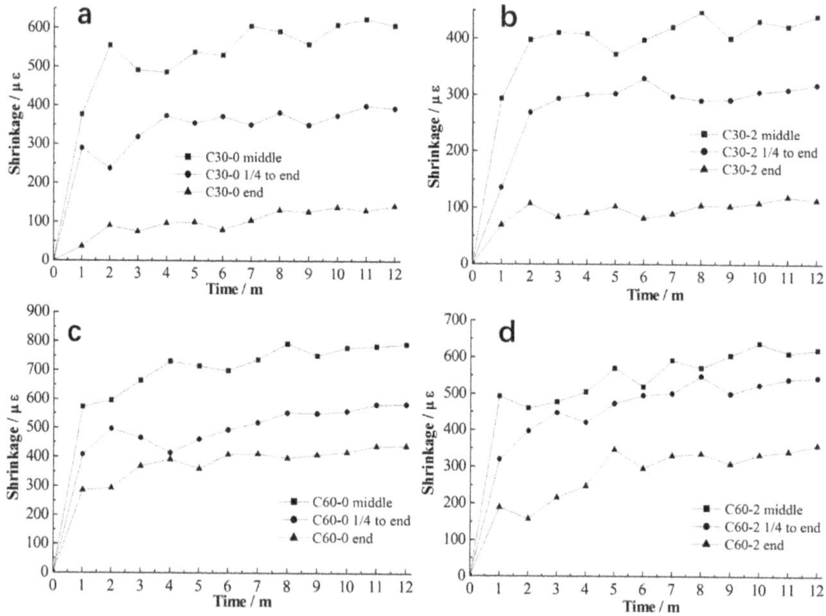

Figure 7. The shrinkage values measured at different positions on the surface of the model beam within 12 months: (**a**) C30-0; (**b**) C30-2; (**c**) C60-0; and (**d**) C60-2.

Moreover, it was found that the shrinkage of C30-2 specimens (Figure 7b) was always less than that of C30-0 (Figure 7a), no matter the data obtained in the middle of the beam or at the end of it. A similar phenomenon was also observed for C60-0 (Figure 7c) and C60-2 specimens (Figure 7d). These experimental results indicated again that the internal curing effect of the saturated ceramsite can significantly reduce the long-term shrinkage of concrete beams, which is consistent with the aforementioned results measured by the embedded strain sensors (Figure 6).

In fact, it is found that this surface-attached-sensor method does not seem to be a good option to monitor the long-term shrinkage of concrete. The main reasons for this are as follows: (1) The crack size of the concrete beams. The length of most strain gauges is about 10 cm, and their attached area are vulnerable to being penetrated by cracks, causing the failure of strain gauges and difficulties in obtaining long-term shrinkage data. (2) The glue used for fixing the sensors on the beam surface has certain aging problems during the long-term service. They usually become hard and brittle, which makes it hard to ensure the coordination between sensors and the concrete deformation process. In some cases, this could lead to the buckling of strain gauges or them even falling off. (3) Multiple strain gauges must be applied to minimize the test error as their locations have a great influence on the results.

3.6. Assessment of the Cracking Risk of Model Beams

The cracking risk of the concrete beam could be estimated by calculating the cracking stresses in the beams. Studies have shown that a greater risk of cracking occurred when the concrete shrinkage stress was ≥70% of its tensile strength [30–32]. The cracking risk of concrete can be defined with Equation (1) [33]. When the cracking risk reaches 0.7 or more, the concrete possesses a high probability of cracking.

$$\zeta = \frac{\sigma_s(t)}{f_t(t)} = \frac{\varepsilon_s(t)E_c}{f_t(t)} \tag{1}$$

where $\sigma_s(t)$—maximum tensile stress at moment t; $\varepsilon_s(t)$—shrinkage value of concrete at moment t; E_c—elastic modulus of concrete; and $f_t(t)$—tensile strength of concrete at moment t.

The elastic modulus of concrete was measured to be 3.4 × 10^4 MPa, 3.25 × 10^4 MPa, 3.92 × 10^4 MPa, and 3.77 × 10^4 MPa for C30-0, C30-2, C60-0, and C60-2, respectively. The tensile strength of concrete was generally approximated to be 0.1 times its compressive strength. According to the preliminary data, the variations in the compressive strength of the concrete samples with time were fitted and their fitted equations are shown in Table 6. Therefore, the tensile strength at moment t could be obtained, and the cracking risk was calculated by combining it with the shrinkage data in Equation (1), as shown in Figure 8.

Table 6. The fitted relationship between compressive strength and time.

Sample	The Fitted Equation
C30-0	F(x) = (48.88x + 0.7329)/(x + 3.516)
C30-2	F(x) = (42.38x − 1.84)/(x + 5.133)
C60-0	F(x) = (91.3x − 2.462)/(x + 2.371)
C60-2	F(x) = (86.92x − 3.41)/(x + 2.581)

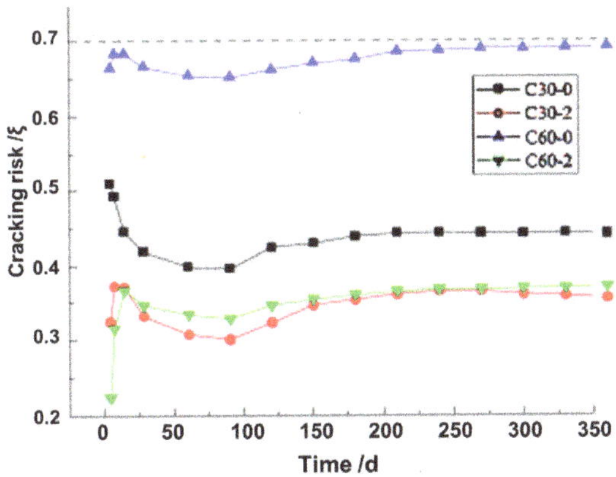

Figure 8. Dependence of the cracking risk of various concrete samples on the time.

As can be seen from Figure 8, concrete generally has a larger cracking risk in the early stage than in the later stage. Particularly, the largest cracking risk of all the four kinds of concretes appeared within the first three days, indicating its great susceptibility to cracking in this period. It is also noted that all the cracking risks of the four concretes are less than 0.7, which was suggested to be the critical cracking risk value of concrete previously [30,32], as concrete has a great possibility of cracking once the value is ≥0.7. These results imply the lower cracking risk of our well-designed concrete systems. The curve for C60-0 is always above that of C30-0, indicating that, for the concretes without the incorporation of saturated ceramsite, the higher the strength level of concrete the greater its cracking risk. However, with the incorporation of saturated ceramsite, the cracking risk values of both the two concretes with different strength levels were reduced significantly, especially for the higher-strength concrete. For instance, the cracking risk of C60-2 after 300 days is about 0.37, which is reduced by more than 46% in comparison with that of C60-0 at the same time. More interestingly, it can be seen in Figure 8 that, besides the shift downwards of the curves for both C30-2 and C60-2 compared to those for C30-0 and C60-0, the curve of C60-2 is almost overlaying that of C30-2, indicating the remarkable internal curing effect of saturated ceramsite on reducing the cracking risk of the high-strength concrete sample (C60-2), which reaches almost the same low cracking risk level of the ordinary concrete sample (C30-2). Furthermore, the valuable internal curing effect of saturated ceramsite could work well for a very long time (at least 1 year), as shown in Figure 8. In fact, it can be seen that the cracking risk value of the internally cured high-strength concrete (C60-2) changes little after 150 days, indicating its long-term stability.

4. Conclusions

The shrinkage and internal temperature development of high-performance internally cured concrete were investigated. The conclusions are given as follows:

(1) The inhibitory effect of saturated ceramsite on the hydration temperature rise of concrete was verified by tests on the model beam. With the incorporation of saturated ceramsite, the hydration temperature rise of concrete was reduced.
(2) The constrained shrinkage behavior of internally cured concrete beams was revealed. Constraints may lead to internal stresses within the concrete beams as well as early-stage cracking. The internal curing effect of saturated ceramsite significantly reduced the confined shrinkage of concrete.
(3) Both early-stage and long-term shrinkage development of the internally cured concrete beams were characterized. Most shrinkage of concrete beams over a one-year period was accomplished within the first month, after which it changed little. The internal curing effect significantly reduced the long-term shrinkage of the concrete beams.
(4) Besides the method of embedded strain sensors, the surface-attached-sensor technique could also be used to monitor the shrinkage of the concrete beam. In this technique, the location of the strain sensor affected the measured shrinkage behavior greatly, and the largest shrinkage values were obtained in the middle of the concrete beam.
(5) The maximum cracking risk of all the concrete beams within 12 months occurred in their early curing stage. For ordinary concretes, the higher the strength level, the greater the cracking risk. The incorporation of saturated ceramsite could significantly reduce the cracking risk of concretes, especially for higher-strength concrete.

In the bridge structure elements, long-term shrinkage of concrete may result in significant pre-stress loss and stiffness reduction. The internal curing technique via the incorporation of saturated ceramsite is effective in reducing the long-term shrinkage of concretes, leading to significant improvement in their volumetric stability and reduction in their cracking risk.

Author Contributions: Conceptualization, P.N. and S.H.; data curation, B.Z. and E.X.; formal analysis, P.N.; investigation, P.N. and B.Z.; methodology, B.Z., S.H. and T.H.; validation, S.H., T.H. and E.X.; visualization, T.H. and E.X.; writing—original draft, P.N.; writing—review and editing, P.N. All authors have read and agreed to the published version of the manuscript.

Funding: This research received no external funding.

Data Availability Statement: The data presented in this study are available upon request from the corresponding author.

Conflicts of Interest: Authors Pengpeng Nan, Bin Zuo, Tao Huang are employed by the China State Construction Engineering Corporation. Author Erpo Xiao is employed by the China Construction Second Engineering Bureau Ltd. The remaining authors declare that the research was conducted in the absence of any commercial or financial relationships that could be construed as a potential conflict of interest.

Nomenclature

SAPs	Super-absorbing polymers
LWA	Lightweight aggregate
C30	Concrete with compressive strength level of 'C30'
C60	Concrete with compressive strength level of 'C60'
ζ	Cracking risk of concrete
σ_s	Maximum tensile stress;
ε_s	Shrinkage value of concrete;
E_c	Elastic Modulus of concrete;
$f_t(t)$	Tensile strength of concrete at moment t

References

1. Huang, X.; Liu, X.; Rong, H.; Yang, X.; Duan, Y.; Ren, T. Effect of Super-Absorbent Polymer (SAP) Incorporation Method on Mechanical and Shrinkage Properties of Internally Cured Concrete. *Materials* **2022**, *15*, 7854. [CrossRef] [PubMed]
2. Lei, Y.; Yang, T.; Wu, D. Effect of biochar on mechanical and shrinkage properties of concrete. *Sichuan Build. Mater.* **2021**, *47*, 15–16.
3. Zhang, G.; Ge, J.; Ding, Q. Preparation and Formation Mechanism of Lightweight Ultra-high Performance Concrete. *J. Chin. Ceram. Soc.* **2021**, *49*, 381–390.
4. Kovler, K.; Souslikov, A.; Bentur, A. Pre-soaked light-weight aggregates as additives for internal curing of high strength concretes. *Cem. Concr. Aggreg.* **2004**, *26*, 131–138.
5. Xia, H.; Zhang, G.; Zhao, X.; Liu, G.; Song, L.; Chen, H. Super-absorbent polymer for concrete internal curing agent and its latest research progress. *Concrete* **2020**, *12*, 85–89.
6. Hu, S.; Zhang, G.; Yang, J. Effect of SAP on the Properties of Low Shrinkage Ultra High Performance Concrete. *Concrete* **2021**, 54–58. [CrossRef]
7. Liu, J.; Farzadnia, N.; Khayat, K.H.; Shi, C. Effects of SAP characteristics on internal curing of UHPC matrix. *Constr. Build. Mater* **2021**, *280*, 122530. [CrossRef]
8. Morinaga, Y.; Akao, Y.; Fukuda, D.; Elakneswaran, Y. Delayed Absorption Superabsorbent Polymer for Strength Development in Concrete. *Materials* **2022**, *15*, 2727. [CrossRef] [PubMed]
9. Lv, X.; Liu, Z.; Dong, Y. Comparative Study of the Effect of Super Absorbent Polymer (SAP) and Ceramsite on the Performance of Concrete. *Bull. Chin. Ceram. Soc.* **2019**, *38*, 246–252.
10. Jóźwiak-Niedźwiedzka, D. Scaling resistance of high performance concretes containing a small portion of pre-wetted lightweight fine aggregate. *Cem. Concr. Compos.* **2005**, *27*, 709–715. [CrossRef]
11. Cheng, S.; Shui, Z.; Yu, R.; Sun, T.; Zhang, X. Multiple influences of internal curing and supplementary cementitious materials on the shrinkage and microstructure development of reefs aggregate concrete. *Constr. Build. Mater.* **2017**, *155*, 522–530. [CrossRef]
12. Henkensiefken, R.; Bentz, D.; Nantung, T.; Weiss, J. Volume change and cracking in internally cured mixtures made with saturated lightweight aggregate under sealed and unsealed conditions. *Cem. Concr. Compos.* **2009**, *31*, 427–437. [CrossRef]
13. Ji, T.; Zheng, D.D.; Chen, X.F.; Lin, X.J.; Wu, H.C. Effect of prewetting degree of ceramsite on the early-age autogenous shrinkage of lightweight aggregate concrete. *Constr. Build. Mater.* **2015**, *98*, 102–111. [CrossRef]
14. Cusson, D.; Hoogeveen, T. Internal curing of high-performance concrete with pre-soaked fine lightweight aggregate for prevention of autogenous shrinkage cracking. *Cem. Concr. Res.* **2008**, *38*, 757–765. [CrossRef]
15. Guo, R.; Li, G.; An, M. Effect of internal curing techniques on the properties of high performance concrete of different strength classes. *China Railw. Sci.* **2015**, *36*, 24–29.
16. Wang, X.F.; Fang, C.; Kuang, W.Q.; Li, D.W.; Han, N.X.; Xing, F. Experimental investigation on the compressive strength and shrinkage of concrete with pre-wetted light-weight aggregates. *Constr. Build. Mater.* **2017**, *155*, 867–879. [CrossRef]
17. Wang, Y.; Qiu, J.; Zeng, C. Shrinkage and strength properties of coal gangue ceramsite lightweight aggregate concrete. *Adv. Mater. Sci. Eng.* **2020**, *2020*, 3575709. [CrossRef]
18. Xie, J.; Liu, J.; Liu, F.; Wang, J.; Huang, P. Investigation of a new lightweight green concrete containing sludge ceramsite and recycled fine aggregates. *J. Clean. Prod.* **2019**, *235*, 1240–1254. [CrossRef]
19. Fan, L.; Zhang, Z.; Yu, Y.; Li, P.; Cosgrove, T. Effect of elevated curing temperature on ceramsite concrete performance. *Constr. Build. Mater.* **2017**, *153*, 423–429. [CrossRef]
20. Gong, J.; Zeng, W.; Zhang, W. Influence of shrinkage reducing agent and polypropylene fiber on shrinkage of ceramsite concrete. *Constr. Build. Mater.* **2018**, *159*, 155–163. [CrossRef]
21. Shen, Y.; Ma, X.; Huang, J.; Hao, F.; Lv, J.; Shen, M. Near-zero restrained shrinkage polymer concrete incorporating ceramsite and waste rubber powder. *Cem. Concr. Compos.* **2020**, *110*, 103584. [CrossRef]
22. Shen, Y.; Huang, J.; Ma, X.; Hao, F.; Lv, J. Study on the free shrinkage of lightweight polymer concrete incorporating waste rubber powder and ceramsite. *Compos. Struct.* **2020**, *242*, 112152. [CrossRef]
23. Lin, J.; Lu, J. Experimental Study on Effect for Internal Curing of Porous Saturated Aggregate on the Crack Resistance and Strength of Concrete. *Fly Ash Compr. Util.* **2020**, *34*, 68–73.

24. Yang, J.; Tang, Y.; Yu, J. Experimental study of cast-in-situ all lightweight foamed concrete used by shale ceramsite. *Ind. Constr.* **2017**, *47*, 133–140.
25. *GB/T50082-2009*; Standard for Test Methods of Long-Term Performance and Durability of Ordinary Concrete. China Architecture & Building Press: Beijing, China, 2009.
26. *GB/T50081-2002*; Standard for Test Method of Mechanical Properties on Ordinary Concrete. China Architecture & Building Press: Beijing, China, 2003.
27. Bazant, Z.P.; Murphy, W.P. Creep and shrinkage prediction model for analysis and design of concrete structures-model B3. *Mater. Struc.* **1995**, *28*, 357–365.
28. Bazant, Z.P.; Baweja, S. Justification and refinements of model B3 for concrete creep and shrinkage 1. statistics and sensitivity. *Mater. Struc.* **1995**, *28*, 415–430. [CrossRef]
29. Liang, T.; Luo, P.; Mao, Z.; Huang, X.; Deng, M.; Tang, M. Effect of Hydration Temperature Rise Inhibitor on the Temperature Rise of Concrete and Its Mechanism. *Materials* **2023**, *16*, 2992. [CrossRef]
30. Li, H.; Liu, J.; Wang, Y.; Yao, T.; Tian, Q.; Li, S. Deformation and cracking modeling for early-age sidewall concrete based on the multi-field coupling mechanism. *Constr. Build. Mater.* **2015**, *88*, 84–93. [CrossRef]
31. Ishida, T.; Maekawa, K.; Kishi, T. Enhanced modeling of moisture equilibrium and transport in cementitious materials under arbitrary temperature and relative humidity history. *Cem. Concr. Res.* **2007**, *4*, 565–578. [CrossRef]
32. Gao, Y.; Zhang, J.; Hou, D. Calculation of moisture induced stress and evaluation of cracking risk in early age concrete. *Eng. Mech.* **2012**, *29*, 121–128.
33. Wang, J.; Yan, P. Analyzing of Cracking Risk of Early-age Concrete Structure. *J. Basic Sci. Eng.* **2006**, *2*, 262–267.

buildings

MDPI

Article

Feasibility of Recycled Aggregate Concrete in a Novel Anchoring Connection for Beam-to-Concrete-Filled Steel Tube Joints

Jianhua Su [1,2], Qian Zhao [1,2], Li'ao Cai [3], Xiaohui Li [3], Hongyin Pu [3], Wei Dai [3], Jian Zhang [3], Deng Lu [4] and Feng Liu [3,*]

1 Guangzhou Research Institute of Construction Industry Co., Ltd., Guangzhou 510520, China
2 Guangzhou Construction Engineering Co., Ltd., Guangzhou 510030, China
3 School of Civil and Transportation Engineering, Guangdong University of Technology, Guangzhou 510006, China
4 China Construction Third Engineering Bureau Group Co., Ltd., Wuhan 430074, China
* Correspondence: fliu@gdut.edu.cn

Abstract: Owing to the substantial benefits in environmental protection and resource saving, recycled aggregate concrete (RAC) is increasingly used in civil engineering; among the different types, RAC-filled steel tubes are an efficient structural form utilizing the advantages of concrete and steel tubes. This paper proposed a novel full-bolted beam-to-concrete-filled steel tube (CFST) joint and investigated the anchoring behavior of the steel plates embedded in RAC-filled steel tubes, which represents the behavior of the tensile zone in this joint, to demonstrate the feasibility of utilizing RAC in composite structures. The specimen consisted of a CFST and a connecting plate embedded in the CFST. In total, 18 specimens were tested to study the effects of concrete type (i.e., recycled aggregate concrete and natural aggregate concrete), anchoring type (i.e., plate with holes, notches, and rebars), and plate thickness on the pullout behavior, such as anchorage strength, load–displacement response, and ductility. Based on experimental results, the aggregate type of the concrete does not affect the pullout behavior obviously but the influence of anchoring type is significant. Among the three anchoring methods, the plate with rebars exhibits the best performance in terms of anchorage strength and ductility, and is recommended for the beam-to-CFST joint. In addition, plate thickness obviously affects the behavior of plates with holes and notches, the bearing area of which is proportional to the thickness, whereas the pullout behavior of the plates with rebars is independent of thickness. Finally, design formulas are proposed to estimate the anchorage strength of the connecting plates, and their reasonability is validated using the experimental results.

Keywords: recycled aggregate concrete; anchoring behavior; pullout test; connecting plates; load–displacement response; theoretical model

Citation: Su, J.; Zhao, Q.; Cai, L.; Li, X.; Pu, H.; Dai, W.; Zhang, J.; Lu, D.; Liu, F. Feasibility of Recycled Aggregate Concrete in a Novel Anchoring Connection for Beam-to-Concrete-Filled Steel Tube Joints. *Buildings* **2024**, *14*, 1178. https://doi.org/10.3390/buildings14041178

Academic Editor: Flavio Stochino

Received: 30 March 2024
Revised: 16 April 2024
Accepted: 18 April 2024
Published: 21 April 2024

1. Introduction

With rapid urbanization, lots of infrastructures are demolished worldwide, resulting in a huge amount of construction waste. This had led to severe impacts on the environment and sustainability [1–7]. To solve this issue, we proposed using crushed waste concrete as the coarse aggregate of concrete, named recycled aggregate concrete (RAC) [8–10]. The utilization of RAC efficiently reduces waste concrete, which mitigates the environment impacts, and solves the resource shortage problem in the construction industry. Currently, extensive experimental and theoretical studies have been conducted on the mechanical properties and mixture design of RAC, as well as on the structural behavior of RAC members and structures [11–14]. In general, the mechanical properties of RAC are somewhat less superior to those of natural aggregate concrete (NAC), but the carbon footprint embedded in RAC is much lower than that of NAC [15,16]. It was found that the replacement ratio of recycled aggregate is the critical parameter that affects the behavior of RAC. As the replacement ratio increases, the mechanical properties of RAC tend to decrease [17,18].

The RAC-filled steel tube is a commonly used structural form in practice, and could utilize the advantages of both steel and concrete [19,20]. The steel tube could provide confinements to RAC to increase the strength and ductility of the concrete [21,22]. Furthermore, the local buckling of steel tube could be postponed or avoided by the in-filled concrete, resulting in an enhancement of the steel tube's local buckling capacity. The RAC-filled steel tube has the advantages of high strength, high ductility, and high energy dissipation capacity [23–26]. Currently, the commonly used type for the I section beam-to-square concrete-filled steel tube (CFST) column connection include joints with interior diaphragms, exterior diaphragms, and through diaphragms, hybrid joints with welds and bolts, joints with vertical stiffeners, and joints with binding bars [27–30]. These joints involve on-site welding, which may increase the construction period and affect the quality of welds. In addition, welds in joints are prone to brittle failure during earthquakes. Therefore, full-bolted-joints are encouraged for beam-to-column connections to improve the performance of the connections.

This study proposed a new type of I beam-to-CFST joint that is fully bolted. As shown in Figure 1, there are three connecting plates, and one end of each is embedded in CFST while the other end is bolted to the flanges and the web of the I-beam. Different measures could be adopted to increase the anchorage of the connecting plates, such as setting steel bars, and cutting holes or notches in the plates. Specially, the configuration of connecting plates with steel bars is similar to that of perfobond rib connectors (PBLs), in which the shear resistance comes from the bearing of the end concrete, the shear of the concrete dowel, and the shear of the rebar [31–33]. Nevertheless, PBL is used in steel–concrete composite beams to maintain the bond between the concrete slab and the steel beam. The anchoring behavior of connecting plates in CFST is different from that of PBLs. In addition, it is more difficult to set rebars in a position for close sections than it is to set them in composite beams. Therefore, it is worthwhile to explore other efficient methods to improve the anchorage of the connection plates in CFSTs and investigate the anchoring behavior for the proposal of design methods.

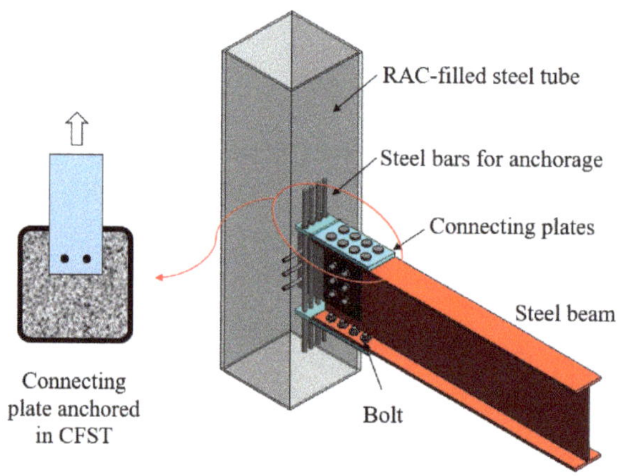

Figure 1. A novel full-bolted beam-to-CFST joint.

Although there are no studies for the proposed configurations, the anchorage behavior of blind bolts in CFSTs have been extensively investigated via pullout tests and may provide some insights for this study. Various types of blind bolts were studied, such as Extended Hollo Bolt [34], T-bolt [35], and the Ajax anchored bolt [36]. By setting anchors, the anchorage strength could be greatly enhanced and the pullout capacity comes mainly from the anchorage of the bolt's anchor and the bearing of the tube face wall [37]. The anchorage strength depends on the embedment depth of the anchor, the bearing area of anchor-to-concrete, and the strength of the concrete. It is noted that this study focused on the anchorage of a steel plate, which is different from bolts with anchors, and the anchorage mechanism is also different.

This study focused on the anchoring behavior of steel plates embedded in square RAC/NAC-filled steel tubes using pullout tests. Three types of anchoring methods, namely plates with circular holes, plates with rectangular notches, and plates with rebars, were proposed and investigated. In total, 18 specimens were tested to investigate the effects of the key parameters, including the replacement ratio of RAC, thickness of the plates, and anchoring methods, on the pullout behavior (e.g., ultimate capacity, load–displacement curves, and ductility). Thereafter, theoretical studies were conducted and formulas were proposed to estimate the ultimate pullout capacities (i.e., the anchorage strength) of the steel plates anchored in the RAC-filled steel tubes.

2. Experimental Program

2.1. Specimens

Pullout specimens were adopted to investigate the tensile behavior of the tension zone of the beam-to-column joint. In total, 18 specimens were tested under monotonic pullout loads. The parameters investigated in this study include the concrete type (i.e., normal concrete and RAC), the thickness of the connecting plate (i.e., 6 mm and 10 mm), and the anchoring type (i.e., the plate with holes, plate with notches, and plate with rebars).

The specimen consisted of a 650 mm long RAC (or NAC)-filled steel tube (CFST) and a 120 mm by 280 mm connecting plate anchored in the CFST. Dimensions of the specimen are shown in Figure 2, in which T is the plate thickness and the embedment depth of the plate is 80 mm. The dimensions of the CFST are 200 mm × 650 mm × 8 mm. Details of the anchoring type are shown in Figure 3, where A1, A2, B1, B2, and C indicate the anchorage type. The size of the rectangular slot on the top surface of the CFST is 122 mm by T + 2 mm, where T is the thickness of the connecting plate.

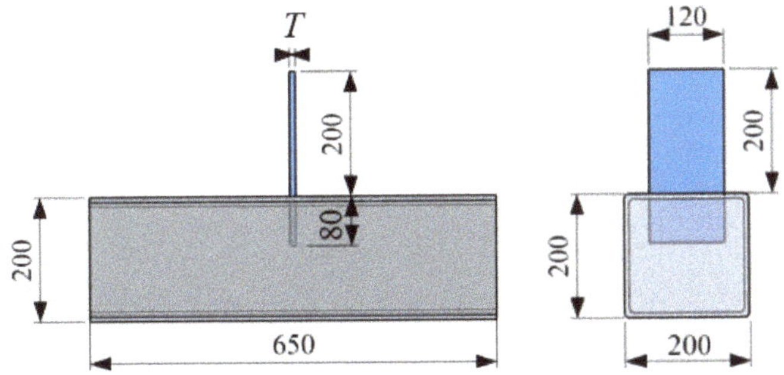

Figure 2. Dimensions of the pullout specimen.

Details of all the tested specimens are listed in Table 1. The label of the specimen consists of the anchorage configuration (A1, A2, B1, B2, and C, as illustrated in Figure 3), rebar diameter ("r" followed by the nominal diameter in mm for a type C specimen), tube thickness ("T" followed by the nominal thickness in mm), and RA replacement ratio ("F" followed by the replacement percentage). For example, C-r8-T6-F100 means that the pullout specimen has a 6 mm thick connecting plate with 8 mm diameter rebars, and the RA replacement ratio is 100%.

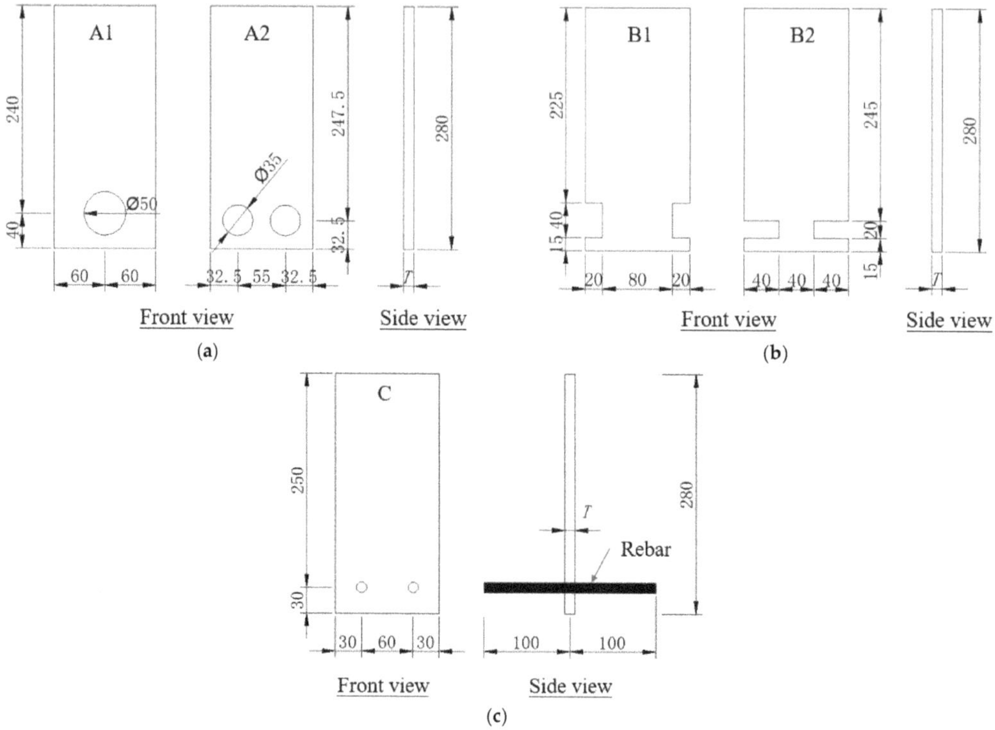

Figure 3. Details of the connecting plates with (**a**) holes; (**b**) notches; (**c**) steel bars.

Table 1. Specimen table.

Specimen	Configuration	T (mm)	RA Replacement (%)	Remark
A1-T6-F100	1 Φ 50 hole	6	100	With holes
A1-T10-F100	1 Φ 50 hole	10	100	
A1-T10-F0	1 Φ 50 hole	10	0	
A2-T6-F100	2 Φ 35 hole	6	100	
A2-T10-F100	2 Φ 35 hole	10	100	
A2-T10-F0	2 Φ 35 hole	10	0	
B1-T6-F100	2□20 × 40 notch	6	100	With notches
B1-T10-F100	2□20 × 40 notch	10	100	
B1-T10-F0	2□20 × 40 notch	10	0	
B2-T6-F100	2□40 × 20 notch	6	100	
B2-T10-F100	2□40 × 20 notch	10	100	
B2-T10-F0	2□40 × 20 notch	10	0	
C-r8-T6-F100	2 Φ 8 rebar	6	100	With rebars
C-r8-T10-F100	2 Φ 8 rebar	10	100	
C-r8-T10-F0	2 Φ 8 rebar	10	0	
C-r12-T6-F100	2 Φ 12 rebar	6	100	
C-r12-T10-F100	2 Φ 12 rebar	10	100	
C-r12-T10-F0	2 Φ 12 rebar	10	0	

The manufacturing procedure of the pullout specimens is as follows: (1) fix the hollow tube on the timber plate with silicon; (2) position the connecting plate and rebars (if applicable) with temporary supports; (3) cast the concrete. Figure 4 shows the pictures of the steel tubes, connecting plates, and specimens before we casted the concrete and casted the specimens. Six concrete cylinders with a diameter of 150 mm and height of 300 mm were also casted to measure the concrete strength. After casting the concrete, all the specimens and cylinders were cured at ambient temperature until the testing date. The actual curing time of the specimens was 28~32 days.

(a) (b) (c) (d)

Figure 4. Preparation of pullout specimens: (**a**) steel tubes; (**b**) connecting plates; (**c**) specimen before casting concrete; (**d**) casted specimens.

2.2. Materials

The designed grade of the concrete is C50 with a water-to-cement ratio of 0.55, and the mixture is shown in Table 2. Both recycled aggregate concrete (RAC) and natural aggregate concrete (NAC) were adopted in this study. The recycled aggregate (RA) was obtained from the demolished concrete of old buildings. The old concrete was crushed into small sizes, screened, washed, and dried (Figure 5a). The mechanical and physical properties of the RA met the requirements in [38]. The NA was fabricated from crushed basalt (Figure 5b). Both RA and NA had continued particle size distributions with a size range of 5~20 mm, which met the requirement indicated in GB/T 25177-2010 [39]. River sand was adopted as the fine aggregate. General-purpose cement (Grade 42.5) was used as the binder, and its initial and final setting time were >45 min and <10 h, respectively. A polycarboxylate-based water reducer was used for the concrete to improve workability. At the time of the specimen test, a compressive test was conducted on the concrete cylinders with a diameter of 150 mm and height of 300 mm to measure the compressive strength (f_c') and Young's modulus (E_c); the average measured values are listed in Table 2. The compressive strengths of RAC and NAC are very close. The main reason is that the RA was crushed from concrete with granite as coarse aggregate, whereas the coarse aggregate for NAC was basalt, the strength and hardness of which are lower than those of granite. In addition, RA has higher water absorption than NA does, resulting in a lower water-to-cement ratio for RAC. Therefore, strength reduction was not observed after replacing NA with RA.

Table 2. Mixture and hardened properties of concrete.

Replacement Ratio of RA (%)	Mass per Unit Volume (kg/m^3)						f_c' (MPa)	E_c' (GPa)
	Cement	Sand	Natural CA	Recycled CA	Water	Water Reducer		
0%	425	762	1018	0	175.5	8.5	50.3	49.4
100%	425	762	0	1018	175.5	8.5	49.3	41.5

Note: CA = coarse aggregate.

(a)

(b)

Figure 5. Coarse aggregates: (**a**) RA; (**b**) NA.

The square steel tubes were cold-formed and purchased from the market. Both the square steel tube and connecting plates were made of Q235 steel, the nominal yield strength of which is 235 MPa. The grade of the deformed steel rebars was HRB400 (nominal yield strength = 400 MPa). Tensile coupon tests were conducted on the steel plates and rebars, and the measured material properties are listed in Table 3. It is necessary to mention that the properties of the steel tube were not measured experimentally, as past similar studies [37,40] indicated that the yield strength of a steel tube is not a governing parameter affecting the anchorage of steel plates.

Table 3. Material properties of steel plates and rebars.

Type	E_s (GPa)	f_y (MPa)	f_u (MPa)
Steel plate	211	328	444
Steel rebar	209	453	672

2.3. Experimental Setup

Figure 6 shows the experimental setup for the pullout test. The connecting plate was gripped by the 500 kN universal testing machine (MTS-322), and the load was applied upward. Two 500 mm long welded rectangular hollow tubes (130 mm × 150 mm × 30 mm) were placed on the CFST, working as the reaction beams to resist the upward movement of the specimen. Each reaction beam was connected to the base plate of the test machine via two high-strength steel (HSS) rods measuring a diameter of 28 mm. The clear distance between the reaction beams was about 350 mm, ensuring that the constraining effect at beam ends did not affect the pullout behavior of the specimen. The load was applied under displacement control with a loading rate of 1.2 mm/min. The experiment was terminated when obvious failure occurred, such as when the pullout of the steel plate, the fracture of the steel plate, or a substantial drop in the applied load was observed.

Two linear variable displacement transducers (LVDTs) were deployed between the actuator and the top surface of the CFST (near the edge) to measure the vertical displacement of the connecting plate (Figure 6). It is necessary to mention that the average value of LVDTs was the same as that for the displacement recorded by the test machine. Seven strain gauges (S1, S2, S3, S4, S5, S6 and S7) were fixed on the top surface of CFST, as shown in Figure 7, to measure the strain distribution of the steel tube. All the data, such as load, displacements, and strains, were recorded simultaneously during the experiment.

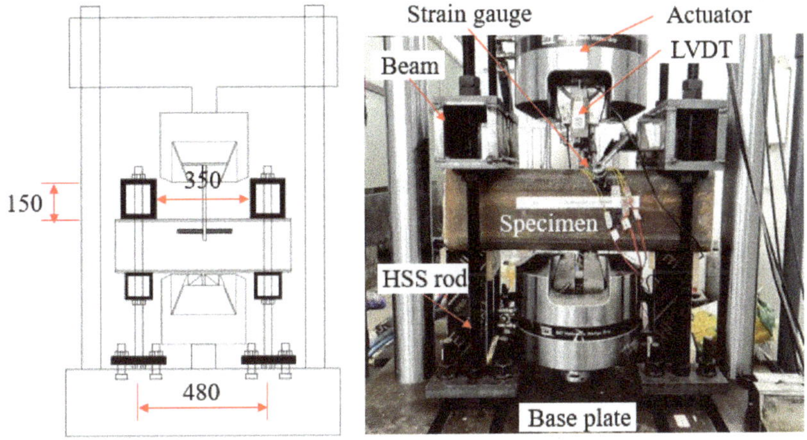

Figure 6. Experimental setup for pullout test.

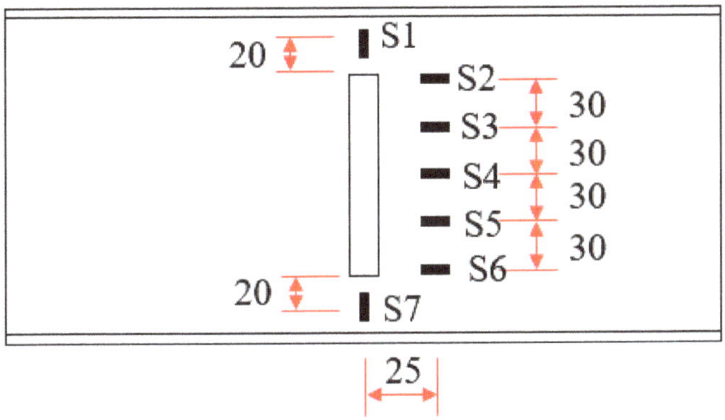

Figure 7. Layout of strain gauges.

3. Experimental Results

3.1. Failure Modes

The deformed shapes of the tested specimens are presented in Figure 8. After each test, the CFSTs were cut off and the concrete was removed to observe the failure modes of the connecting plates, as shown in Figure 8. In general, owing to the different anchorage mechanisms, the failure modes of the tested specimens (plates with holes, notches, and rebars) varied significantly.

As shown in Figure 8a, the deformation of the steel tube was insignificant during the test, indicating that only a small portion of concrete was involved in resisting the applied force. Furthermore, the bearing deformation of the steel plate was not obvious. Localized failure happened in the concrete within the holes. This was likely due to the shear failure and bearing failure (concrete-to-steel) of the concrete. For specimens with notches (Figure 8b), the deformation of the CFST was not obvious. However, significant deformation was observed in steel plate's notches. The bottom edge of the notch could be regarded as a cantilever beam in contact with the filled concrete. As shown in Figure 8b, with the increase in notch width (from 20 mm to 40 mm), the governing failure mode of the notch edge changed from shearing-governed to bending-governed due to the increase in the length-to-depth ratio of the cantilever beam. Excessive deformation of CFST face wall was observed for specimens with rebars (Figure 8c). Shear failure of the rebars occurred for the plates with thickness of 10 mm. A combination of the shear and bending failure of rebars and the tearing of the steel plate was observed for the 6 mm thick plate. The deformation of the rebar was mainly induced by the shear and bending effects, which could be regarded as a beam on continuous springs (i.e., surrounded concrete) and it is under laterally distributed loads. Su et al. [41] and He et al. [42] observed the similar failure mode and defined it as the tension-shear failure.

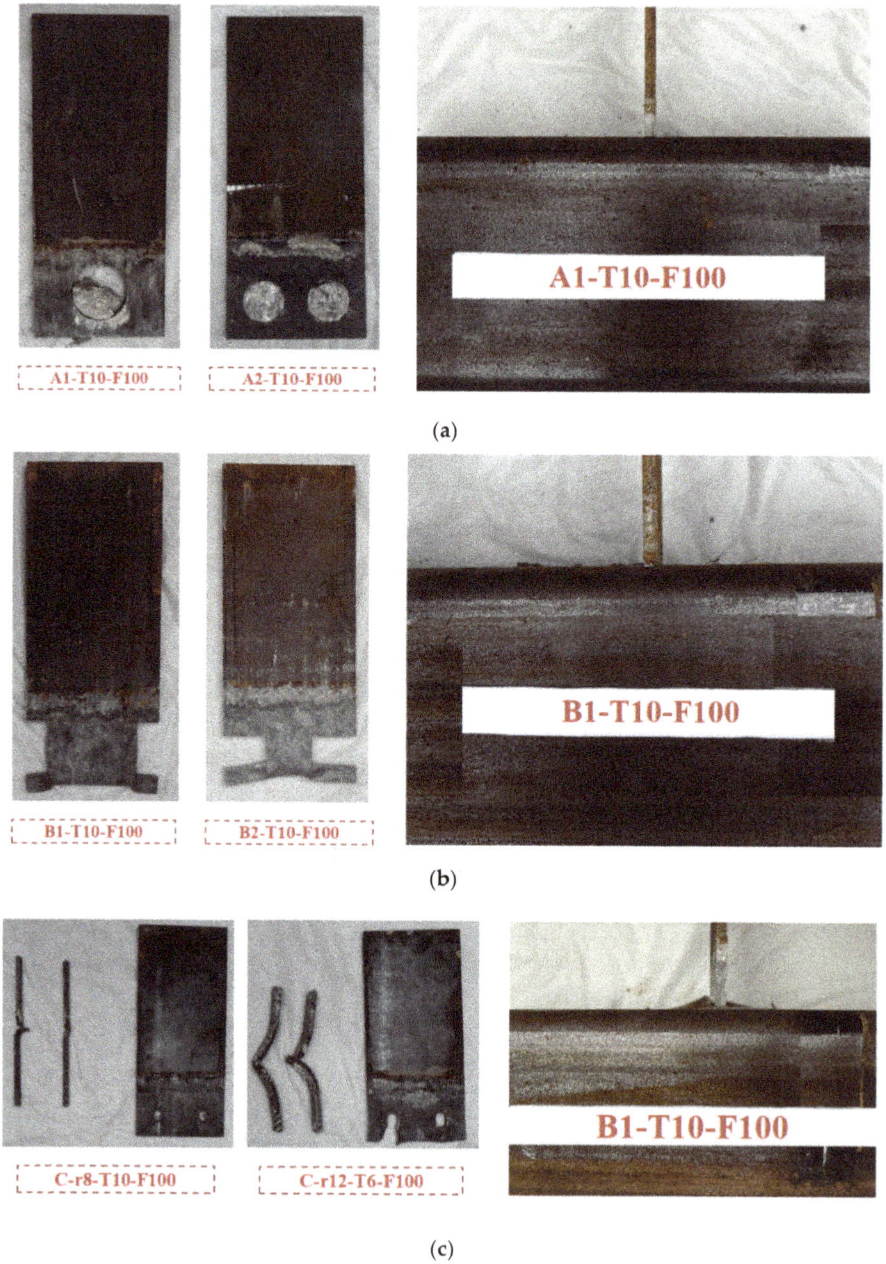

Figure 8. Failure modes: (**a**) plate with holes; (**b**) plate with notches; (**c**) plate with rebars.

3.2. Load–Displacement Curves

Load–displacement curves of all the tested specimens are shown in Figure 9, in which the vertical displacement was recorded by the test machine. In general, the peak load and the shapes of the curves varies greatly depending on the anchoring methods and configurations.

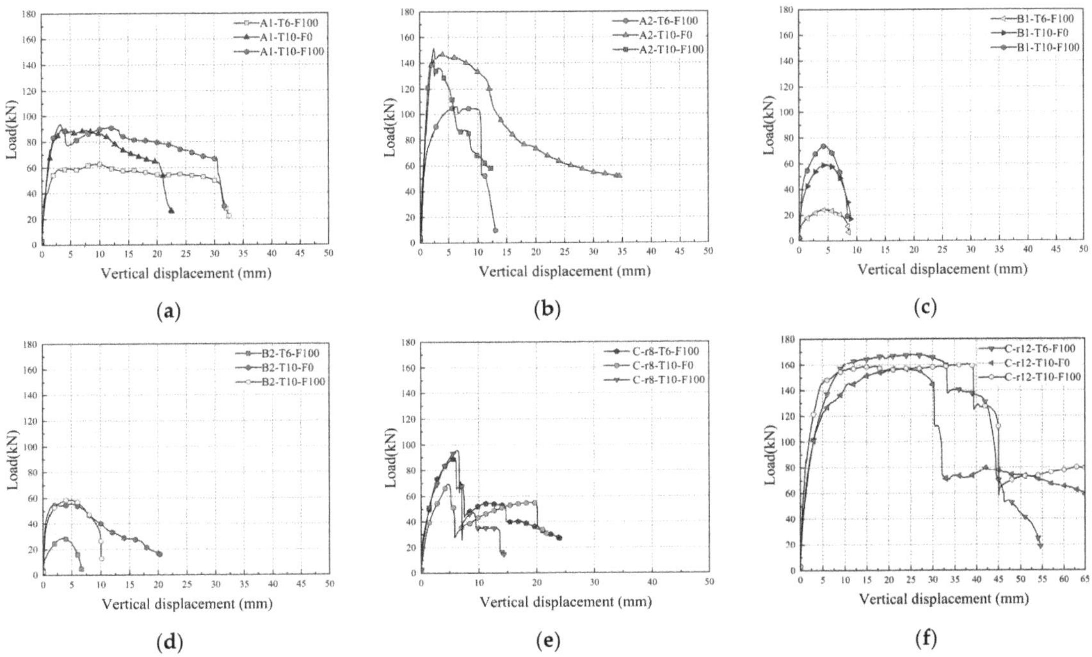

Figure 9. Load–displacement curves of tested specimens. (**a**) Plates with one hole (A1), (**b**) plates with two holes (A2), (**c**) plates with B1-type notches, (**d**) plates with B2-type notches, (**e**) plates with 8 mm diameter bars, (**f**) plates with 12 mm diameter bars.

For connecting plates with holes, the anchorage strength increases by about 50% if the thickness is increased from 6 mm to 10 mm (Figure 9a). Although the total void areas are the same, plates with one Φ50 hole have a lower anchorage strength than those with two Φ35 holes (Figure 9b), which is due to the larger bearing area of the latter. After reaching the peak load, plates with one hole (A1) can sustain the applied load without an obvious load drop, indicating that this configuration (with one larger hole) has better ductility than the A2 configuration (with two smaller holes). The reason is that the bearing deformation in the A1 plate is more significant than that in the A2 plate and that the bearing failure is usually more ductile than the shear failure of the concrete cylinder within the hole. For the plates with holes, the bearing area is proportional to the hole diameter whereas the shear area is proportional to the square of hole diameter. By decreasing the diameter (e.g., from 50 mm to 35 mm), the failure mode may change from bearing-dominated to shear-dominated, and the latter failure will be more brittle. Effects of concrete type (NAC vs. RAC) on the anchorage strength and curve shape are not obvious, except that specimen A2-T10-F100 shows a faster load drop after the peak load than its natural aggregate counterpart does. In addition, the displacements corresponding to the anchorage strength are about 2~4 mm, confirming the experimental observation that the deformation of CFST was not obvious.

Figure 9c,d show the load–displacement curves of the plates with notch dimensions (width × height) of 20 mm × 40 mm (B1) and 40 mm × 20 mm (B2), respectively. As expected, concrete type does not influence the pullout behavior obviously. Nevertheless, the plate thickness greatly affects the anchorage strength as the bearing area is proportional to the thickness. In general, effects of the notch dimensions on the peak load are not obvious whereas the B2 group shows a slightly slower load drop than that of the B1 group.

For plates with rebars (Figure 9e,f), the effects of plate thickness and concrete type on the pullout behavior are insignificant. Nevertheless, the anchorage strength and ductility are greatly enhanced by an increase in the rebar diameter from 8 mm to 12 mm, which is caused by the increase in the bearing area in concrete. After a dramatic load drop, a strength re-gain is observed for the plates with Φ6 rebars (Figure 9e). In general, specimens with Φ12 rebars exhibit the best performance among the tested specimens.

3.3. Key Experimental Results

Based on the load–displacement curves, key experimental results, including the yield capacity (P_y), ultimate capacity (P_u), displacement at P_y (Δ_y), displacement at P_u (Δ_u), and ductility index (DI) could be determined. The P_u, also called anchorage strength in this study, is the peak load of the load–displacement curve. As there is no obvious yielding plateau of the load–displacement curves, the method in [43] was adopted to specify the P_y and Δ_y. As illustrated in Figure 10, the procedures for determining P_y are as follow: (1) draw the line OA, the slope of which is the initial stiffness; (2) draw a vertical line at A to obtain intersection B; (3) draw the line OB and find intersection C; (4) draw a vertical line at C to obtain the yield point D (Δ_y, P_y). In this paper, the ductility index is defined as the ratio of the area covered by the load–displacement curve of OBDEF to OBD (Figure 10), in which the load at F is equal to $0.8P_u$. All the values of P_u, P_y, Δ_u, Δ_y, and DI are summarized in Table 4 and will be used for the discussions in Section 4.

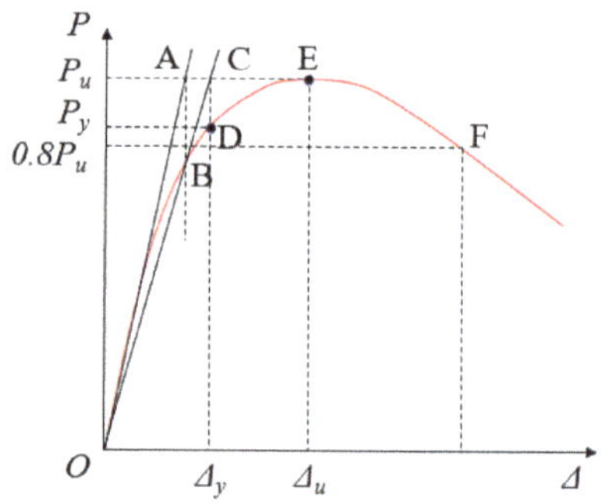

Figure 10. Determination of the yield point and the ductility index.

Table 4. Experimental results and predictions.

Specimen	P_y (kN)	Δ_y (mm)	P_u (kN)	Δ_u (mm)	DI	N_p (kN)	N_p/P_u
A1-T6-F100	36.4	1.14	62.6	9.3	82.5	60.6	1.03
A1-T10-F100	54.5	1.57	93.6	3.3	45.2	101.0	1.08
A1-T10-F0	70.4	1.51	89.8	3.4	22.3	103.2	1.15
A2-T6-F100	62.3	1.56	105.9	6.4	20.1	84.8	0.80
A2-T10-F100	102.4	1.53	142.7	2.5	8.0	141.4	0.99
A2-T10-F0	109.8	2.01	151.5	2.5	14.0	144.4	0.95
B1-T6-F100	11.7	0.67	24.4	4.4	38.8	28.0	1.15
B1-T10-F100	48.0	1.29	73.8	4.4	13.0	60.3	0.82
B1-T10-F0	31.6	1.15	59.2	4.6	19.5	61.6	1.04
B2-T6-F100	12.6	0.76	28.4	3.6	25.8	28.0	0.99
B2-T10-F100	37.2	0.85	58.7	4.3	26.4	60.3	1.03
B2-T10-F0	36.3	0.55	55.7	4.8	43.9	61.6	1.11
C-r8-T6-F100	58.0	1.80	89.4	5.9	7.9	71.9	0.80
C-r8-T10-F100	63.6	2.27	95.6	6.3	6.4	71.9	0.75
C-r8-T10-F0	26.9	2.02	69.4	4.8	9.4	72.0	1.04
C-r12-T6-F100	110.1	3.58	168.0	25.2	31.1	160.0	0.95
C-r12-T10-F100	110.8	3.64	160.4	38.0	30.1	160.0	1.00
C-r12-T10-F0	107.6	5.10	157.0	21.8	15.2	160.1	1.02

3.4. Strain Distributions of CFST

The strain distributions on the CFST's face wall of the specimens with RAC and a plate thickness of 10 mm are shown in Figure 11, which includes the strain data corresponding to the loads of $0.2P_u$, $0.4P_u$, $0.6P_u$, and $0.8P_u$ before reaching the peak load. The layout of the strain gauges is illustrated in Figure 7. The strains in the corner (S2 and S6) are positive (in tension), whereas the strains in other locations (S1, S3~S5, and S7) are negative (in compression). The strain data were mainly used for assessing the deformation of the CFST and verifying the finite element models that will be constructed in a future study.

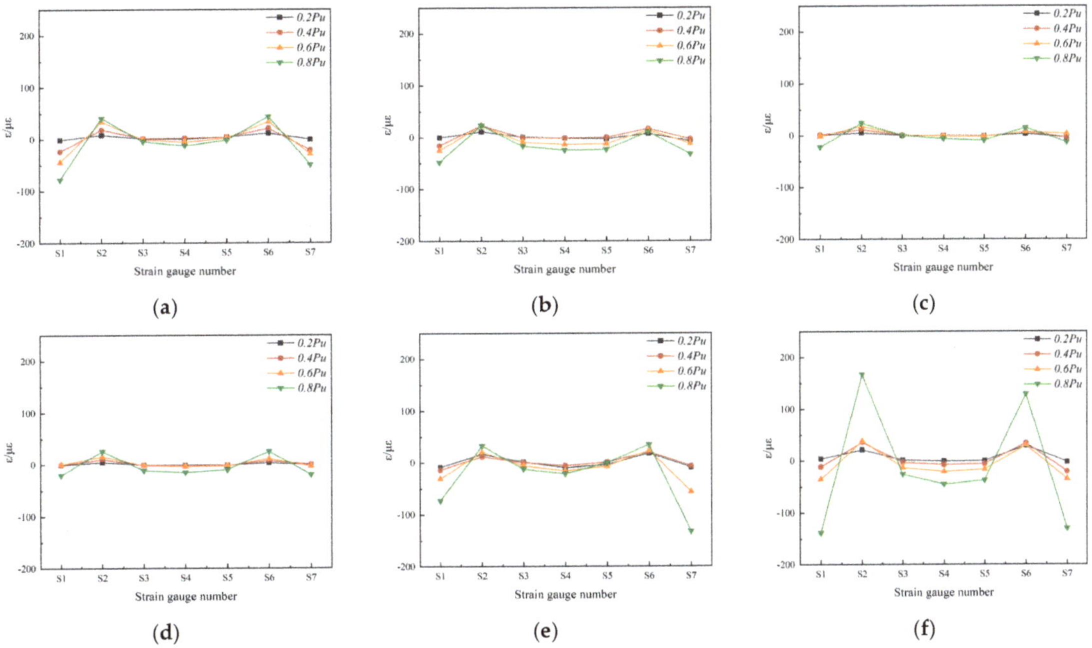

Figure 11. Strain distributions. (**a**) A1-T10-F100, (**b**) A2-T10-F100, (**c**) B1-T10-F100, (**d**) B2-T10-F100, (**e**) C-r8-T10-F100, and (**f**) C-r12-T10-F100.

As shown in Figure 11, the absolute strain values are quite small, and are less than 50 με, except for specimen C-r12-T10-F100, indicating that the deformation of the steel tube was negligible. It can be concluded that the deformation of the connecting plate and the CFST can be ignored during service load. Furthermore, in the determination of the initial stiffness of a beam-to-column joint carried out by using the well-known component method, the stiffness of CFST and connecting plates could be assumed as infinity. A higher value of the strain indicates larger deformation of the tube face wall, and this was induced by the expansion and movement of the in-filled concrete. Therefore, more concrete was involved in anchoring the connecting plates if the strains were larger. As shown in Figure 11, specimen C-r12-T10-F100 has the most concrete involved and would have the highest anchorage strength, which is in agreement with the experimental results in Table 4. Plates with holes and notches had a lower anchorage strength as only a small portion of concrete could participate in resisting the applied load. In addition, the strain in the middle (SG4) was larger than the strains in the locations near the side walls (i.e., SG3 and SG5). This phenomenon could be explained by simplifying the tube face wall as a fixed supported beam (i.e., larger strain at mid-span).

4. Discussions

4.1. Effect of Anchoring Type and Configurations

The effects of the key parameters, such as concrete type, anchoring type, anchoring configurations (e.g., size and number of holes, direction of notches, and diameter of rebars), and plate thickness on the pullout behavior of the connecting plate are discussed in this section. The pullout behavior mainly includes the anchorage strength, load–displacement response, and ductility index, which were defined in Figure 10.

As shown in Table 4 and Figure 9, the anchoring type (hole, notch or rebar) has a significant effect on the anchorage strength, ductility index, and pullout load–displacement curves, owing to the different anchorage mechanisms. The plates with rebars or holes had much higher anchorage strength than plates with notches did. For plates with similar total void areas, the configuration of smaller holes led to a higher anchorage strength (by about 70%) but lower ductility. With the same notch area, the effect of notch direction (see Figure 3b) on the anchorage strength was not obvious. The plate with wider notches exhibited slightly better ductility (i.e., a slower load drop and a higher ductility index) than that with narrow notches. By increasing the diameter of the rebars, the anchorage strength and ductility were greatly improved. The connecting plates with Φ12 rebars performed the best among the tested specimens. Therefore, the anchoring type of using a rebar is recommended for the connecting plates in beam-to-column joints. Nevertheless, it is necessary to mention that the construction of this configuration is less convenient than that of connecting plates with holes or notches.

4.2. Effect of Plate Thickness

As shown in Table 4 and Figure 12, the anchorage strength of the 10 mm thick connecting plates with holes is 35~50% larger than that of the 6 mm thick plates, whereas the enhancement is 100~170% for the plates with notches. Plate thickness has a significant impact on anchorage strength as the anchorage strength is mainly contributed by the bearing effect, the bearing area of which depends on the plate thickness. Nevertheless, the ductility index shows a decreasing trend (about 50%) with the increase in plate thickness. For plates with rebars, the effects of thickness on the pullout behavior are not obvious. The reason is that the applied load is mainly resisted by the bearing of the rebars, which is not related to the plate thickness.

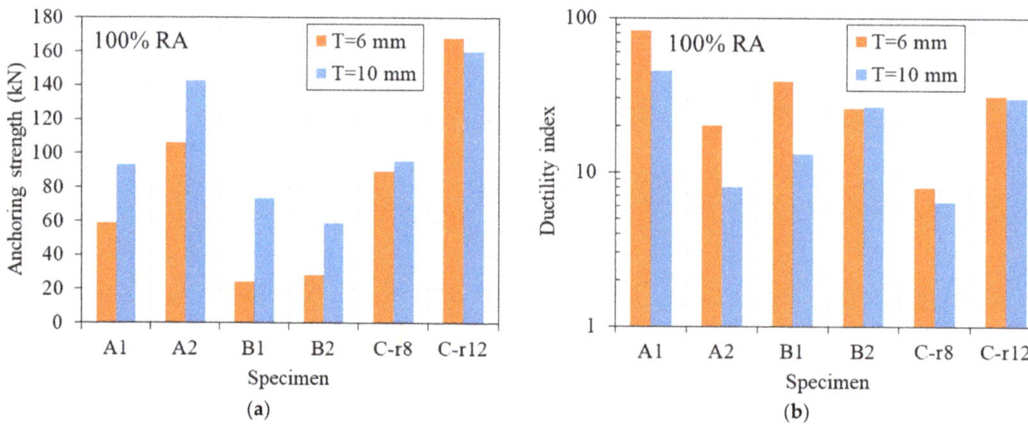

Figure 12. Effect of plate thickness on the pullout behavior. (**a**) Anchorage strength and (**b**) ductility index.

4.3. Effect of Concrete Type

One of the aims of this study was to investigate whether or not recycled aggregate has an obvious effect on the anchoring behavior of plates. As reported in Section 2.2, with the same mix proportions, RAC has a similar compressive strength but slightly lower Young's modulus compared with NAC, probably because of the lower modulus of recycled aggregates than that of natural aggregates. As shown in Figures 9 and 13, the effects of the RA content on the anchorage strengths, and load–displacement curves are insignificant. NAC specimens showed slightly higher ductility than RAC specimens did, as shown in Figure 13b, likely due to the higher modulus of NAC. Nevertheless, this observation is only valid for four of the six groups (i.e., A2, B1, B2, and C-r6) and more research is needed. Therefore, it could be concluded that RAC has similar anchoring behavior to that of NAC, provided that the compressive strengths are similar.

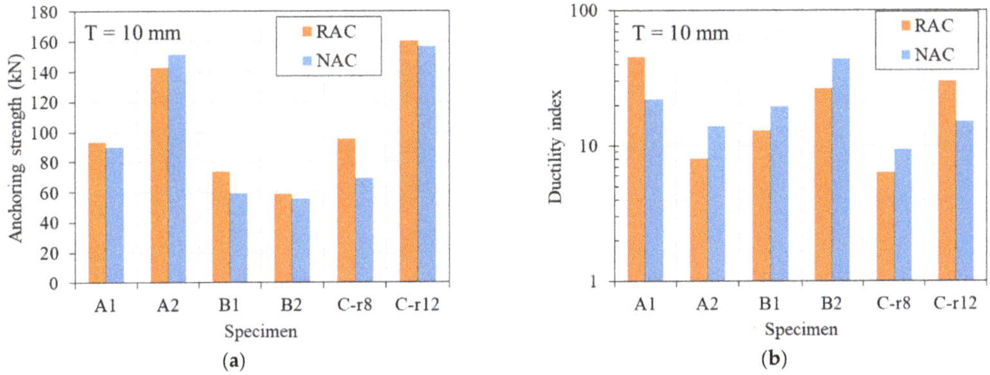

Figure 13. Effect of concrete type on the pullout behavior. (**a**) Anchorage strength and (**b**) ductility index.

5. Prediction for the Anchorage Strength of the Connecting Plate

5.1. Connecting Plate with Holes

The possible failure modes of the connecting plates with holes are the bearing failure of the hole, the shear failure of the concrete cylinder within the hole, and the bearing failure of the concrete within the hole. The anchorage strength (N_p) is the lower strength of the bearing strength of steel (N_{bs}), shear strength of concrete (N_{sc}), and bearing strength of concrete (N_{bc}).

$$N_p = \min(N_{bs}, N_{sc}, N_{bc}) \quad (1)$$

Based on the design methods for bolted connections, Equation (2) from GB50017 [44] was adopted to predict the bearing strength of the steel plate:

$$N_{bs} = nd_h t f_{bs} \quad (2)$$

where N_{bs} is the bearing strength of steel plate, d_h is the diameter of the hole, n is the number of the holes, t is the thickness of the plate, and f_{bs} is the bearing strength of the steel plate, taking this as 405 MPa for Q235 steel [39] or considering it equal to $0.67(f_y + f_u)$, as suggested by the HK code [45], where f_y and f_u are the yield strength and ultimate strength, respectively.

The current design method for determining the shear strength of a perfobond rib connector in JTG [46] is adopted to calculate the shear strength of concrete within the holes (N_{sc}):

$$N_{sc} = 1.4d_h^2 f_c' \quad (3)$$

where f_c' is the compressive strength of concrete.

As indicated by the experimental observations, the failure mode of the tested specimens is the localized bearing failure of the concrete within the hole. Equation (4) is proposed to estimate the bearing strength of the concrete:

$$N_{bc} = \eta nd_h t f_c' \quad (4)$$

where f_c' is the compressive strength of concrete, and η is an empirical coefficient accounting for the bearing strength enhancement of the localized concrete induced by confinement. The value for η is obtained from the regression analysis of the experimental data, and it is recommended to be 4.1. Therefore, the anchorage strength could then be determined via Equation (1).

5.2. Connecting Plate with Notches

By analyzing the bearing stress in the notch, Equation (5) is proposed to predict the anchorage strength of the plate with notches:

$$N_p = \chi n b_n t f_c' \tag{5}$$

where n is the number of notches, b_n is the width of the notch, t is the plate thickness, f_c' is the concrete compressive strength, and χ is an empirical coefficient that considers the effect of notch size on anchorage strength.

Based on the experimental results, there are several important findings regarding the anchorage strength and failure modes: (1) the anchorage strength is not proportional to the thickness of the plate; (2) Although the bearing areas are the same, the anchorage strength differs somewhat for different notch widths (b_n); (3) the failure of the connecting plate can be regarded as a combination of the bearing failure of concrete and the bending failure of the cantilever deep beam with a cross-section of h_n by t_n and a length of b_n (Figure 14). Therefore, χ is assumed to be related to the size of the notch, and governs the bending and shear behavior of the notch. Via regression analysis, as shown in Figure 15, Equation (6) is proposed to estimate the value of χ:

$$\chi = \frac{5}{\sqrt{b_n^2/(h_n t)}} \tag{6}$$

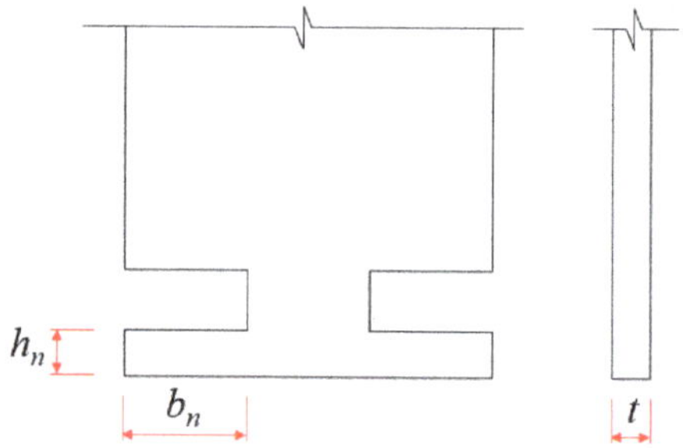

Figure 14. Dimensions of the notch.

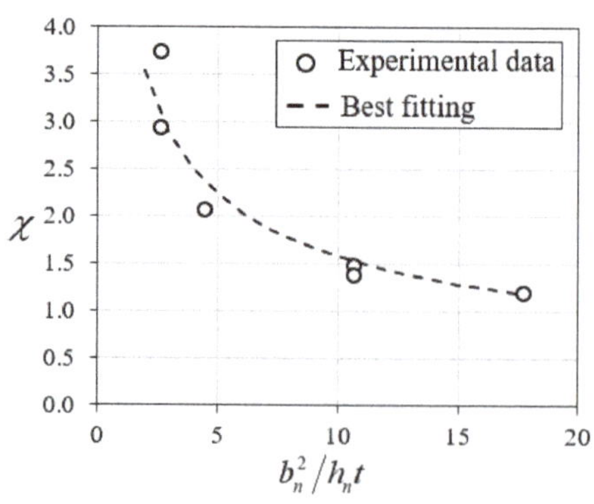

Figure 15. Determination of the empirical coefficient, χ, in Equation (5).

5.3. Connecting Plate with Rebars

The experimental results indicated that the anchoring mechanism of the plate with rebars is similar to that of perfobond rib connectors in resisting shear load. The anchorage strength is proposed to be estimated via the existing design method for perfobond rib connectors in JTG [46]:

$$N_p = 1.4n(d_h^2 - d_s^2)f_c' + 1.2nd_s^2 f_y \tag{7}$$

where n is the number of rebars, d_h is the diameter of the hole, d_s is the diameter of the rebar, $f_c{}'$ is the compressive strength of the concrete, and f_y is the yield strength of the rebar.

5.4. Verifications

A comparison of the experimental and predicted anchorage strengths of the tested specimens is summarized in Table 4 and Figure 16. The average experimental-to-predicted ratios are 1.00, 1.02, and 0.93, whereas the COVs (coefficient of variations) are 0.11, 0.10, and 0.12, for the connecting plates with holes, notches, and rebars, respectively. Therefore, the proposed methods could reasonably estimate the anchorage strength. Nevertheless, it is necessary to mention that the coefficients in Equations (4) and (6) are determined by regressing the experimental data of this study. If more data are available in the future, those formulas could be further refined. The prediction of JTG [46] on the anchorage strength of the steel plate with rebars is conservative. The main reason would be that the concrete in this study is confined by the square steel tube but the confinement effect does not exist in perfobond rib connectors.

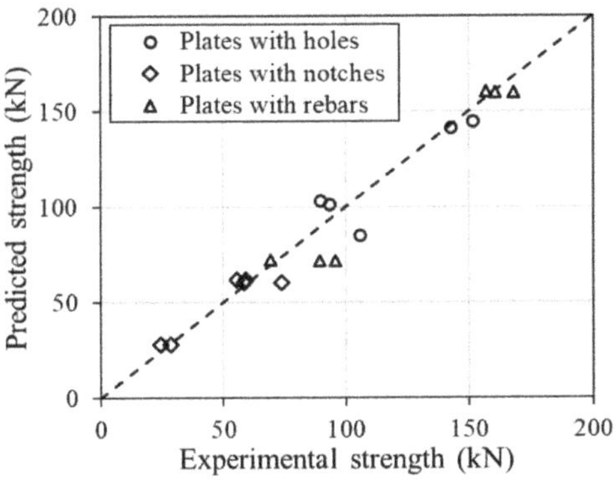

Figure 16. Comparison of the experimental and predicted anchorage strengths.

6. Conclusions

This paper presents an experimental and theoretical study on the pullout behavior of connecting plates embedded in recycled/natural aggregate concrete-filled square tubes (CFSTs), which could simulate the behavior of the tension zone in a novel full-bolted steel beam-to-CFST joint. Effects of the recycled aggregate replacement, anchoring type, and plate thickness on the pullout behavior (e.g., anchorage strength, load–displacement curves, and ductility) are discussed. Theoretical models are then proposed to estimate the anchorage strength of the connecting plates. The following conclusions could be drawn:

(1) Mechanical properties of recycled aggregate concrete are similar to those of natural aggregate concrete with the same mixtures. Using recycled aggregate for a CFST does not obviously affect the pullout behavior of plates, demonstrating the feasibility of using RAC for composite structures.

(2) The failure modes of the specimens greatly depend on the anchoring types and configurations. Plates with rebars exhibit higher strength and ductility than the plates with holes or notches, so it is a recommended anchoring method for beam-to-CFST joints.

(3) With the same void area, plates with two small holes have higher strength than plates with one large hole do, likely due to the increase in bearing area. The effect of the notch dimension (20 mm × 40 mm vs. 40 mm × 20 mm, same area but different direction) on the anchorage strength is not significant, whereas the ductility is slightly influenced. The diameter of the rebars greatly affects the anchorage strength and ductility of the connecting plates.
(4) Because the bearing area is proportional to the plate thickness, thickness has a significant influence on the pullout behavior of the connecting plates with holes and notches. However, the behavior of the plates with rebars is not affected by the plate thickness.
(5) Deformations of the connecting plate and CFST are negligible, and their stiffness could be assumed to be infinity in the determination of the initial stiffness of a beam-to-column joint using the component method.
(6) The proposed formulas could reasonably predict the anchorage strength of the connecting plates with holes and notches. The current method for perfobond rib connectors somewhat underestimates the anchorage strength of the connecting plates with rebars, likely due to the lack of consideration of the confining effect in CFSTs, which is a major difference compared with that in perfobond rib connectors.

Author Contributions: Conceptualization, J.S. and Q.Z.; methodology, J.S.; software, J.Z.; validation, X.L. and D.L.; investigation, L.C., H.P. and W.D.; resources, Q.Z. and D.L.; writing—original draft, L.C.; writing—review and editing, X.L.; visualization, H.P., W.D. and J.Z.; software, J.Z.; Supervision, F.L.; project administration, F.L.; funding acquisition, J.S. and F.L. All authors have read and agreed to the published version of the manuscript.

Funding: The authors gratefully acknowledge the financial support provided by the Special Foundation for Assembly Construction of Guangzhou Construction Engineering Co., Ltd. ([2019]-KJ002), the National Natural Science Foundation of China (12072080).

Data Availability Statement: The original contributions presented in the study are included in the article, further inquiries can be directed to the corresponding authors.

Conflicts of Interest: Authors Jianhua Su and Qian Zhao were employed by the company Guangzhou Research Institute of Construction Industry Co., Ltd. Authors Jianhua Su and Qian Zhao were employed by the company Guangzhou Construction Engineering Co., Ltd. Author Deng Lu was employed by the company China Construction Third Engineering Bureau Group Co., Ltd. The remaining authors declare that the research was conducted in the absence of any commercial or financial relationships that could be construed as a potential conflict of interest.

References

1. Lai, Y.-Y.; Yeh, L.-H.; Chen, P.-F.; Sung, P.-H.; Lee, Y.-M. Management and Recycling of Construction Waste in Taiwan. *Procedia Environ. Sci.* **2016**, *35*, 723–730. [CrossRef]
2. Zhen, H.; Xiong, Z.; Song, Y.; Li, L.; Qiu, Y.; Zou, X.; Chen, B.; Chen, D.; Liu, F.; Ji, Y. Early mechanical performance of glass fibre-reinforced manufactured sand concrete. *J. Build. Eng.* **2024**, *83*, 108440. [CrossRef]
3. Zhu, H.; Xiong, Z.; Song, Y.; Zhou, K.; Su, Y. Effect of Expansion Agent and Glass Fiber on the Dynamic Splitting Tensile Properties of Seawater–Sea-Sand Concrete. *Buildings* **2024**, *14*, 217. [CrossRef]
4. Lin, J.-X.; Luo, R.-H.; Su, J.-Y.; Guo, Y.-C.; Chen, W.-S. Coarse synthetic fibers (PP and POM) as a replacement to steel fibers in UHPC: Tensile behavior, environmental and economic assessment. *Constr. Build. Mater.* **2024**, *412*, 134654. [CrossRef]
5. Chen, Z.; Yu, J.; Nong, Y.; Yang, Y.; Zhang, H.; Tang, Y. Beyond time: Enhancing corrosion resistance of geopolymer concrete and BFRP bars in seawater. *Compos. Struct.* **2023**, *322*, 117439. [CrossRef]
6. Pan, Z.; Liu, F.; Li, H.; Li, X.; Wang, D.; Ling, Z.; Zhu, H.; Zhu, Y. Performance Evaluation of Thermal Insulation Rubberized Mortar Modified by Fly Ash and Glass Fiber. *Buildings* **2024**, *14*, 221. [CrossRef]
7. Xiong, Z.; Mai, G.; Pan, Z.; Chen, Z.; Jian, J.; Wang, D.; Ling, Z.; Li, L. Synergistic effect of expansive agents and glass fibres on fatigue bending performance of seawater sea sand concrete. *Constr. Build. Mater.* **2024**, *421*, 135665. [CrossRef]
8. Ding, T.; Xiao, J.; Tam, V.W.Y. A closed-loop life cycle assessment of recycled aggregate concrete utilization in China. *Waste Manag.* **2016**, *56*, 367–375. [CrossRef] [PubMed]
9. Chen, X.; Xiao, X.; Wu, Q.; Cheng, Z.; Xu, X.; Cheng, S.; Zhao, R. Effect of magnesium phosphate cement on the mechanical properties and microstructure of recycled aggregate and recycled aggregate concrete. *J. Build. Eng.* **2021**, *46*, 103611. [CrossRef]

10. Superti, V.; Houmani, C.; Hansmann, R.; Baur, I.; Binder, C.R. Strategies for a Circular Economy in the Construction and Demolition Sector: Identifying the Factors Affecting the Recommendation of Recycled Concrete. *Sustainability* **2021**, *13*, 4113. [CrossRef]
11. Zhang, X.; Kuang, X.; Wang, F.; Wang, S. Strength indices and conversion relations for basalt fiber-reinforced recycled aggregate concrete. *Dyna-Bilbao* **2019**, *94*, 82–87. [CrossRef]
12. Kurda, R.; de Brito, J.; Silvestre, J.D. Combined influence of recycled concrete aggregates and high contents of fly ash on concrete properties. *Constr. Build. Mater.* **2017**, *157*, 554–572. [CrossRef]
13. Papavasileiou, G.S.; Charmpis, D.C. Earthquake-resistant buildings with steel or composite columns: Comparative assessment using structural optimization. *J. Build. Eng.* **2020**, *27*, 100988. [CrossRef]
14. Ma, H.; Liu, F.; Wu, Y.; Xin, X.; Zhao, Y. Axial compression tests and numerical simulation of steel reinforced recycled concrete short columns confined by carbon fiber reinforced plastics strips. *Front. Struct. Civ. Eng.* **2022**, *16*, 817–842. [CrossRef]
15. Duan, Z.; Li, B.; Xiao, J.; Guo, W. Optimizing mix proportion of recycled aggregate concrete by readjusting the aggregate gradation. *Struct. Concr.* **2020**, *22*, E22–E32. [CrossRef]
16. Liang, C.; Liu, T.; Xiao, J.; Zou, D.; Yang, Q. Effect of Stress Amplitude on the Damping of Recycled Aggregate Concrete. *Materials* **2015**, *8*, 5298–5312. [CrossRef]
17. Wang, Y.; Deng, Z.; Xiao, J.; Li, T.; Li, J. Mechanical properties of recycled aggregate concrete under compression-shear stress state. *Constr. Build. Mater.* **2021**, *271*, 121894. [CrossRef]
18. Xiao, J.; Zhang, K.; Akbarnezhad, A. Variability of stress-strain relationship for recycled aggregate concrete under uniaxial compression loading. *J. Clean. Prod.* **2018**, *181*, 753–771. [CrossRef]
19. Tam, V.W.Y.; Xiao, J.; Liu, S.; Chen, Z. Behaviors of recycled aggregate concrete-filled steel tubular columns under eccentric loadings. *Front. Struct. Civ. Eng.* **2018**, *13*, 628–639. [CrossRef]
20. Yang, Y.-F.; Han, L.-H.; Zhu, L.-T. Experimental Performance of Recycled Aggregate Concrete-Filled Circular Steel Tubular Columns Subjected to Cyclic Flexural Loadings. *Adv. Struct. Eng.* **2009**, *12*, 183–194. [CrossRef]
21. Ma, H.; Xue, J.; Zhang, X.; Luo, D. Seismic performance of steel-reinforced recycled concrete columns under low cyclic loads. *Constr. Build. Mater.* **2013**, *48*, 229–237. [CrossRef]
22. Ma, H.; Xue, J.; Liu, Y.; Dong, J. Numerical analysis and horizontal bearing capacity of steel reinforced recycled concrete columns. *Steel Compos. Struct.* **2016**, *22*, 797–820. [CrossRef]
23. Ma, H.; Qiang, J.; Xi, J.; Zhao, Y. Cyclic loading tests and horizontal bearing capacity of recycled concrete filled circular steel tube and profile steel composite columns. *J. Constr. Steel Res.* **2022**, *199*, 107572. [CrossRef]
24. Zhang, W.-H.; Wang, R.; Zhao, H.; Lam, D.; Chen, P. Axial-load response of CFST stub columns with external stainless steel and recycled aggregate concrete: Testing, mechanism analysis and design. *Eng. Struct.* **2022**, *256*, 113968. [CrossRef]
25. Lyu, W.-Q.; Han, L.-H. Investigation on bond strength between recycled aggregate concrete (RAC) and steel tube in RAC-filled steel tubes. *J. Constr. Steel Res.* **2019**, *155*, 438–459. [CrossRef]
26. Zhao, H.; Zhang, W.-H.; Wang, R.; Hou, C.-C.; Lam, D. Axial compression behaviour of round-ended recycled aggregate concrete-filled steel tube stub columns (RE-RACFST): Experiment, numerical modeling and design. *Eng. Struct.* **2023**, *276*, 115376. [CrossRef]
27. Choi, S.-M.; Park, S.-H.; Yun, Y.-S.; Kim, J.-H. A study on the seismic performance of concrete-filled square steel tube column-to-beam connections reinforced with asymmetric lower diaphragms. *J. Constr. Steel Res.* **2010**, *66*, 962–970. [CrossRef]
28. Qin, Y.; Chen, Z.; Wang, X. Experimental investigation of new internal-diaphragm connections to CFT columns under cyclic loading. *J. Constr. Steel Res.* **2014**, *98*, 35–44. [CrossRef]
29. Qin, Y.; Chen, Z.; Wang, X. Elastoplastic behavior of through-diaphragm connections to concrete-filled rectangular steel tubular columns. *J. Constr. Steel Res.* **2014**, *93*, 88–96. [CrossRef]
30. Yang, J.; Sheehan, T.; Dai, X.; Lam, D. Experimental study of beam to concrete-filled elliptical steel tubular column connections. *Thin-Walled Struct.* **2015**, *95*, 16–23. [CrossRef]
31. Vianna, J.d.C.; de Andrade, S.; Vellasco, P.d.S.; Costa-Neves, L. Experimental study of Perfobond shear connectors in composite construction. *J. Constr. Steel Res.* **2013**, *81*, 62–75. [CrossRef]
32. Fang, Z.; Wu, J.; Xian, B.; Zhao, G.; Fang, S.; Jiang, H.; Ma, Y. Shear Performance and Design Recommendations of Single Embedded Nut Bolted Shear Connectors in Prefabricated Steel-UHPC Composite Beams. *Steel Compos. Struct.* **2024**, *50*, 319–336.
33. Fang, Z.; Wu, J.; Xu, X.; Ma, Y.; Fang, S.; Zhao, G.; Jiang, H. Grouped rubber-sleeved studs–UHPC pocket connections in prefabricated steel–UHPC composite beams: Shear performance under monotonic and cyclic loadings. *Eng. Struct.* **2024**, *305*, 117781. [CrossRef]
34. Tizani, W.; Cabrera, M.; Mahmood, M.; Ninic, J.; Wang, F. The behaviour of anchored extended blind bolts in concrete-filled tubes. *Steel Constr.* **2022**, *15*, 51–58. [CrossRef]
35. Ng, W.H.; Kong, S.Y.; Chua, Y.S.; Bai, Y. Tensile behaviour of innovative one-sided bolts in concrete-filled steel tubular connections. *J. Constr. Steel Res.* **2022**, *191*, 107165. [CrossRef]
36. Yao, H.; Goldsworthy, H.; Gad, E. Experimental and numerical investigation of the tensile behavior of blind-bolted T-stub connections to concrete-filled circular columns. *J. Struct. Eng.* **2008**, *134*, 198–208. [CrossRef]
37. Li, Y.; Chan, T.; Zhao, X. Review on blind bolted connections to concrete-filled steel tubes. *Thin-Walled Struct.* **2023**, *183*, 110444. [CrossRef]

38. *GB 175*; Common Portland Cement. China Standards Press: Beijing, China, 2020. (In Chinese)
39. *GB/T25177*; Recycled Coarse Aggregate for Concrete. China Construction Industry Press: Beijing, China, 2010. (In Chinese)
40. Oktavianus, Y.; Chang, H.; Goldsworthy, H.M.; Gad, E.F. Component model for pull-out behaviour of headed anchored blind bolt within concrete filled circular hollow section. *Eng. Struct.* **2017**, *148*, 210–224. [CrossRef]
41. Su, Q.; Yang, G.; Bradford, M.A. Bearing Capacity of Perfobond Rib Shear Connectors in Composite Girder Bridges. *J. Bridg. Eng.* **2016**, *21*, 06015009. [CrossRef]
42. He, S.; Fang, Z.; Fang, Y.; Liu, M.; Liu, L.; Mosallam, A.S. Experimental study on perfobond strip connector in steel–concrete joints of hybrid bridges. *J. Constr. Steel Res.* **2016**, *118*, 169–179. [CrossRef]
43. Guo, Z.; Shi, X. *Theory and Analysis Ofreinforced Concrete*; Tsinghua University Press: Beijing, China, 2003; pp. 163–337. (In Chinese)
44. *GB 50017*; Code for Design of Steel Structures. China Construction Industry Press: Beijing, China, 2017. (In Chinese)
45. *Code of Practice for the Structural Use of Steel 2011*; Buildings Department: Hong Kong, China, 2023.
46. *JTG D64*; Specifications for design of highway steel bridge. China Communications Press: Beijing, China, 2015. (In Chinese)

buildings

MDPI

Article

Bond–Slip Performance of Steel–Fiber-Reinforced Polymer Composite Bars (SFCBs) and Glass Fiber with Expansion-Agent-Reinforced Seawater Sea-Sand Concrete (GF-EA-SSSC) under Freezing–Thawing Environment

Yufu Li [1], Jiayu Jian [2], Yuying Song [2], Wei Wei [1,2,*], Yilin Zhang [2], Gangliang Li [3], Huanyu Zhu [2], Jiawei Lin [2] and Zhe Xiong [2]

1 School of Civil Engineering, Guangdong Construction Polytechnic, Guangzhou 510440, China; liyufu@gdcvi.edu.cn
2 School of Civil and Transportation Engineering, Guangdong University of Technology, Guangzhou 510006, China; 2112109107@mail2.gdut.edu.cn (J.J.); 1112309005@mail2.gdut.edu.cn (Y.S.); 2112109063@mail2.gdut.edu.cn (Y.Z.); 3121002912@mail2.gdut.edu.cn (H.Z.); 3120002981@mail2.gdut.edu.cn (J.L.); gdgyxz263@gdut.edu.cn (Z.X.)
3 Guangdong Yonghe Construction Group Co., Ltd., Guangzhou 510330, China; ligl@gdyhjs.cn
* Correspondence: weiwei@gdcvi.edu.cn

Abstract: The combined application of steel–FRP composite bars (SFCBs) and seawater sea-sand concrete (SSSC) in marine engineering not only solves the problem of resource scarcity and reduces the construction cost but also avoids the problems of chloride corrosion of steel reinforcement in seawater sea-sand concrete and the lack of ductility of FRP bars. At the same time, the addition of glass fiber (GF) and expansion agent (EA) in appropriate amounts improves the crack resistance and seepage resistance of concrete. However, the durability of SFCB with GF- and EA-reinforced SSSC in freezing–thawing environment remains unclear, which limits its potential application in cryogenic marine engineering. This study investigates the bonding properties between SFCB and GF-EA-SSSC interfaces using eccentric pullout experiments under different thicknesses of concrete protective cover and a number of freezing–thawing cycles. The results showed that the compressive strength and dynamic elastic modulus of SSSC decrease, while the mass loss increases with an increasing number of freezing–thawing cycles. Additionally, the bond strength and stiffness between SFCB and SSSC decrease, leading to an increase in relative slip. However, the rate of bond strength and stiffness loss decreases with an increase in the thickness of the concrete protective cover. Furthermore, formulas for bond strength, relative slip, and bond stiffness are established to quantify the effects of the thickness of the concrete protective cover and the number of freezing–thawing cycles. The experimental values obtained verify the accuracy of these formulas, with a relative error of less than 5%. Moreover, a bond stress–slip constitutive model is developed for SFCB and GF-EA-SSSC, and the fitting results closely resemble the experimental values, demonstrating a high level of model fit.

Keywords: seawater sea-sand concrete; glass fiber; expansion agent; steel–FRP composite bars; freezing–thawing environment; bond–slip performance

Citation: Li, Y.; Jian, J.; Song, Y.; Wei, W.; Zhang, Y.; Li, G.; Zhu, H.; Lin, J.; Xiong, Z. Bond–Slip Performance of Steel–Fiber-Reinforced Polymer Composite Bars (SFCBs) and Glass Fiber with Expansion-Agent-Reinforced Seawater Sea-Sand Concrete (GF-EA-SSSC) under Freezing–Thawing Environment. *Buildings* **2024**, *14*, 1121. https://doi.org/10.3390/buildings14041121

Academic Editor: Dan Bompa

Received: 6 March 2024
Revised: 8 April 2024
Accepted: 15 April 2024
Published: 17 April 2024

1. Introduction

In recent years, the rapid development of infrastructure construction in coastal cities and island areas, coupled with the acceleration of urbanization, has led to an increasing demand for concrete [1,2]. However, the production process of concrete requires the consumption of a significant amount of natural resources, such as fresh water and river sand [3–5]. The excessive exploitation and use of these resources have resulted in a severe shortage and significant damage to the natural environment. Moreover, the transportation cost of traditional building materials is higher in construction projects located in coastal and island areas. In response to these challenges, researchers have proposed the use of seawater and sea sand as alternatives to freshwater and river sand in the production of concrete structures [6–8]. Sea sand has the advantages of large resource reserves and wide distribution, and the physical properties are similar to river sand, but the grading is slightly different. Existing research indicates that the mechanical properties of seawater sea-sand concrete (SSSC) are comparable to those of ordinary concrete [9,10]. Consequently, SSSC presents not only substantial economic benefits but also satisfactory working performance, making it highly promising for widespread application. However, the high concentration of chloride ions in seawater and sea sand introduces a significant challenge, as it accelerates the corrosion of steel reinforcement, leading to potentially severe safety hazards. Therefore, there are difficulties in popularizing the application of SSSC.

In order to promote the application of SSSC and address the issue of steel bar corrosion, researchers have proposed the utilization of fiber-reinforced polymer bar (FRP bar) as a replacement for steel bars, in combination with SSSC in building structures [11–15]. With advancements in technology and process refinement, the use of FRP bars has become more streamlined, cost-effective, and widespread in building and bridge structures. This has garnered increasing attention from scholars, who have begun to focus on the research regarding the application of FRP reinforcement [16–20]. FRP bars are used in concrete as reinforcing steel bars, which have the advantages of chloride ion corrosion resistance, high tensile strength, lightweight properties, and fatigue resistance [21–24]. Therefore, the adoption of FRP bars instead of traditional steel bars plays an important role in resolving the issue of chloride corrosion in seawater sea sand. However, FRP bars tend to have low stiffness in concrete structures, large structural deflection in the use phase, easy to brittle damage, as well as high price and other shortcomings, making it difficult to be widely promoted in engineering applications. To address these challenges, steel–FRP composite bars (SFCBs) are a new type of reinforced composite material with steel reinforcement as the core and fiber layer as the outer layer. SFCB combines the favorable characteristics of both steel bars and FRP bars, including a high modulus of elasticity, stable secondary stiffness, excellent corrosion resistance, and a lower price point [25,26]. The application of SFCB in SSSC offers a theoretical solution to the problems associated with steel reinforcement erosion caused by chloride ions and the lack of ductility observed in FRP bars.

The substitution of SFCB for steel addresses two key problems, steel corrosion and the lack of ductility in FRP bars. A large number of previous studies have demonstrated the favorable bonding properties between FRP bars and concrete, resulting in a synergistic effect that enhances the integrity and safety of FRP-reinforced concrete [27–29]. However, it is imperative to investigate whether SFCB exhibits similar bonding properties with concrete and creates a beneficial synergy. Moreover, building structures must not only perform well in comfortable environments but also withstand the impact of aggressive conditions such as freezing and thawing, high temperatures, seawater erosion, and wet–dry cycles. Scholars have already conducted relevant studies on the bonding performance between SFCB and concrete [25,30]. However, these studies have mostly focused on conventional environments, wet–dry cycles, and seawater immersion, with fewer investigations conducted on bond performance during freezing–thawing cycles, which present extreme conditions [31–33]. Freezing–thawing environments have diverse and multifaceted effects on building structures. They can induce spalling, expansion cracking of concrete, interface cracking between different materials, accelerated carbonation of concrete, and increased sulfate corrosion, all of which contribute to a decline in structural performance [34–36]. Therefore, ensuring the durability of the interfacial bond is crucial for maintaining the overall integrity of a building structure over a prolonged period under freezing–thawing conditions. Investigating the interfacial bonding properties between SFCB and SSSC under freezing–thawing cycles is imperative for the successful application of SFCB in such environments.

A substantial body of research has explored the durability of concrete in freezing–thawing environments, with scholars proposing various theories regarding the deterioration mechanism of concrete under such conditions [37–39]. Freezing–thawing damage is commonly divided into two stages. The initial stage involves the solidification and expansion of internal liquid water, leading to crack formation and characterized by a loss of dynamic modulus and concrete strength. The subsequent stage involves the disintegration of the cement paste body, resulting in the shedding of surface mortar. This stage is characterized by mass loss and partial or complete destruction of the concrete structure. Considering this two-stage perspective, improving the cracking and permeability resistance of concrete can enhance its ability to withstand freezing–thawing losses, consequently improving the durability of the bond between reinforcement and concrete in freezing–thawing environments. To enhance the frost resistance of concrete, glass fiber plays a crucial role by limiting the extension of vertical cracks and promoting the bonding of parallel cracks. This optimization of the internal structure of concrete improves its crack resistance [40]. The incorporation of an expansion agent in concrete contributes to the formation of expansion mineral crystals through a reaction with $Ca(OH)_2$, a product of cement hydration. This reaction slows down the cement hydration rate, generating expansion stress within the concrete. This stress helps resist part of the tensile stress resulting from drying contraction, inhibiting concrete deformation, and reducing cracks. Furthermore, an appropriate amount of expansion agent can enhance the seepage resistance of concrete. According to our previous studies [41,42], there is evidence of a synergistic effect when combining an appropriate amount of expansion agent with glass fiber. The concrete mixed with glass fiber and expansion agent can not only inhibit crack development but also improve the concrete compressive strength, flexural strength and toughness, and other aspects of performance [8,41–43]. Therefore, in this study, the incorporation of a suitable amount of glass fiber and expansion agent aims to enhance the crack resistance and seepage resistance of SSSC, ultimately improving its ability to resist freezing and thawing losses.

To ensure the frost resistance of building structures, it is crucial to examine the durability of the bond between reinforcement and concrete in freezing–thawing environments. Previous studies have focused on the bond durability between FRP bars and concrete, primarily through axial pullout tests [25,33]. However, these tests were conducted with a concrete cover thickness greater than 75 mm and a ratio of concrete cover thickness c to FRP reinforcement diameter d greater than 5, which does not align with the recommendation in ACI 440.1R-2015 stating that c/d should not exceed 3.5 [44]. Consequently, the use of axial pullout tests may result in specimens with an inconsistent concrete protective layer thickness, leading to an overestimation of the bond durability between stressed reinforcement and concrete under freezing–thawing conditions. To address this issue, an eccentric pullout test is employed in this study to investigate the bond performance of the SFCB and SSSC interface under varying concrete protective layer thicknesses and freezing–thawing cycles. This approach allows for the consideration of the impact of concrete protective layer thickness on the frost durability of the bonded interface, while also being more representative of the actual engineering situation.

This study addresses the issues of chloride corrosion in traditional steel bars submerged in seawater sea-sand concrete and the limited ductility of FRP bars by introducing a novel composite steel bar known as SFCB. To enhance the crack resistance and seepage resistance of SSSC, an appropriate amount of glass fiber and expansion agent are incorporated. Consequently, the frost resistance of the concrete is improved. To assess the impact of the concrete protective layer thickness on the frost durability of the bonded interface, eccentric pullout experiments were conducted to replicate the actual engineering situation. Furthermore, the bonding performance of SFCB and GF-EA-SSSC interfaces was examined under varying concrete protective layer thicknesses and a number of freezing–thawing cycles. These investigations aim to promote the advancement and widespread application of seawater sea-sand concrete.

2. Experimental Program

2.1. Materials

Each raw material utilized in preparing seawater sea-sand concrete is shown in Figure 1. Ordinary silicate cement (P.O 42.5R, with an apparent density of 3114.6 kg/m^3) served as the cementitious material for concrete mixing, the chemical properties as shown in Table 1. The artificial seawater, formulated in accordance with ASTM D1141 [45], had a chemical composition comprising 16.26 g/L Cl^-, 1.88 g/L SO_4^{2-}, 1.83 g/L Na^+, 0.21 g/L Mg^{2+}, 0.069 g/L K^+, and 0.124 g/L Ca^{2+}. Sea sand, characterized by an apparent density of 2617 kg/m^3, water content of 1.55%, water absorption of 3.26%, mud content of 0.98%, and a fineness modulus of 2.41, was employed as the fine aggregate. Granite gravel, with a particle size distribution of 5–10 mm and an apparent density of 2689 kg/m^3, served as the coarse aggregate. The granite gravel had a water content of 1.56%, water absorption of 1.84%, and mud content of 1.55%. The particle size distribution of both fine and coarse aggregates is illustrated in Figure 2. The glass fiber used in the study had a length of 18 mm, a monofilament diameter of 13 μm, a density of 2730 kg/m^3, a tensile strength of 1.7 GPa, a modulus of elasticity of 75 GPa, an elongation at break of 2.1%, and a water content of less than 0.1%. The calcium sulfoaluminate–calcium oxide composite expansion agent employed had a magnesium oxide content of 4.8%, water content of 2.5%, total alkali content of 0.7%, chloride ion content of 0.04%, and a specific area of 250 m^2/kg. Additionally, a water-reducing agent was chosen as the aqueous solution of a polycarboxylic acid-based mixture.

The SFCB employed in this study possessed a nominal diameter of 12 mm, an inner core diameter of 8 mm, and a strength grade of HRB400-threaded steel bar. The outer layer featured a wrapped GFRP with a thickness of 4 mm, adopting a threaded surface form, as shown in Figure 3. To assess mechanical properties, five randomly selected SFCB samples underwent testing according to ASTM D7205/D7205M [46]. The equivalent diameter of SFCB was determined using the drainage method. The average equivalent diameter of the SFCB was measured to be 11.6 mm. The average tensile strength and modulus of elasticity were found to be 816.93 MPa and 72.7 GPa, as shown in Table 2.

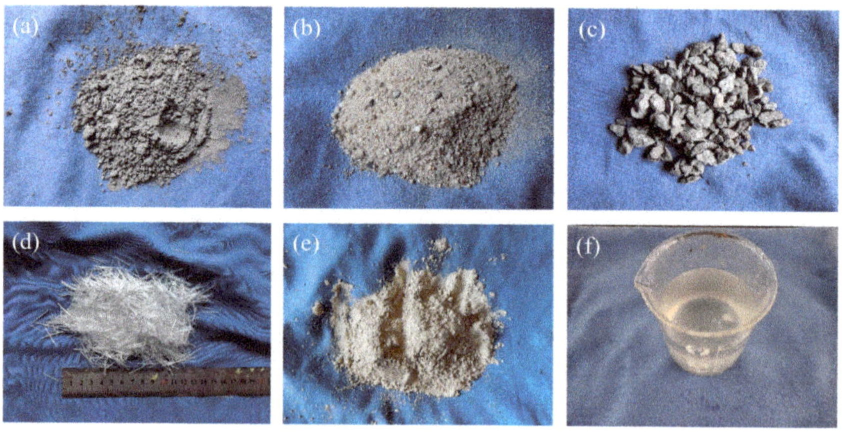

Figure 1. Raw materials of seawater sea-sand concrete: (**a**) cement, (**b**) sea sand, (**c**) crushed granites, (**d**) glass fiber, (**e**) expansion agent, and (**f**) superplasticizer.

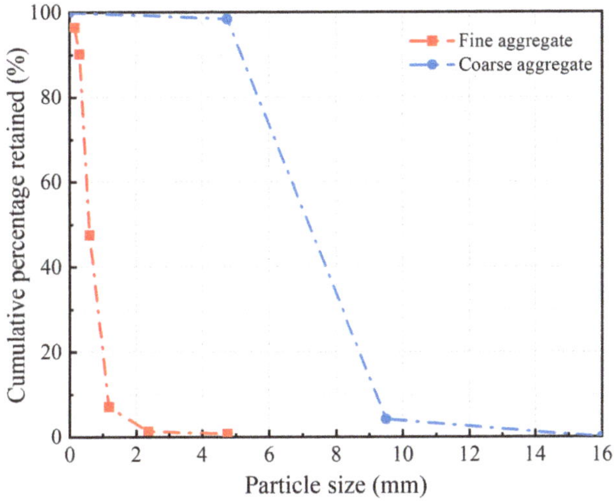

Figure 2. Particle size distribution diagram of aggregate.

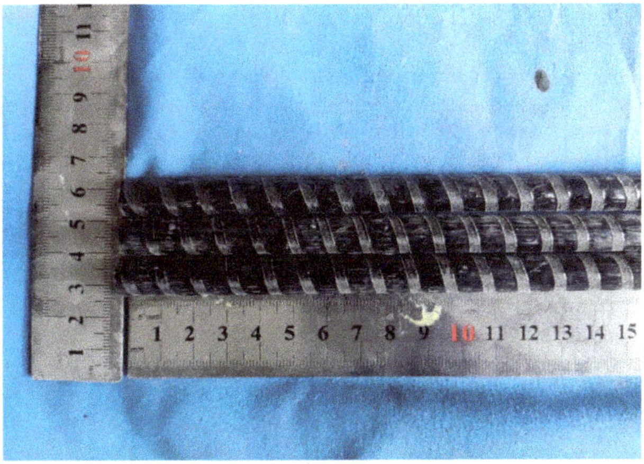

Figure 3. Steel–FRP composite bars.

Table 1. Chemical properties of cement.

Chemical Component	Na_2O	MgO	Al_2O_3	SiO_2	SO_3	K_2O	CaO	Fe_2O_3	LOI
Cement	0.04	1.57	4.34	20.01	1.33	1.02	66.05	2.87	2.77

Table 2. Mechanical properties of SFCB.

Nominal Diameter (mm)	Core Diameter (mm)	Thickness of FRP (mm)	Equivalent Diameter (mm)	Average Tensile Strength (MPa)	Average Elastic Modulus (GPa)
12	8	4	11.6	816.39	72.7

2.2. Specimen Preparation

In this study, glass fiber and expansion agent were mixed into seawater sea-sand concrete to investigate their impact on the frost resistance of the concrete. The mixture components of the seawater sea-sand concrete are represented by F-EA, where F0 and F0.3

denote the glass fiber admixture of 0 and 0.3%, respectively (referring to the volume of glass fiber as a percentage of the total volume of sea-sand). Similarly, EA0 and EA1 represent the expansion agent admixture of 0 and 1%, respectively (referring to the volume of the expansion agent as a percentage of the total volume). The specified proportions included a water–cement ratio of 0.42, a slurry volume ratio of 40%, and a fine aggregate to coarse aggregate volume ratio of 1, as shown in Table 3.

Five different numbers of freezing–thawing cycles T (0, 50, 100, 150, and 200) and four different thicknesses of the concrete protective layer c (30 mm, 35 mm, 40 mm, and 45 mm) were used as the two research variables. This resulted in a total of 13 experimental groups, with three samples for each group, totaling 39 eccentric drawing specimens. The experimental protocol is detailed in Table 4. The eccentric drawing specimens were composed of 150 mm × 150 mm × 150 mm concrete cubes and 500 mm long SFCB. The bond length was set at 60 mm, equivalent to 5 times the nominal diameter.

In Figure 4a, the mold of the eccentric drawn specimen is illustrated, with the length of the bonded area controlled by the PVC pipe and cap. Following the molding process, the specimens underwent a standard curing procedure in a chamber with a temperature of 20 ± 2 °C and relative humidity exceeding 95% for a duration of 24 days, as shown in Figure 4b. To protect the specimen inside the PVC pipe from direct erosion by freezing–thawing damage, the PVC pipe was filled with paraffin. Subsequently, the specimen was immersed in water at a temperature of 20 ± 5 °C for 4 days, ensuring the water surface was 20 mm higher than the specimen. This step aimed to achieve a heavily saturated state for the SSSC. After completing the freezing–thawing cycles, the loading zone of the specimens was anchored using seamless steel pipes.

Fifteen cylindrical specimens with dimensions Ø 150 mm × 300 mm were prepared to investigate the impact of the number of freezing–thawing cycles on compressive strength. Additionally, 15 rectangular specimens with dimensions 100 mm × 100 mm × 400 mm were prepared specifically to study the effect of the number of freezing–thawing cycles on the dynamic modulus of elasticity.

Table 3. Mix proportion of SSSC.

Number	Cement (kg/m^3)	Sea-Sand (kg/m^3)	Crushed Granites (kg/m^3)	Seawater (kg/m^3)	GF (kg/m^3)	EA (kg/m^3)	SP (kg/m^3)
F0EA0	540.46	773.89	787.89	234.36	0.00	0.00	2.70
F0.3EA1	537.36	766.87	780.74	233.08	8.04	5.37	5.43

Table 4. Grouping and numbering of pullout specimens.

Number	Concrete Cover Thickness (mm)	Freezing–Thawing Cyclic Number
F0EA0-T0-c40	40	0
F0EA0-T50-c40	40	50
F0.3EA1-T0-c30	30	0
F0.3EA1-T0-c35	35	0
F0.3EA1-T0-c40	40	0
F0.3EA1-T0-c45	45	0
F0.3EA1-T50-c30	30	50
F0.3EA1-T50-c35	35	50
F0.3EA1-T50-c40	40	50
F0.3EA1-T50-c45	45	50
F0.3EA1-T100-c40	40	100
F0.3EA1-T150-c40	40	150
F0.3EA1-T200-c40	40	200

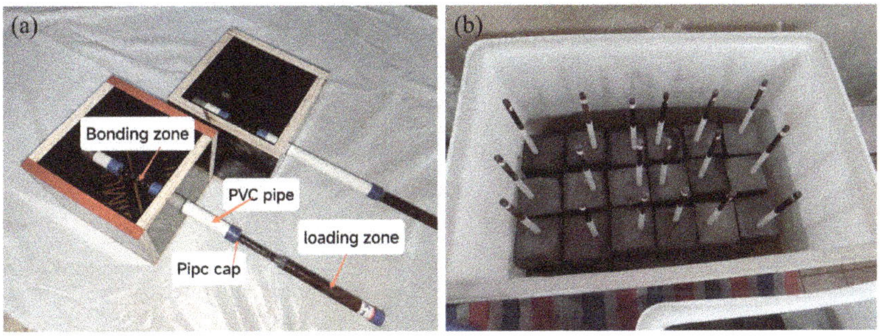

Figure 4. Eccentric pullout specimen preparation: (**a**) mold and (**b**) curing condition.

2.3. Test Methods

The freeze–thaw test in this study employed the rapid freeze–thaw method. As shown in Figure 5, the rapid freeze–thaw testing machine TDRF-II was utilized to subject the drawing specimen to freeze–thaw cycle actions. The thawing temperature was maintained at 5 ± 2 °C, and the freezing temperature was set at −18 ± 2 °C. Every single freezing–thawing cycle lasted for 3–4 h, with the cooling time (the process of the inner core of specimen temperature decreasing from 5 °C to −18 °C) being greater than the freezing time by 1/2. The warming time (the process of the inner core of specimen temperature increasing from −18 °C to 5 °C) was also greater than 1/2 of the thawing time. The freezing and thawing conversion time was kept below 10 min. The temperature–time curve for the freezing and thawing cycle process is illustrated in Figure 6.

The axial compression test of SSSC was conducted using the MATEST universal testing machine. The deformation in the axial compression region of the cylindrical specimen was measured using collars and linear displacement testers (LVDTs), as depicted in Figure 7. The dynamic modulus of elasticity of SSSC was carried out using a DT-20 dynamic modulus of elasticity tester, and the fundamental vibration frequency was measured and determined by the resonance method as shown in Figure 8. Furthermore, the eccentric pullout test was carried out using the MTS322 testing machine with displacement loading, as illustrated in Figure 9.

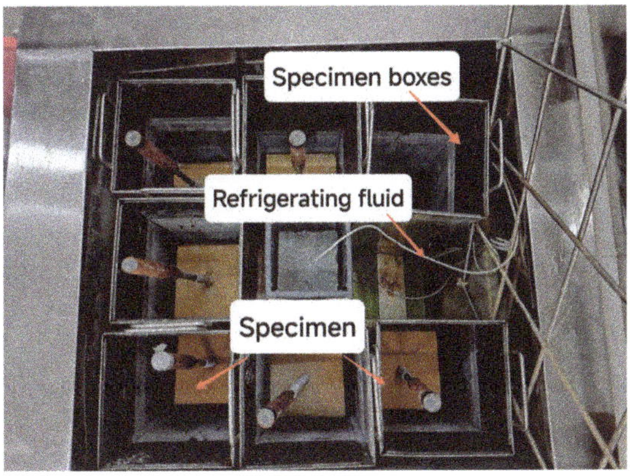

Figure 5. TDRF-II type concrete rapid freezing–thawing test machine.

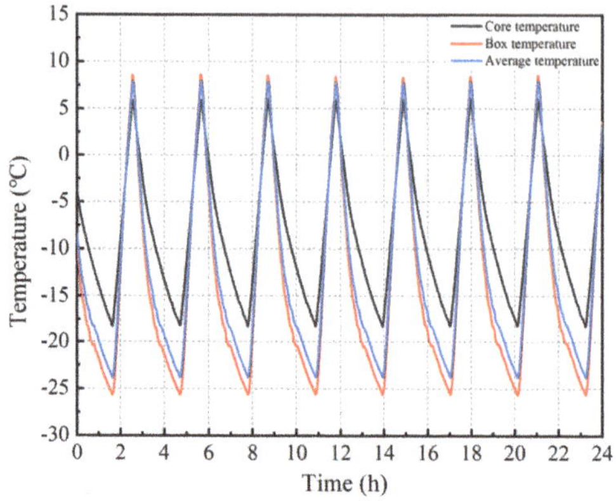

Figure 6. Temperature–time curve of freezing–thawing cycles.

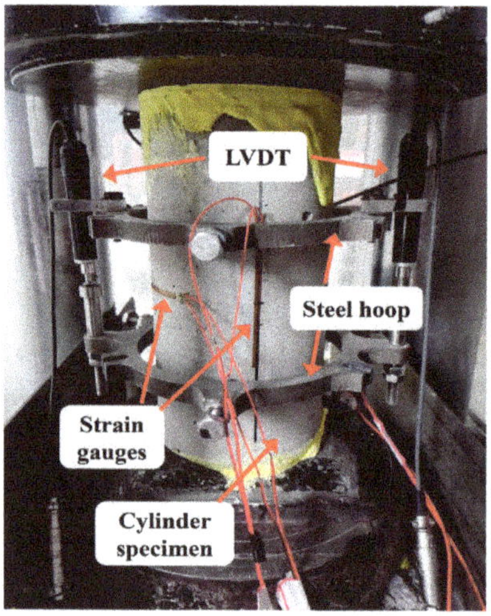

Figure 7. SSSC compressive test setup.

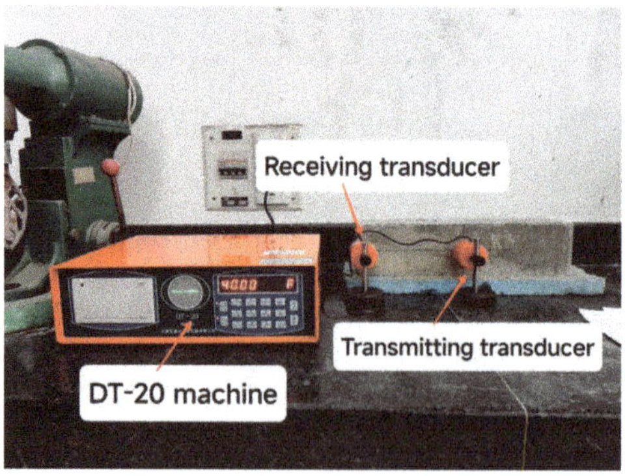

Figure 8. DT-20 dynamic elastic modulus tester.

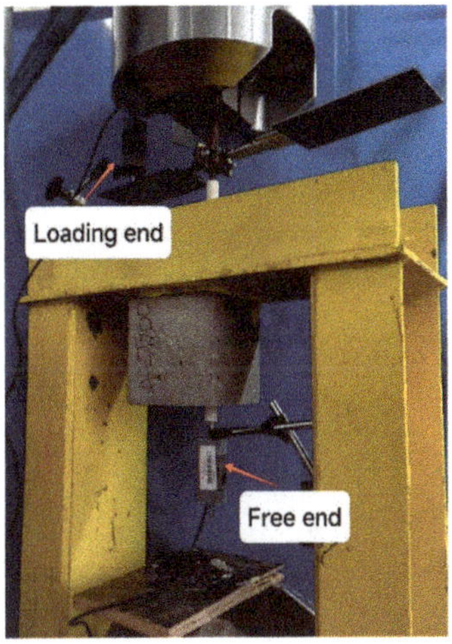

Figure 9. Eccentric pullout test setup.

3. Results and Discussion

3.1. Mechanical Properties of SSSC in Freezing–Thawing Environment

3.1.1. Failure Modes

Figure 10 shows the surface morphology of the specimens subjected to varying numbers of freezing–thawing cycles. In F0.3EA1 after 50 freezing–thawing cycles, there is no discernible change in the morphology, implying that the freezing and thawing effects are limited at this stage, and the concrete surface remains unaffected. However, after 100 freezing–thawing cycles, the surface mortar begins to detach, and microcracks become evident. Subsequently, after 150 freezing–thawing cycles, extensive spalling of the surface mortar and exposure of coarse aggregates occur. Furthermore, after 200 freezing–thawing cycles, noticeable spalling of the coarse aggregates is observed, resulting in evident surface damage. The freezing–thawing damage initiates from the outer layer and gradually progresses toward the inner layers. Under the repeated influence of hydrostatic pressure and osmotic pressure, the cracks and pores in the specimen progressively propagate from the exterior to the interior. As concrete exhibits internal structure heterogeneity, freezing–thawing damage initially affects the weakest points, leading to crack development [38]. As the cracks and pores continue to propagate and accumulate, spalling of the concrete surface commences. Consequently, the freezing–thawing damage expands from the weak points, ultimately resulting in an enlargement of the spalled concrete area. Compared to seawater sea-sand concrete, the incorporation of glass fiber and expansion agents enhances the ability of concrete to resist freezing–thawing damage.

Compressive tests were conducted on the specimens after 50–200 freezing–thawing cycles, and the damage pattern is illustrated in Figure 11. As the number of freezing–thawing cycles increased, the development of cracks and pores persisted, resulting in a notable deterioration of the mechanical properties of the concrete. Notably, within the range of 50–100 freezing–thawing cycles, the trend of crack deterioration exhibited a more gradual increase. Initially, cracks emerged from weak areas of the specimens, extending in narrow strips toward the center, gradually enlarging and generating new microcracks. Concrete blocks were predominantly dislodged and broken. Following 150 cycles of freezing and thawing, severe damage due to freezing and thawing was observed at both ends of the specimen, with significant exposure of coarse aggregate and spalling of fine aggregate. The unevenness on the surface of both ends necessitated multiple applications of gypsum leveling to rectify. The incremental trend of crack deterioration continued to increase, dislodged concrete debris accumulated, the dislodged area expanded, and broken glass fibers were visible within the cracks. After 200 freezing–thawing cycles, the trend of crack deterioration escalated sharply, accompanied by a substantial increase in the volume of dislodged concrete pieces. The specimen as a whole exhibited diagonal cut damage, with exposed aggregates and a substantial amount of dislodged glass fibers and aggregates. In the freezing–thawing environment, the inclusion of admixed glass fiber and expansion agent demonstrated the ability to delay crack expansion during the compression process of SSSC. However, as the number of freezing–thawing cycles increased, the effectiveness of the glass fiber and expansion agent gradually diminished, and the damage mode shifted from tensile damage to diagonal cut damage.

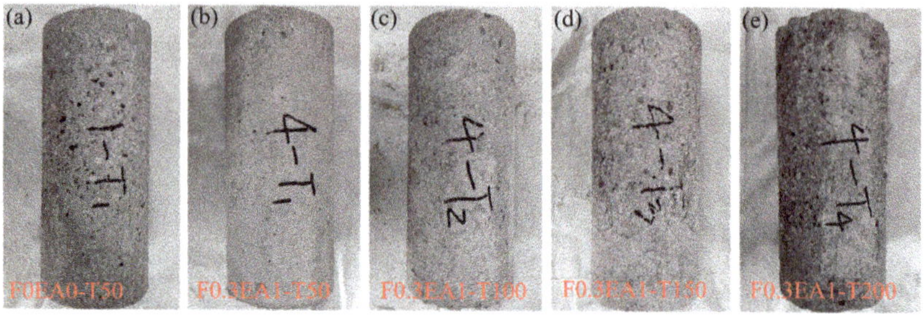

Figure 10. Surface appearances of specimens after 50–200 freezing–thawing cycles: (**a**) F0EA0-T50, (**b**) F0.3EA1-T50, (**c**) F0.3EA1-T100, (**d**) F0.3EA1-T150, and (**e**) F0.3EA1-T200.

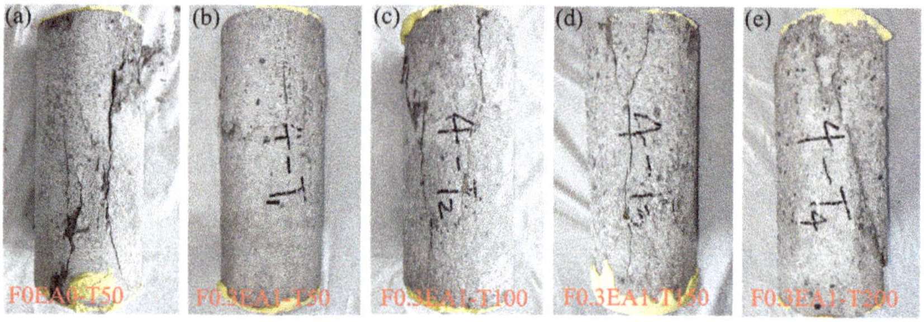

Figure 11. Failure modes of specimens after 50–200 freezing–thawing cycles: (**a**) F0EA0-T50, (**b**) F0.3EA1-T50, (**c**) F0.3EA1-T100, (**d**) F0.3EA1-T150, and (**e**) F0.3EA1-T200.

3.1.2. Compressive Strength Degradation

The uniaxial compressive strength plays a crucial role in assessing the mechanical properties of concrete, serving as an indicator of its degradation under freezing–thawing cycles. Figure 12 shows the variations in compressive strength of SSSC as a function of the number of freezing–thawing cycles. With the increase in the number of freezing and thawing cycles, the repeated freezing and thawing action of water in the internal pores and cracks of concrete makes the number of internal pores and cracks of concrete increase, the diameter increases, and the mechanical properties continue to decrease. Initially, during the first 50 freezing–thawing cycles, a slight reduction in compressive strength was observed. This decrease primarily stemmed from the freezing of liquid water within the pore medium, resulting in a volumetric expansion of 9% [47]. Nonetheless, independent closed pores exist within the SSC, which partially alleviate the expansion stress induced by water freezing, thereby slowing down crack formation and growth. Consequently, when the number of freezing–thawing cycles is low, the detrimental impact of microcracks on compressive strength remains negligible.

As the number of freezing–thawing cycles increases, the number of microcracks and pores caused by freezing–thawing losses also increases. During the thawing phase, these microcracks and pores are filled with water again. During the freezing stage, the volume expansion caused by water freezing is greater, and the microcracks in the SSSC start connecting with each other, leading to the formation of macrocracks and a decreasing trend in compressive strength. When the number of freezing–thawing cycles reaches more than 100 times, the freezing–thawing damage accumulates, causing an increase in pore crack size, a stable morphology, and a change in the internal structure of concrete from dense to loose. This results in a continuous decrease in compressive strength, which is consistent with the results obtained by previous studies [37]. At 200 freezing–thawing cycles, the loss of compressive strength reaches 55.05%.

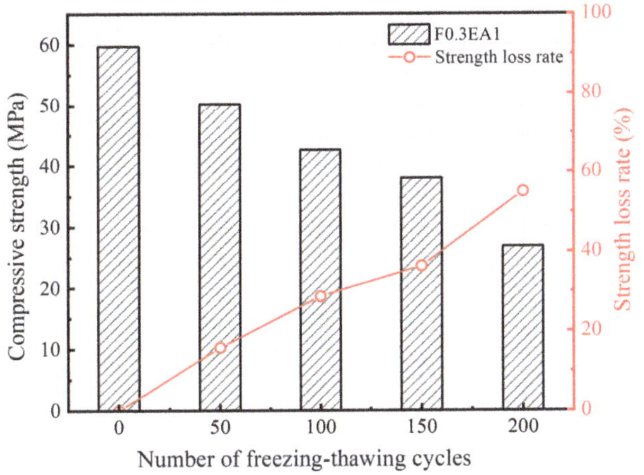

Figure 12. The changing trend of compressive strength under freezing–thawing cycles.

3.1.3. Stress–Strain Relationships

The stress–strain curves for all specimens exhibited a similar pattern, consisting of a linear rising stage, a nonlinear rising stage, and a falling stage. Specifically, the stress–strain curves for the specimens in group F0.3EA1 are displayed in Figure 13. In the absence of freezing–thawing cycles, the descending phase curves appeared relatively smooth and exhibited good ductility. However, as the number of freezing–thawing cycles increased, several notable changes were observed. Firstly, the compressive strength of the specimens decreased, accompanied by an increase and subsequent decrease in peak strain. Additionally, the stress–strain curves exhibited a sharp decline after reaching the peak point. These findings suggest that freezing–thawing cycles contribute to the brittleness of the SSSC [48]. Moreover, the ductility of the SSSC, as represented by the envelope area of the curve, decreased with an increasing number of freezing–thawing cycles [49].

During the initial 100 freezing–thawing cycles, the water in the concrete internal pores and cracks underwent repeated icing expansion. However, the independent closed pores within the concrete partially alleviated the expansion stress caused by water icing, thereby slowing down the formation and progression of cracks. Consequently, the damage stage of the curve appeared relatively smooth and exhibited ductile characteristics. Nevertheless, when the number of freezing–thawing cycles reached 150–200, the concrete experienced an increased number and size of pores and cracks, resulting in significant inhomogeneous damage. This result is consistent with the results obtained by previous studies [49]. Consequently, the destruction stage of the curve appeared steeper, highlighting the brittle characteristics of the concrete.

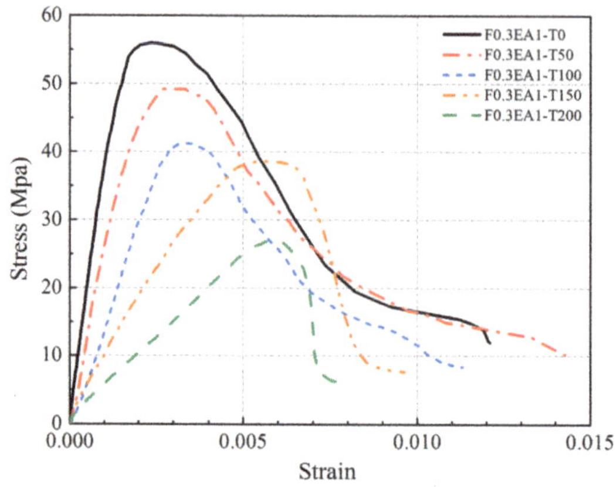

Figure 13. Stress–strain curve of F0.3EA1 under freezing–thawing cycles.

3.1.4. Dynamic Elastic Modulus

The dynamic modulus of elasticity refers to the modulus of elasticity when subjected to dynamic loading conditions. It is closely linked to the level of internal structure compactness within the specimen and serves as an indicator of crack and pore development within the concrete. According to GB/T 50082-2009 [50], the dynamic modulus of elasticity can be calculated by Equation (1).

$$E_d = \frac{13.244 \times 10^{-4} \times WL^3 f^2}{a^4} \tag{1}$$

where E_d is the modulus of dynamic elasticity of concrete (GPa), a is the side length of the square cross-section of the specimen (mm), L is the length of the specimen (mm), W is the mass of the specimen (kg, to the nearest 0.01 kg), and f is the fundamental frequency of vibration of the specimen during transverse vibration (Hz).

Figure 14 illustrates the variation in the dynamic elastic modulus of the F0.3EA1 group across different numbers of freezing–thawing cycles. It can be observed that the dynamic elastic modulus exhibits a decline as the number of freezing–thawing cycles increases. Initially, during the early stages of freezing–thawing cycles, the kinetic elastic modulus experiences a gradual decrease. However, as the number of freezing–thawing cycles continues to rise, the rate of decline in the kinetic elastic modulus accelerates. This phenomenon can be attributed to the expansion stresses generated throughout the repeated freezing process, causing the cracks and pores within the SSC to expand. It is worth noting that the impact of freezing–thawing cycles on concrete damage gradually progresses from the outer layers toward the inner layers. Consequently, the extent of freezing–thawing damage within the specimen is directly related to the volume of freezing–thawing damaged concrete present.

In the initial 100 freezing–thawing cycles, the freezing–thawing damage primarily impacts the outermost layer of the concrete specimen. Cracks and pore development predominantly occur in this outer layer, although their influence on the overall mechanical properties of the specimen is minor. As the number of freezing–thawing cycles increases to 150–200, the freezing–thawing damage progressively penetrates from the outer concrete toward the inner layers. The cracks and pores within the concrete gradually deepen, while the inner concrete exerts a more pronounced effect on the mechanical properties of the entire specimen. This effect manifests as a reduction in compressive strength and a decrease in the basic vibration frequency of the specimen, ultimately resulting in a decline in the dynamic modulus of elasticity.

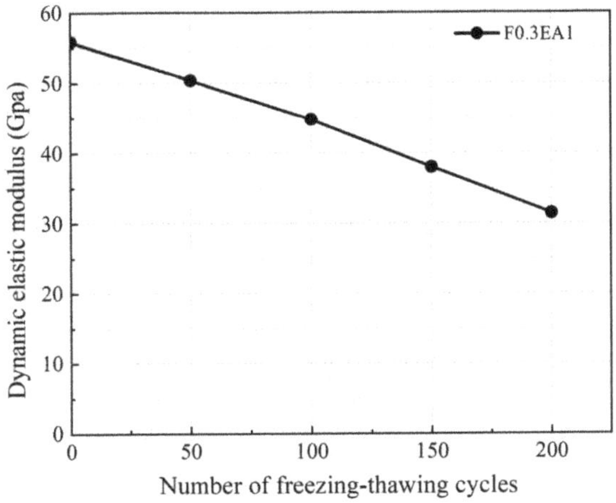

Figure 14. Dynamic elastic modulus under a number of freezing–thawing cycles.

3.1.5. Mass Loss

Mass loss is an important parameter to characterize the performance of concrete under a freezing–thawing environment. The surface morphology of specimens with varying numbers of freezing–thawing cycles is illustrated in Figure 15. In the absence of any freezing–thawing cycles, the specimen exhibits a smooth and intact surface. However, after 50 freezing–thawing cycles, some portions of the mortar begin to flake off, although the overall integrity of the specimen remains uncompromised. With the progression of 100 freezing–thawing cycles, a significant amount of mortar flakes off, resulting in a roughened surface that gradually exposes the outline of the coarse aggregate. Subsequent to 150–200 freezing–thawing cycles, the surface mortar experiences spalling, commencing from the inner layers and extending towards the outer surface. This process leads to complete spalling of the surface mortar, accompanied by substantial loss of mortar. Additionally, the surface coarse aggregate becomes distinctly visible, displaying signs of peeling off. A considerable number of fractured glass fibers can be observed on the surface of the specimen. The mass loss rate of SSSC can be calculated according to Equation (2).

$$M = \frac{M_0 - M_n}{M_0} \times 100\% \tag{2}$$

where M is the mass loss rate, M_0 is the initial mass of the concrete specimen before the freezing–thawing cycle, and M_n is the mass of the concrete specimen after n freezing–thawing cycles.

Figure 16 shows the mass loss rate of specimens in group F0.3EA1 at varying numbers of freezing–thawing cycles. As the number of freezing–thawing cycles increases, the mass loss rate of the specimens also increases. This can be attributed to the expansion stresses generated during the repeated freezing process, which leads to the formation of microcracks in the SSSC. Moisture continuously infiltrates these microcracks, causing further development and expansion of cracks and pores within the concrete. Consequently, the concrete spalls from the outer surface to the inner layers, resulting in a decrease in concrete quality. The SSSC becomes loose and prone to peeling off, and the overall structure gradually transitions from dense to porous.

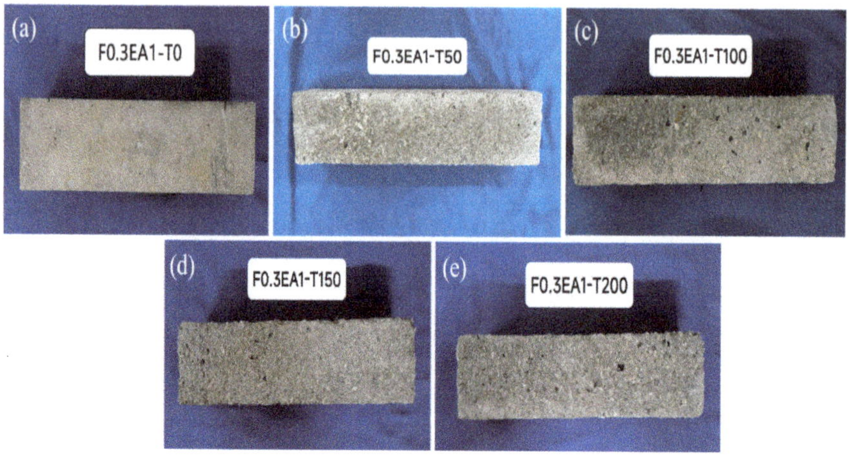

Figure 15. Surface changes of freezing–thawing cycles: (**a**) F0.3EA1-T0, (**b**) F0.3EA1-T50, (**c**) F0.3EA1-T100, (**d**) F0.3EA1-T150, and (**e**) F0.3EA1-T200.

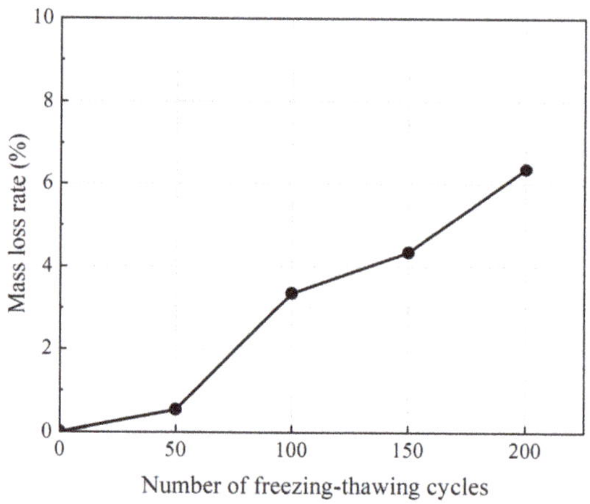

Figure 16. Mass loss rate under a number of freezing–thawing cycles.

3.2. SFCB-SSSC Bonding Behavior

3.2.1. Bond Interfaces

In this study, the pullout tests revealed two primary modes of damage under freezing–thawing cycles, rebar pullout damage and concrete splitting damage [35]. The majority of pullout specimens experienced rebar pullout damage during the test. However, in cases where the thickness of the concrete protective layer was small, concrete splitting damage occurred. The bond between SFCB and SSSC was primarily composed of chemical adhesion and friction when there was no or minimal slip. When a slip occurred, the mechanical occlusal force and friction force became the dominant factors in the bond. Figure 17 illustrates the interaction force between SFCB and SSSC. The mechanical occlusion force acting on the SFCB can be further broken down into axial and radial stresses along the SFCB. These radial stresses then produce circumferential tensile stresses in the concrete surrounding the SFCB [51]. In cases where the thickness of the concrete protective layer is thin and the circumferential tensile stress exceeds the tensile strength of the concrete, cracking occurs from the inside, progressing from the surface of the SFCB to the surface of the specimen. Therefore, the thickness of the concrete protective layer has a significant impact on the durability of the freeze bond. To investigate this effect, an eccentric pullout test was conducted in this study to examine the bond interface of the specimens in the freezing–thawing environment with varying thicknesses of the concrete protective layer.

To observe the damaged sample of the bonded interface, after the pullout test, the pullout specimen was split and the interface damage was observed. As shown in Figure 18, different phenomena were observed under varying protective layer thicknesses and different numbers of freezing–thawing cycles. The increase in the thickness of the concrete protective layer was found to offer effective confinement for SFCB. Therefore, there was an observable increase in the wear of the concrete interface and the presence of adherent flocculent fibers on the concrete surface with the increase in the protective layer thickness. Under unfrozen and thawed conditions, intact ribs were observed at the c30 concrete interface and the entangled fibers fractured. This observation indicates that adequate mechanical anchorage between SSSC and SFCB did not occur when relative slip occurred between them. The surface damage pattern of SFCB revealed a concentration of force at the bond interface during the pullout process. As the thickness of the concrete protective layer increased, the damage pattern of the entangled fibers transitioned from fracture to spalling, and then from spalling to abrasion, resulting in a flocculent shape. The rib surface at the concrete interface suffered severe abrasion, with a significant number of entangled fibers attached. This observation suggests that the concrete exerts a certain binding force on the reinforcement, and increasing the thickness of the concrete protective layer not only enhances the interfacial bond strength but also influences the damage pattern at the bond interface. In the freezing–thawing environment, the attached flocculent fibers at the concrete interface decreased, and the damage to the SFCB entangled fibers slowed down, leading to different damage patterns. This indicates a reduction in bond strength in the freezing–thawing environment, accompanied by alterations in bond behavior. Following freezing–thawing cycles, complete ribs were observed at the concrete interface of c30, while partial traces of ribs were noted at c35, and fiber remnants were still present at the concrete interfaces of c40 and c45. This suggests that the protective concrete layer thickness of c40 and c45 provides effective protection for the bond interface.

When the thickness of the protective layer remained constant (c = 40 mm), the damage to the bond interface of F0.3EA1 was compared under different freezing–thawing conditions. As shown in Figure 18c, a large number of flocculent fibers were observed attached to the concrete interface when it was not subjected to freezing–thawing cycles. Moreover, there was evident abrasion on the rib and reinforcement surfaces of the concrete. As the number of freezing–thawing cycles increased, the residue of fibers at the concrete interface gradually decreased, and the extent of wear on the fiber layer at the SFCB interface diminished. Among them, after 50 freezing–thawing cycles, the internal wear of the fiber layer transitioned to external wear. Subsequently, after 100 freezing–thawing cycles, the fiber layer experienced partial wear instead of complete wear. After 150 freezing–thawing cycles, the surface of the fiber layer appeared more intact. As the number of freezing–thawing cycles increased, the damage at the bond interface gradually diminished, with the concrete interface shifting from bond–slip damage to freezing–thawing damage, and minimal damage occurred on the surface of the steel reinforcement caused by freezing and thawing.

As can be seen in Figure 19, before freeze–thawing, the F0EA0 specimen exhibited a minimal amount of fiber residue at the concrete interface, with distinct rib traces. After 50 freezing–thawing cycles, complete ribs were visible at the concrete interface, indicating poor bonding performance and insufficient adhesion between the SFCB and concrete. In contrast, the F0.3EA1 specimen, with the inclusion of glass fiber and expansion agent, showed the highest fiber residue at the concrete interface prior to freezing–thawing. The internal fiber layer at the SFCB interface exhibited filamentation damage. Following 50 freezing–thawing cycles, the amount of fiber residue at the concrete interface decreased, and slight rib traces could be observed. This suggests that the presence of GF and EA can enhance the bond performance between SFCB and SSSC, with a more pronounced synergistic effect.

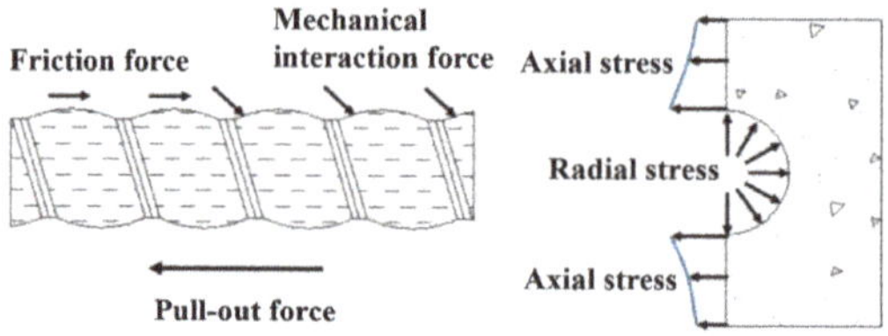

Figure 17. The interaction force of reinforcement and concrete.

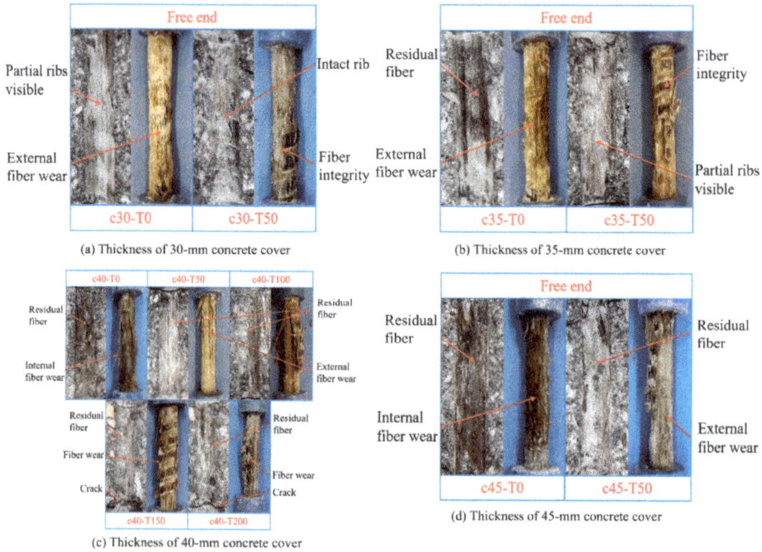

Figure 18. Bonding interfaces of F0.3EA1 under different thicknesses of concrete cover.

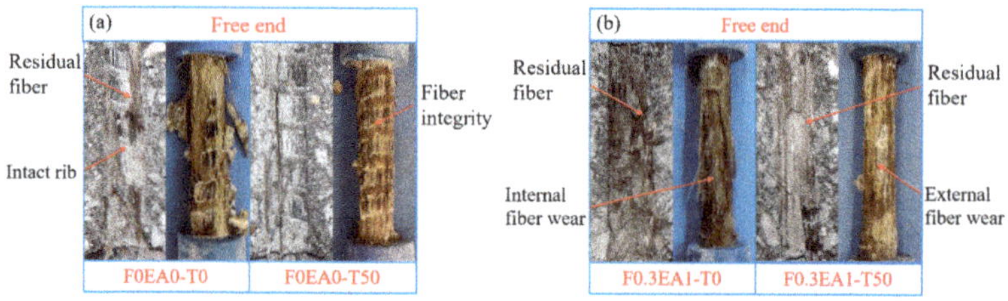

Figure 19. Bonding interfaces of F0EA0 and F0.3EA1 under thicknesses of 40 mm concrete cover: (**a**) F0EA0, (**b**) F0.3EA1.

3.2.2. Bond Stress–Slip Relationships

Figure 20 illustrates bond stress–slip curves for specimens featuring various concrete protective layer thicknesses and different numbers of freezing–thawing cycles. The bond stress–slip relationship can be delineated into two distinct types, one for split-damaged specimens and the other for pullout-damaged specimens [52]. In this study, the bond stress–slip curve pertains to specimens exhibiting pullout damage. These curves can be primarily segmented into three stages, the first stage (ascending section), the second stage (descending section), and the third stage (residual section).

The first stage of the bond stress–slip curve is characterized by linear growth during the initial loading phase. In this phase, the curve ascends with a consistent slope, displaying elastic characteristics in the bond stress between SFCB and concrete. As slip increases, damage becomes evident in the ribs connecting SFCB and concrete, leading to a decline in bonding effectiveness. Simultaneously, circumferential tensile stress rises with the growing load. Once this stress surpasses the tensile strength of the protective concrete layer, the curve transitions to a descending stage. During this descending stage, the interconnected ribs sustain damage, resulting in a decreasing trend in bond stress. However, stress unconcentration introduces variations in the descending segments of the curve. As the slip continues to rise, the SFCB at the free end engages with the bond interface, generating a new mechanical anchorage force. Consequently, the bond stress begins to rise again, marking the curve entry into the residual stage. With the increase in slip, the curve attains a new stress peak. Nevertheless, the pullout process inflicts damage upon the SFCB and concrete, leading to cumulative damage. Therefore, the peak stress in the residual stage is significantly smaller than that observed in the first stage.

The bond stress–slip curves for SFCB and SSSC exhibited multiple peaks before and after freezing and thawing. The peaks gradually increased with the thickness of the concrete protective layer, indicating an enhancement in the bond between concrete and steel reinforcement. Conversely, an increase in the number of freezing–thawing cycles resulted in a decrease in the peak value of the curve, signaling a decline in the bond between concrete and reinforcement. The second peak was consistently smaller than the first peak. This discrepancy arises from the destruction of the bond interface during the ascending stage of bond stress. This observation underscores the ductile nature of pullout damage, revealing its resilience in the face of structural stress.

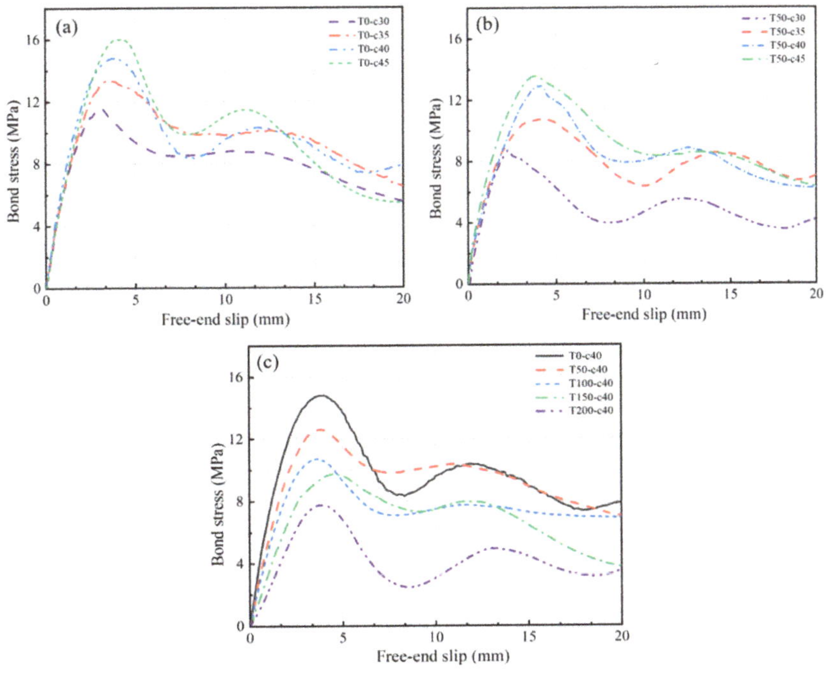

Figure 20. Bond stress–slip relationship of F0.3EA1: (**a**) different concrete cover (T = 0), (**b**) different concrete cover (T = 50), and (**c**) different freezing–thawing cyclic numbers (c = 40).

3.2.3. Bond Strength

In this study, bond strength τ_u, relative slip s_u, and bond stiffness k were chosen to evaluate bond performance. Pullout damage was observed as the prevalent damage mode in all groups, with complete damage occurring between bond interfaces, as detailed in Table 4. Figure 21 illustrates the evolving trend of interfacial bond strength between SFCB and SSSC under varying concrete protective layer thicknesses and freezing–thawing cycles. In Figure 21a, the F0.3EA1 group, incorporating GF and EA, demonstrated a significant enhancement in the anti-freezing bond durability of SSSC compared to the F0EA0 group. The bond strength loss rate decreased from 22.68% to 9.07% after 50 freezing–thawing cycles. In Figure 21b, the bond strength exhibited an upward trajectory with increasing concrete protective layer thickness. In case of an increase in the thickness of the protective layer, the ability of concrete to restrain the reinforcement is strengthened, and the reinforcement is elongated by tensile force; hence, the bond strength increases, which is similar to the conclusion reached by the previous authors [53,54]. However, at a certain thickness of the concrete protective layer, the bond strength of the specimens all decreased after freezing–thawing cycles.

Following 50 freezing–thawing cycles, the specimens exhibited varying degrees of bond strength degradation. After 200 freezing–thawing cycles, the bond strength of F0.3EA1 decreased by 42.31% compared to the unfrozen condition. The decline in bond strength correlates with the damage status of the bond interface, which deteriorates as freezing–thawing cycles increase. The freezing–thawing cycle action induces loosening and spalling of SSSC, accompanied by the appearance of cracks and pores at the concrete interface. Internal cracks expand over time, and the freezing–thawing damage shifts progressively from the external to the internal structure. These factors collectively contribute to the observed decrease in bond strength. Additionally, the higher coefficient of thermal expansion in SFCB than in SSSC, especially at lower temperatures, causes greater contraction in SFCB. This contraction creates gaps between the interfaces, resulting in a loss of bond performance.

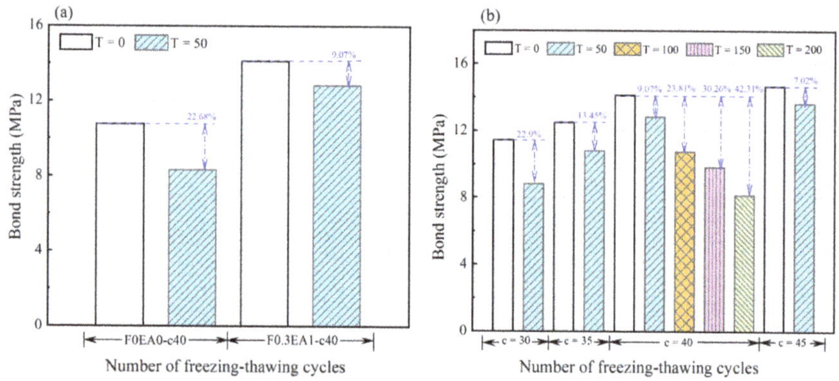

Figure 21. Bond strength–freezing and thawing cyclic number relationships: (**a**) concrete cover (T = 40) of F0EA0 and F0.3EA1 and (**b**) different concrete cover of F0.3EA1.

However, a large number of previous studies have suggested that the freezing–thawing cycling action minimally impacts the bond strength between stressed reinforcement and concrete [55,56]. These conclusions may be nuanced, as the experiments in those studies utilized center-drawn specimens, resulting in a thicker concrete protective layer than typically found in the actual project. This deviation from actual project conditions prompted our study to employ eccentric drawing tests. In this study, the bond strength of F0.3EA1-c40 significantly degraded after 200 freezing–thawing cycles. The larger the thickness of the concrete protective layer, the smaller the rate of bond strength loss under the same number of freezing–thawing cycles. This suggests a notable influence of concrete protective layer thickness on bond durability between SFCB and concrete in freezing–thawing environments. The existing studies may overestimate the bond durability between the stressed reinforcement and concrete under a freezing–thawing environment. Thus, considering both the number of freezing–thawing cycles and the concrete protective layer thickness, we propose the following formula for τ_u:

$$\tau_u = A + B \tag{3}$$

$$A = -5.8731 + 0.8024c - 0.1087n \tag{4}$$

$$B = -0.0077c^2 + 2.766 \times 10^{-6}n^2 + 0.00198cn \tag{5}$$

where A is primarily related to the concrete protective layer thickness (c) and the number of freezing–thawing cycles (n), while B is a secondary formula considering the same parameters. Table 5 presents a comparison between theoretical values and experimental data, with the relative error (RE) staying within 5%.

Analysis of Table 5 reveals a substantial decrease in relative slip (s_u) with an increasing number of freezing–thawing cycles, aligning with findings from previous studies [57,58]. To account for the impacts of freezing–thawing cycles and concrete protective layer thickness on relative slip, we propose the following formula for s_u:

$$s_u = C + D \tag{6}$$

$$C = -4.0755 + 0.3625c - 0.0357n \tag{7}$$

$$D = -0.00386c^2 + 7.651 \times 10^{-6}n^2 + 7.058 \times 10^{-4}cn \tag{8}$$

where C is the primary formula connected to concrete protective layer thickness (c) and freeze–thaw cycles (n), while D is a secondary formula considering the same parameters. The comparison of theoretical values with test data in Table 4 indicates a relative error (RE) within 5%.

Table 5. Analysis of the bond performance parameters of F0.3EA1.

Number	τ_u (MPa)			s_u (mm)			k (MPa/mm)		
	Test	Eq.	RE	Test	Eq.	RE	Test	Eq.	RE
T0-c30	11.44	11.30	1.2%	3.33	3.32	0.2%	6.29	6.20	1.4%
T50-c30	8.82	8.83	0.2%	2.61	2.61	0.1%	5.59	5.69	1.8%
T0-c35	12.49	12.81	2.6%	3.91	3.88	0.7%	6.62	6.55	1.1%
T50-c35	10.81	10.85	0.4%	3.30	3.34	1.4%	6.01	6.03	0.4%
T0-c40	14.11	13.95	1.1%	4.22	4.24	0.5%	6.72	6.79	1.2%
T50-c40	12.83	12.48	2.7%	3.83	3.88	1.5%	6.21	6.28	1.2%
T100-c40	10.75	11.03	2.6%	3.65	3.57	2.2%	5.63	5.59	0.7%
T150-c40	9.84	9.59	2.5%	3.34	3.29	1.5%	4.95	4.72	4.7%
T200-c40	8.14	8.17	0.4%	3.01	3.05	1.4%	3.67	3.66	0.2%
T0-c45	14.67	14.70	0.2%	4.46	4.41	1.0%	6.96	6.96	0%
T50-c45	13.64	13.73	0.7%	4.18	4.23	1.3%	6.46	6.45	0.2%

3.2.4. Bond Stiffness

Figure 22 illustrates the impact of varying concrete protective layer thicknesses and freezing–thawing cycles on bond stiffness. According to the previous study [59], bond stiffness (k) is defined as the slope of the cut line in the bond stress–slip curve at a slip of 0.5 mm.

From Figure 22a, it can be found that the specimens of the F0.3EA1 group have an inhibitory effect on the freezing–thawing degradation of bond stiffness due to the incorporation of GF and EA, in contrast to the F0EA0 group. At room temperature, the bond stiffness of the GF- and EA-doped specimens closely resembles that of the F0EA0 group. However, after 50 freezing–thawing cycles, the bond stiffness loss rate increases from 28.15% to 7.59%. From Figure 22b, an observable trend is the increase in bond stiffness with the augmentation of the protective layer thickness. A thicker protective layer in concrete enhances the restraining effect on SGFCB bar pullout, leading to a reduced rate of bond stiffness loss with increasing layer thickness. Specifically, when the protective layer thickness escalates from 30 mm to 45 mm, the bond stiffness loss rate decreases from 11.13% to 7.18% after 50 freezing–thawing cycles. It is noteworthy that bond stiffness decreases as the number of freezing–thawing cycles increases, holding the thickness of the concrete protective layer constant.

During the initial 50 freezing–thawing cycles, freezing–thawing damage primarily accumulated on the thinner side of the concrete, resulting in a relatively small bond stiffness loss rate in the specimens. However, with a growing number of freezing–thawing cycles, the damage extended from the outer concrete layer of SFCB to the inner layers, leading to a gradual increase in the bond stiffness loss rate. After 200 freezing–thawing cycles, this rate can escalate to 45.39%. The mechanism behind the bond stiffness loss parallels that of bond strength. The freezing–thawing cycle action causes the expansion of pores and cracks in SSSC, elevating porosity. This contributes to a reduction in the clamping force of SSSC on SFCB, making it easier for SFCB to be pulled out.

To account for the impact of freezing–thawing cycles and the thickness of the protective concrete layer on bond stiffness (k), we establish the following formula for k:

$$k = 2.2493 + 0.1861c - 0.00844n - 0.00181c^2 - 3.362 \times 10^{-5}n^2 \quad (9)$$

where c is the thickness of the protective layer with concrete, and n is the number of freezing–thawing cycles. The comparison of theoretical values with experimental data is shown in Table 5, with the relative error (RE) found to be within 5%.

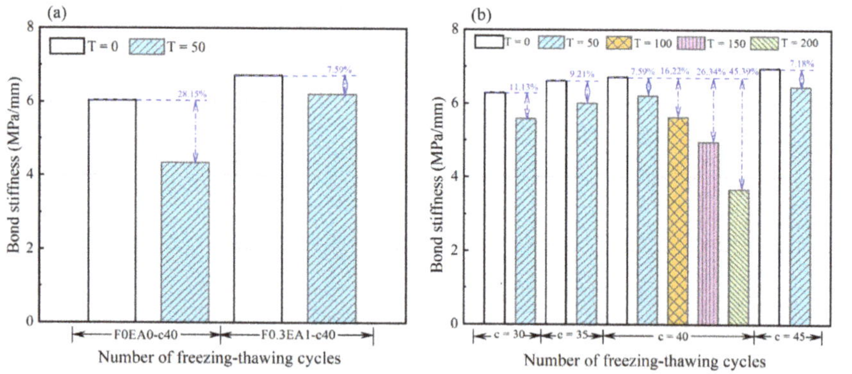

Figure 22. Bond stiffness–freezing and thawing cyclic number relationships: (**a**) concrete cover (T = 40) of F0EA0 and F0.3EA1 and (**b**) different concrete cover of F0.3EA1.

4. Calibration of Constitutive Models

The bond stress–slip principal model, as the key to bond research, offers a comprehensive depiction of the bond interaction between reinforcement and concrete. Currently, a large number of models reflecting the bond–slip behavior of reinforced concrete have been developed, leading scholars to establish FRP bar-concrete stress–slip constitutive models, primarily the BPE model and CMR model [60–62]. However, the ontological model for the stress–slip relationship between SFCB and concrete remains imperfect. Notably, existing models do not account for the effects of the freezing–thawing environment and the thickness of the concrete protective layer on bond performance. Therefore, it is necessary to extend the bond stress–slip ontological model of SFCB with concrete to more accurately depict the bonding performance between SFCB and SSSC. Given the enhanced frost durability of SSSC with the incorporation of GF and EA, the effects of the number of freezing–thawing cycles and the thickness of the concrete protective layer are also considered. In this study, the commonly used modified Bertero–Eligehausen–Popov (mBPE) model and Cosenza–Manfredi–Realfonzo (CMR) model are used as references to establish the bond stress–slip ontological model about SFCB and GF-EA-SSSC.

Among them, the bond stress–slip relationship of the mBPE model can be expressed as follows:

$$\tau = \begin{cases} \tau_u (s/s_u)^a & 0 \leq s \leq s_u \\ \tau_u \left[1 - p\left(\frac{s}{s_u} - 1\right)\right] & s_u < s \leq s_r \\ \tau_r & s > s_r \end{cases} \quad (10)$$

where a and p are empirical parameters, and τ_r and s_r are the bond stress and slip at the intersection of the descending and residual stages, respectively.

However, the mBPE model employs a horizontal straight line to approximate the residual segment curve, resulting in an inadequate prediction of the residual stage. In this study, only the ascending and descending segments of the mBPE model were fitted, disregarding the residual stage.

On the other hand, the CMR model primarily focuses on describing the ascending phase of the curve and does not fully capture the entire bond stress–slip curve. The bond stress–slip relationship in the CMR model is expressed as follows:

$$\tau = \tau_u \left[1 - \exp\left(-\frac{s}{\alpha}\right)\right]^{\beta} \ (0 \leq s \leq s_u) \quad (11)$$

where α and β are empirical parameters.

In comparison, the CMR model exhibits a better fit for the ascending phase than the mBPE model. The fitting results of the improved intrinsic model match well with the experimental data, as shown in Figure 23. The specific empirical parameter values are given in Table 6.

Table 6. Empirical parameters of mBPE model and CMR model.

Number	mBPE Model		CMR Model	
	a	p	α	β
T0-c30	0.50599	0.24459	0.79414	1.83709
T50-c30	0.68258	0.24511	0.69323	1.78494
T0-c35	0.53014	0.20189	0.9558	1.71732
T50-c35	0.40923	0.31301	1.08141	1.15897
T0-c40	0.45585	0.46078	0.95284	1.71102
T50-c40	0.57138	0.25634	1.00502	1.7262
T100-c40	0.57394	0.3414	0.95719	1.72866
T150-c40	0.55914	0.30071	1.09304	2.05805
T200-c40	0.78326	0.6089	0.95291	2.91302
T0-c45	0.58459	0.50329	1.12851	1.74061
T50-c45	0.50843	0.24175	1.21384	1.16974

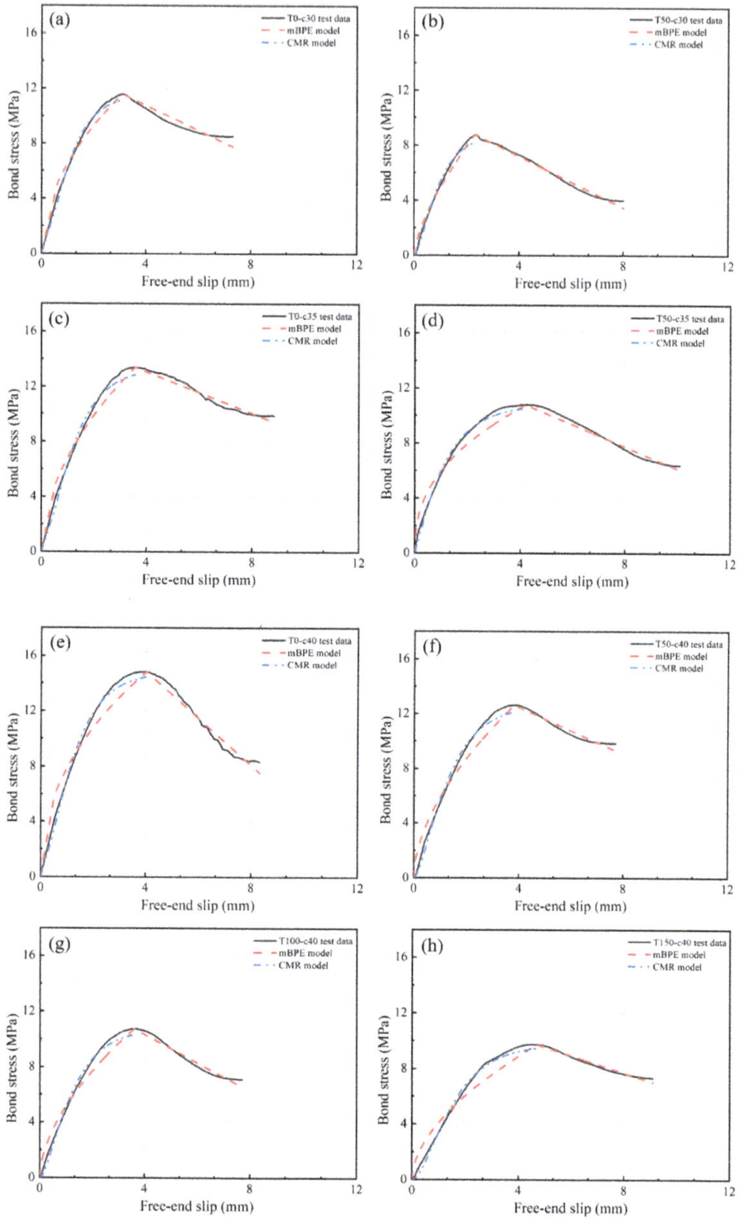

Figure 23. *Cont.*

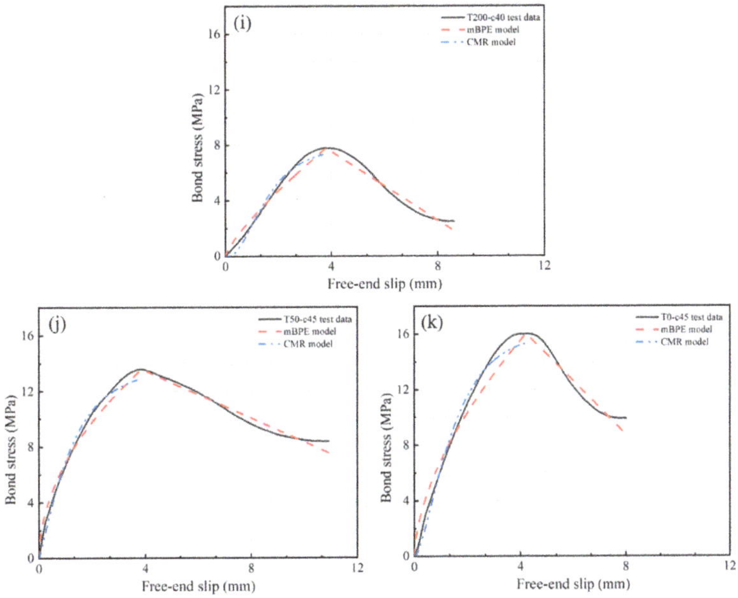

Figure 23. Comparison between constitutive models and test data of F0.3EA1: (**a**) T0-c30, (**b**) T50-c30, (**c**) T0-c35, (**d**) T50-c35, (**e**) T0-c40, (**f**) T50-c40, (**g**) T100-c40, (**h**) T150-c40, (**i**) T200-c40, (**j**) T0-c45, and (**k**) T50-c45.

5. Conclusions

In this study, 13 concrete mix proportions were devised, and 39 specimens of SFCB embedded in SSSC underwent eccentric pullout tests. The investigation focused on assessing the bond durability of SFCB and GF-EA-SSSC under freezing–thawing conditions. The study considered five sets of freezing–thawing cycles (T) and four varying thicknesses of the concrete protective layer (c) as the independent variables. The primary conclusions drawn from this research are as follows:

(1) The SSSC undergoes a loosening and brittleness transformation under freezing–thawing cycles. Incorporating glass fiber and expansion agents can enhance the ability of concrete to resist freezing–thawing damage. Initially, cracks emerge from the weakest point of the concrete. As the number of freezing–thawing cycles increases, damage accumulates from the exterior to the interior, leading to a gradual escalation in both crack size and quantity. Upon reaching 150–200 freezing–thawing cycles, significant spalling occurs on the surface of the specimen, with an expanding spalling area. The damage pattern shifts from tensile damage to diagonal cut damage as freezing–thawing cycles progress.

(2) As the number of freezing–thawing cycles increases, there is a concurrent decrease in compressive strength and an increase in mass loss. In the initial stages of the freezing–thawing cycle, the decrease in compressive strength is gradual, and mass loss is minimal. Microcracks have a negligible impact on compressive strength, and the specimen retains its structural integrity. After surpassing 100 freezing–thawing cycles, concrete damage accumulates significantly. The internal structure transitions from dense to loose, leading to extensive spalling from the exterior to the interior. Compressive strength exhibits a noticeable decline, accompanied by an increase in quality loss. By the time 200 freezing–thawing cycles are reached, the loss of compressive strength amounts to 55.05%, and the quality loss rate reaches 6.36%.

(3) Analyzing the stress–strain curve reveals that freezing–thawing action not only impacts the compressive strength of concrete but also influences the characteristics of the concrete damage stage. With an escalation in the number of freezing–thawing cycles,

compressive strength diminishes, the peak strain initially rises and then decreases, and the stress–strain curve sharply descends after reaching the peak point. This phenomenon indicates an increase in brittleness and a decrease in ductility for SSSC. The freezing–thawing cycles induce heterogeneous damage within the concrete interior. Stress concentration during the damage stage results in the manifestation of brittle damage.

(4) The dynamic elastic modulus experiences a decline as the number of freezing–thawing cycles increases. In the initial phase of the freezing–thawing cycle, the decrease in dynamic elastic modulus is gradual. However, as the number of freezing–thawing cycles advances, the impact of freezing–thawing damage intensifies, progressing from the outer layer to the inner layers of concrete. Consequently, the development of cracks and pores within the concrete becomes increasingly profound, leading to an accelerated rate of decrease in dynamic elastic modulus.

(5) Following freezing–thawing cycles, the majority of specimens subjected to the eccentric pullout test exhibited rebar pullout damage. The bond performance at the bonded interface of the pulled specimens diminished, and there was a continuous decrease in the quantity of residual fibers on the concrete interface, along with wear on the SFCB interface. Maintaining a constant number of freezing–thawing cycles, an increase in the thickness of the concrete protective layer altered the damage mode of entangled fibers. The shift occurred from fracture to spalling and subsequently from spalling to abrasion. Notably, residual fibers remained at the interface, and the SSSC has a better restraining effect on the SFCB. With a fixed concrete protective layer thickness, an escalation in the number of freezing–thawing cycles resulted in a gradual reduction in bond interface damage. The damage at the concrete interface transitioned from bond–slip damage to freezing–thawing damage.

(6) The bond stress–slip curves for specimens exhibiting pullout damage vary with different thicknesses of the concrete protective layer and the number of freezing–thawing cycles. In the initial loading stage, the bond stress between SFCB and SSSC demonstrates elastic characteristics. Once the curve reaches its peak stress value, it transitions into the descending and residual sections. The bond stress–slip curve exhibits multiple peaks before and after freezing and thawing. As the thickness of the concrete protective layer increases, the peak value of the curve gradually rises, indicating enhanced bond stress between SFCB and SSSC. Conversely, with an increase in the number of freezing–thawing cycles, the peak of the curve decreases, signifying a reduction in the bond stress between SFCB and SSSC. The second peak is generally much smaller than the first peak, indicating that the pullout damage is ductile in nature.

(7) Under various thicknesses of the concrete protective layer and differing numbers of freezing–thawing cycles, the trends observed in bond strength, relative slip, and bond stiffness exhibited similarities. With a constant concrete protective layer thickness, an increase in the number of freezing–thawing cycles resulted in a decrease in both bond strength and bond stiffness, accompanied by an increase in relative slip. After 200 freezing–thawing cycles, the bond strength decreased by 42.31% relative to the unfrozen-thawed condition, and the loss of bond stiffness could reach 45.39%. The restraining effect on SFCB pullout was found to improve with a thicker concrete protective layer. Additionally, the rate of loss in bond strength and bond stiffness reduced as the protective layer thickness increased. Formulas for bond strength, relative slip, and bond stiffness were established, taking into account the effects of the concrete protective layer thickness and the number of freezing–thawing cycles in the eccentric pullout test. The relative error between theoretical and experimental values was within 5%.

(8) In this study, GF and EA were introduced to enhance the frost durability of SSSC, and the mBPE model and CMR model were refined to account for the influence of the number of freezing and thawing cycles as well as the thickness of the concrete protective layer. These modifications aimed to establish a bond stress–slip constitutive model for the interaction between SFCB and the composite material of GF-EA-SSSC. The fitting results of

the improved constitutive model closely resembled the experimental values, demonstrating a high level of accuracy in the predictive capabilities of the model.

(9) This paper investigates the bonding properties between steel–FRP composite bar (SFCB) and glass fiber with expansion-agent-reinforced seawater sea-sand concrete (GF-EA-SSSC) interface using eccentric pullout experiments under different thicknesses of concrete protective cover and a number of freezing–thawing cycles. The eccentric pullout test in this study utilized a single SFCB, without considering the impact of spacing multiple SFCBs on the bond strength of GF-EA-SSSC after freezing–thawing cycles. This divergence from practical engineering standards suggests the potential for refining the test methodology. Following 200 freezing–thawing cycles, the specimens underwent nearly complete freezing, an occurrence rarely observed in practical engineering scenarios. Future research could delve into variations in dynamic elastic modulus and ultrasonic testing post freezing–thawing damage to specific cross-section areas, and the influence of the bonding performance of the interface between steel and concrete, so as to propose a more reliable evaluation method for reinforced concrete members after freezing–thawing damage in practical engineering.

Author Contributions: Conceptualization, Y.L. and W.W.; Methodology, Y.L. and G.L.; Software, J.J.; Validation, Y.S.; Formal analysis, Y.Z. and J.L.; Investigation, Y.Z., J.L. and H.Z.; Resources, G.L.; Data curation, J.L.; Writing—original draft, J.J. and Y.S.; Writing—review & editing, H.Z.; Supervision, W.W.; Project administration, Z.X.; Funding acquisition, W.W. and Z.X. All authors have read and agreed to the published version of the manuscript.

Funding: The authors gratefully acknowledge the financial support provided by the Department of Education Characteristic Innovation Project of Guangdong Province under No. 2021KTSCX211, Guangdong Construction Vocational and Technical College High-level Talent Research Project under No. GCC2021-02, Special Foundation for Scientific and Technological Innovation Strategy of Guangdong Province under Grant No. pdjh2024a137 (in China), Guangdong Yonghe Construction Group Limited–Guangdong University of Technology Industry–Research Cooperation Project under No. 23HK0171, National Innovation and Entrepreneurship Training Program for College Students under Grant No. 202311845030 (in China).

Data Availability Statement: The original data are available upon request. The data are not publicly available due to project privacy.

Conflicts of Interest: Author Gangliang Li was employed by the company Guangdong Yonghe Construction Group Co., Ltd. The remaining authors declare that the research was conducted in the absence of any commercial or financial relationships that could be construed as a potential conflict of interest.

References

1. Huang, Y.; Chen, T.; Hu, D.; Lin, T.; Zhu, W.; Zhang, G.; Xue, X. Spatiotemporal patterns and influencing factors of urban ecological space availability in coastal cities of China during rapid urbanization. *Ecol. Indic.* **2023**, *154*, 110757. [CrossRef]
2. Zhao, T.; Zhang, Y.; Li, K.; Wang, J.; Cai, E.; Lin, G.; Li, Z. A copula approach in the probabilistic evaluation of chloride ingress process for coastal concrete structures. *Structures* **2023**, *56*, 104918. [CrossRef]
3. Ren, Q.; Tao, Y.; Jiang, Z.; De Schutter, G. Adhesive and rheological properties of fresh mortar with manufactured sand as replacement of river sand. *Constr. Build. Mater.* **2024**, *411*, 134303. [CrossRef]
4. Xiang, J.; Qiu, J.; Wu, P.; Zhang, Q.; Song, Y.; Yang, L. Autolytic capsules incorporating alkali-activated slag system for self-healing cementitious composites. *Cem. Concr. Compos.* **2024**, *147*, 105439. [CrossRef]
5. Xiang, J.; Qiu, J.; Zhao, Y.; Zheng, P.; Peng, H.; Fei, X. Rheology, mechanical properties, and hydration of synergistically activated coal gasification slag with three typical solid wastes. *Cem. Concr. Compos.* **2023**, *147*, 105418. [CrossRef]
6. Wang, X.; Dong, C.; Xu, S.; Song, Q.; Ren, J.; Zhu, J. Influence of seawater and sea sand on early-age performance and cracking sensitivity of concrete. *J. Build. Eng.* **2023**, *79*, 107811. [CrossRef]
7. Zhen, H.; Xiong, Z.; Song, Y.; Li, L.; Qiu, Y.; Zou, X.; Chen, B.; Chen, D.; Liu, F.; Ji, Y. Early mechanical performance of glass fibre-reinforced manufactured sand concrete. *J. Build. Eng.* **2024**, *83*, 108440. [CrossRef]
8. Xiong, Z.; Su, Y.; Hu, Z.; Huang, Y.; Wang, Z.; Chen, J.; Lao, W.; Li, L.; Zhou, K.; Kuang, J. Dynamic compression study of seawater sea sand concrete incorporated with expansive agents and glass fibre. *J. Build. Eng.* **2023**, *79*, 107942. [CrossRef]
9. Guan, H.; Hao, B.; Zhang, G. Mechanical properties of concrete prepared using seawater, sea sand and spontaneous combustion coal gangue. *Structures* **2023**, *48*, 172–181. [CrossRef]

10. Wang, F.; Sun, Y.; Xue, X.; Wang, N.; Zhou, J.; Hua, J. Mechanical properties of modified coral aggregate seawater sea-sand concrete: Experimental study and constitutive model. *Case Stud. Constr. Mater.* **2023**, *18*, e02095. [CrossRef]
11. Li, S.; Zhu, D.; Guo, S.; Yi, Y.; Feng, G. Experimental study on the effects of dynamic loading and seawater attack on bond behavior between basalt fiber reinforced polymer bars and seawater sea-sand concrete. *J. Build. Eng.* **2023**, *72*, 106505. [CrossRef]
12. Xiong, Z.; Mai, G.; Qiao, S.; He, S.; Zhang, B.; Wang, H.; Zhou, K.; Li, L. Fatigue bond behaviour between basalt fibre-reinforced polymer bars and seawater sea-sand concrete. *Ocean Coast. Manag.* **2022**, *218*, 106038. [CrossRef]
13. Peng, K.-D.; Zeng, J.-J.; Huang, B.-T.; Huang, J.-Q.; Zhuge, Y.; Dai, J.-G. Bond performance of FRP bars in plain and fiber-reinforced geopolymer under pull-out loading. *J. Build. Eng.* **2022**, *57*, 104893. [CrossRef]
14. Hu, K.; Chen, Z.; Kang, H.; Tang, Y. 3D vision technologies for a self-developed structural external crack damage recognition robot. *Autom. Constr.* **2024**, *159*, 105262. [CrossRef]
15. Chen, Z.; Yu, J.; Nong, Y.; Yang, Y.; Zhang, H.; Tang, Y. Beyond time: Enhancing corrosion resistance of geopolymer concrete and BFRP bars in seawater. *Compos. Struct.* **2023**, *322*, 117439. [CrossRef]
16. Bujotzek, L.; Beck, D.; Apostolidi, E.; Waldmann, D. Experimental and statistical investigations of the material properties of FRP reinforcement in compression. *Constr. Build. Mater.* **2024**, *414*, 134782. [CrossRef]
17. Zeng, J.-J.; Li, P.-L.; Yan, Z.-T.; Zhou, J.-K.; Quach, W.-M.; Zhuge, Y. Behavior of 3D-printed HPC plates with FRP grid reinforcement under bending. *Eng. Struct.* **2023**, *294*, 116578. [CrossRef]
18. Hakeem, I.; Hosen, M.A.; Alyami, M.; Qaidi, S.; Özkılıc, Y. Influence of Heat–Cool Cyclic Exposure on the Performance of Fiber-Reinforced High-Strength Concrete. *Sustainability* **2023**, *15*, 1433. [CrossRef]
19. Çelik, A.İ.; Özkılıç, Y.O.; Bahrami, A.; Hakeem, I.Y. Effects of glass fiber on recycled fly ash and basalt powder based geopolymer concrete. *Case Stud. Constr. Mater.* **2023**, *19*, e02659. [CrossRef]
20. Gemi, L.; Madenci, E.; Özkılıç, Y.O. Experimental, analytical and numerical investigation of pultruded GFRP composite beams infilled with hybrid FRP reinforced concrete. *Eng. Struct.* **2021**, *244*, 112790. [CrossRef]
21. Rafieizonooz, M.; Kim, J.-H.J.; Varaee, H.; Nam, Y.; Khankhaje, E. Testing methods and design specifications for FRP-prestressed concrete members: A review of current practices and case studies. *J. Build. Eng.* **2023**, *73*, 106723. [CrossRef]
22. Zeng, J.-J.; Chen, S.-P.; Feng, P.; Zhuge, Y.; Peng, K.-D.; Dai, J.-G.; Fan, T.-H. FRP bar-reinforced ultra-high-performance concrete plates with a grouting sleeve connection: Development and flexural behavior. *Eng. Struct.* **2023**, *287*, 116164. [CrossRef]
23. Zaki, M.A.; Rasheed, H.A. Behavior of reinforced concrete beams strengthened using CFRP sheets with innovative anchorage devices. *Eng. Struct.* **2020**, *215*, 110689. [CrossRef]
24. Smith, S.T.; Rasheed, H.A.; Kim, S.J. Full-range load-deflection response of FRP-strengthened RC flexural members anchored with FRP anchors. *Compos. Struct.* **2017**, *167*, 207–218. [CrossRef]
25. Xu, L.; Wang, X.; Pan, J.; Zhou, J.; Liu, W. Bond behavior between steel-FRP composite bars and engineered cementitious composites in pullout conditions. *Eng. Struct.* **2024**, *299*, 117086. [CrossRef]
26. Han, S.; Zhou, A.; Fan, C.; Xiao, G.; Ou, J. Axial-flexural performance of columns reinforced by steel-FRP composite bars and FRP ties. *Compos. Struct.* **2023**, *322*, 117435. [CrossRef]
27. Zhang, P.-F.; Iqbal, M.; Zhang, D.; Zhao, X.-L.; Zhao, Q. Bond strength prediction of FRP bars to seawater sea sand concrete based on ensemble learning models. *Eng. Struct.* **2024**, *302*, 117382. [CrossRef]
28. Wang, Q.; Zhu, H.; Zhang, B.; Zhao, Y.; Li, F. Bond enhancement for BFRP bar in concrete by using a resin-filled FRP tube anchorage. *Structures* **2022**, *39*, 1107–1117. [CrossRef]
29. Wang, J.; Xiao, F.; Yang, J. Experimental study on bond behavior between epoxy-coated reinforcement (ECR) and seawater sea-sand concrete (SSC) under FRP-steel confinement. *Constr. Build. Mater.* **2023**, *385*, 131426. [CrossRef]
30. Xu, L.; Ma, X.; Pan, J.; Zhou, J.; Wang, X.; Liu, W. Investigation on bond behavior of steel-FRP composite bars in engineered cementitious composites by spliced beam tests. *Constr. Build. Mater.* **2023**, *407*, 133480. [CrossRef]
31. Dong, Z.; Wu, G.; Xu, Y. Experimental study on the bond durability between steel-FRP composite bars (SFCBs) and sea sand concrete in ocean environment. *Constr. Build. Mater.* **2016**, *115*, 277–284. [CrossRef]
32. Chang, Y.; Wang, Y.; Wang, M.; Zhou, Z.; Ou, J. Bond durability and degradation mechanism of GFRP bars in seawater sea-sand concrete under the coupling effect of seawater immersion and sustained load. *Constr. Build. Mater.* **2021**, *307*, 124878. [CrossRef]
33. Li, J.; Gravina, R.J.; Smith, S.T.; Visintin, P. Bond strength and bond stress-slip analysis of FRP bar to concrete incorporating environmental durability. *Constr. Build. Mater.* **2020**, *261*, 119860. [CrossRef]
34. Lu, Z.; Zhao, C.; Zhao, J.; Shi, C.; Xie, J. Bond durability of FRP bars and seawater–sea sand–geopolymer concrete: Coupled effects of seawater immersion and sustained load. *Constr. Build. Mater.* **2023**, *400*, 132667. [CrossRef]
35. Shakiba, M.; Bazli, M.; Karamloo, M.; Mortazavi, S.M.R. Bond-slip performance of GFRP and steel reinforced beams under wet-dry and freeze-thaw cycles: The effect of concrete type. *Constr. Build. Mater.* **2022**, *342*, 127916. [CrossRef]
36. Cao, Q.; Xu, J.; Wu, Z.; Zhang, L. Performance of damaged concrete stub columns strengthened with post-tensioned CFRP under freezing-thawing environment. *Structures* **2023**, *47*, 544–556. [CrossRef]
37. Mai, G.; Li, L.; Lin, J.; Wei, W.; He, S.; Zhong, R.; Xiong, Z. Bond durability between BFRP bars and recycled aggregate seawater sea-sand concrete in freezing-thawing environment. *J. Build. Eng.* **2023**, *70*, 106422. [CrossRef]
38. Jin, M.; Li, W.; Wang, X.; Tang, J.; Teng, L.; Ma, Y.; Zeng, H. Investigating the dual influence of freezing-thawing cycles and chloride ion penetration on GGBS-AEA concrete deterioration. *J. Build. Eng.* **2023**, *78*, 107759. [CrossRef]

39. Qiu, J.; Li, L.; Li, L.; Luan, X.; Guan, X.; Niu, G. Study on the deterioration characteristics and mechanisms of recycled brick-concrete aggregate concrete under load-freeze-thaw coupling. *Constr. Build. Mater.* **2024**, *413*, 134817. [CrossRef]
40. Xiang, J.; Qu, L.; Fei, X.; Qiu, J.; Kong, X. Depth-dependent self-healing capacity and mechanism of cracked fiber-reinforced concrete by bacterial community. *J. Build. Eng.* **2024**, *84*, 108485. [CrossRef]
41. Xiong, Z.; He, S.H.; Kwan, A.K.H.; Li, L.G.; Zeng, Y. Compressive behaviour of seawater sea-sand concrete containing glass fibres and expansive agents. *Constr. Build. Mater.* **2021**, *292*, 123309. [CrossRef]
42. Zhu, H.; Xiong, Z.; Song, Y.; Zhou, K.; Su, Y. Effect of expansion agent and glass fiber on the dynamic splitting tensile properties of seawater–sea-sand concrete. *Buildings* **2024**, *14*, 217. [CrossRef]
43. Lin, M.; He, S.; Qiao, S.; Xiong, Z.; Qiu, Y.; Zhang, J.; Li, L. Combined effects of expansive agents and glass fibres on the fracture performance of seawater and sea-sand concrete. *J. Mater. Res. Technol.* **2022**, *20*, 1839–1859. [CrossRef]
44. Busel, J.P.; Shield, C.K. *440.1R-15 Guide for the Design and Construction of Structural Concrete Reinforced with Fiber-Reinforced Polymer (FRP) Bars, Fiber Reinforced Polymers*; ACI: Farmington Hills, MI, USA, 2015.
45. *ASTM D1141*; Standard Specification for Substitute Ocean Water. American Society for Testing and Materials: West Conshohocken, PA, USA, 1975.
46. *ASTM D7705/D7705M-12*; D/DM, Standard Test Method for Alkali Resistance of Fiber Reinforced Polymer (FRP) Matrix Composite Bars used in Concrete Construction. ASTM International: Washington, DC, USA, 2019.
47. Lu, J.; Liu, J.; Yang, H.; Wan, X.; Gao, J.; Zhang, J.; Li, P. Experimental investigation on the mechanical properties and pore structure deterioration of fiber-reinforced concrete in different freeze-thaw media. *Constr. Build. Mater.* **2022**, *350*, 128887. [CrossRef]
48. Wang, Y.; Xie, M.; Zhang, J. Mechanical properties and damage model of modified recycled concrete under freeze-thaw cycles. *J. Build. Eng.* **2023**, *78*, 107680. [CrossRef]
49. Wu, J.; Jing, X.; Wang, Z. Uni-axial compressive stress-strain relation of recycled coarse aggregate concrete after freezing and thawing cycles. *Constr. Build. Mater.* **2017**, *134*, 210–219. [CrossRef]
50. *GB/T50082-2009*; Introduction of Revised Standard for Test Methods of Long-Term Performance and Durability of Ordinary Concrete. Ministry of Housing and Urban-Rural Development of the People's Republic of China: Beijing, China, 2010.
51. Akbas, T.; Celik, O.; Yalcin, C.; Ilki, A. Monotonic and Cyclic Bond Behavior of Deformed CFRP Bars in High Strength Concrete. *Polymers* **2016**, *8*, 211. [CrossRef] [PubMed]
52. Zhang, B.; Zhu, H.; Chen, J. Bond durability between BFRP bars and seawater coral aggregate concrete under seawater corrosion environments. *Constr. Build. Mater.* **2023**, *379*, 131274. [CrossRef]
53. Yan, F.; Lin, Z.; Zhang, D.; Gao, Z.; Li, M. Experimental study on bond durability of glass fiber reinforced polymer bars in concrete exposed to harsh environmental agents: Freeze-thaw cycles and alkaline-saline solution. *Compos. Part B Eng.* **2017**, *116*, 406–421. [CrossRef]
54. Majain, N.; Rahman, A.B.A.; Adnan, A.; Mohamed, R.N. Bond behaviour of deformed steel bars in steel fibre high-strength self-compacting concrete. *Constr. Build. Mater.* **2022**, *318*, 125906. [CrossRef]
55. Khanfour, M.A.; El Refai, A. Effect of freeze-thaw cycles on concrete reinforced with basalt fiber-reinforced polymers (BFRP) bars. *Constr. Build. Mater.* **2017**, *145*, 135–146. [CrossRef]
56. Wang, W.; Wang, Y.; Chen, Q.; Liu, Y.; Zhang, Y.; Ma, G.; Duan, P. Bond properties of basalt fiber reinforced polymer (BFRP) bars in recycled aggregate thermal insulation concrete under freeze–thaw cycles. *Constr. Build. Mater.* **2022**, *329*, 127197. [CrossRef]
57. Alves, J.; El-Ragaby, A.; El-Salakawy, E. Effect of fatigue loads on the bond strength of glass FRP bars to concrete. In *Advances in FRP Composites in Civil Engineering: Proceedings of the 5th International Conference on FRP Composites in Civil Engineering (CICE 2010), Beijing, China, 27–29 September 2010*; Springer: Berlin/Heidelberg, Germany, 2010; pp. 368–371.
58. Veljkovic, A.; Carvelli, V.; Haffke, M.M.; Pahn, M. Concrete cover effect on the bond of GFRP bar and concrete under static loading. *Compos. Part B Eng.* **2017**, *124*, 40–53. [CrossRef]
59. Xiong, Z.; Wei, W.; He, S.; Liu, F.; Luo, H.; Li, L. Dynamic bond behaviour of fibre-wrapped basalt fibre-reinforced polymer bars embedded in sea sand and recycled aggregate concrete under high-strain rate pull-out tests. *Constr. Build. Mater.* **2021**, *276*, 122195. [CrossRef]
60. Eligehausen, R.; Popov, E.P.; Bertero, V.V. *Local Bond Stress-Slip Relationships of Deformed Bars under Generalized Excitations: Experimental Results and Analytical Model*; University of California: Berkeley, CA, USA, 2016.
61. Li, H.; Liu, F.; Pan, Z.; Li, H.; Wu, Z.; Li, L.; Xiong, Z. Use of supplementary cementitious materials in seawater–sea sand concrete: State-of-the-art review. *Constr. Build. Mater.* **2024**, *425*, 136009. [CrossRef]
62. Cosenza, E.; Manfredi, G.; Realfonzo, R. Behavior and Modeling of Bond of FRP Rebars to Concrete. *J. Compos. Constr.* **1997**, *1*, 40–51. [CrossRef]

buildings

MDPI

Article

Mechanical and Thermal Properties of an Energy-Efficient Cement Composite Incorporating Silica Aerogel

Tatiana Aleksandrovna Koriakovtseva [1,*], Anna Evgenyevna Dontsova [1] and Darya Viktorovna Nemova [2]

1 School of Hydrotechnical and Power Engineering, Institute of Civil Engineering, Peter the Great St. Petersburg Polytechnic University, 195251 Saint Petersburg, Russia; dontsova_ae@spbstu.ru
2 Laboratory of Protected and Modular Structures, Peter the Great St. Petersburg Polytechnic University, 195251 Saint Petersburg, Russia; nemova_dv@spbstu.ru
* Correspondence: musorina_t@spbstu.ru

Abstract: The thermal performance of the building envelope is significant in energy-efficient construction. Because concrete is widely used in civil engineering, options to reduce its R-value should be considered. This study explores the thermal and structural properties of aerogel-enhanced concrete. Silica aerogel powder was mixed with fine-grained concrete at 15 vol.%. Two series of samples were prepared to identify the preferred technology. The first series of samples were mixed without isopropyl alcohol; for the second series, the alcohol was mixed with silica aerogel before mixing into the dry mix. The thermal conductivity, compressive strength, and bending resistance of the specimens were measured. The presence of silica aerogel admix resulted in a decrease in the compressive strength of the specimens by 30% compared with that of the reference samples and a reduction in the bending strength of the samples by 9% compared with that of the reference samples. For the first and second series of samples, the K-values of the aerogel-enhanced specimens varied in the range from 0.83 W/(m·K) to 1.13 W/(m·K), respectively. To further decrease the thermal conductivity, gypsum putty was then added to the specimens, resulting in the K-values further decreasing to 0.64 W/(m·K) and 0.84 W/(m·K), respectively. The calculation of heat losses through 1 m^2 of the aerogel-enhanced concrete wall was performed. It has been shown that energy expenses for heating can be lowered by 30%. The calculation of the greenhouse gas emissions from the combustion of fuel required for heating was also considered. The emissions decreased by 30.2% compared with the reference sample. Microscopic examination of the face and section surfaces of the sample revealed a significant number of pores compared with conventional fine-grain concrete.

Keywords: concrete; aerogel; experimental study; compressive strength; three-point bending test; thermal conductivity

Citation: Koriakovtseva, T.A.; Dontsova, A.E.; Nemova, D.V. Mechanical and Thermal Properties of an Energy-Efficient Cement Composite Incorporating Silica Aerogel. *Buildings* **2024**, *14*, 1034. https://doi.org/10.3390/buildings14041034

Academic Editor: Yun Gao

Received: 29 February 2024
Revised: 25 March 2024
Accepted: 3 April 2024
Published: 8 April 2024

1. Introduction

Energy efficiency improvement is an important challenge in civil engineering. Heating systems consume the largest share of energy. It is necessary to use powerful heating equipment to maintain comfortable indoor climate parameters in colder regions. The main energy expense item is heat loss through the enclosing structures. Thus, the thermal insulation requirements are being toughened. Manufacturers are encouraged to improve insulating materials, which can be used in energy-efficient construction projects.

The multilayer wall structure allows for the improvement of heat resistance. However, minimizing the energy consumption of the building while providing a comfortable indoor climate remains a challenge. Energy-saving technologies, materials, and measures should not only lead to a reduction in energy consumption but also be cost-effective. Unfortunately, investment performance in energy-saving measures is rarely considered [1,2]. Nowadays, there is a growing interest and demand for energy savings, reducing the load on material resources, and reducing pollution. The important condition for solving these complex problems is the development of new building materials with high strength and durability that are safe for human health and the natural environment [3]. The central task of energy-efficient design is to reduce heating costs while maintaining the conditions for human comfort [4]. Although new energy-efficient materials and innovative technologies are constantly emerging in the construction industry, the search for even more promising solutions does not end. New ideas must first meet requirements such as practical significance, operability, economic rationality, and environmental and social safety.

An option to comply with the heat resistance requirements in civil engineering is to decrease the thermal conductivity of the most popular building materials. The use of additives to create heat-insulating concrete is considered. Traditionally, the thermal performance of concrete is improved with a lower density or higher porosity of the composite. For this purpose, plant-based or synthesized aggregates are used. Sometimes, processed raw materials can also serve this purpose. The resulting concrete composite is durable and strong, even though it is more porous.

One of the most promising and revolutionary insulating materials is a nanostructured aerogel based on silicon dioxide, which was first created by scientists in 1931. The first to obtain aerogels in Russia were the staff of the Federal Research Center Boreskov Institute of Catalysis (Novosibirsk, Russia) and the Joint Institute for Nuclear Research (Dubna, Russia) [5–7]. In this study, silica aerogel powder was used as an aggregate to reduce the thermal conductivity of concrete. Cement-based aerogel mixes were first presented in 2008 [8] as heat-resistant concrete. The properties of the novel composite have shown its potential in engineering [9–15]. The 'high-performance aerogel concrete' made of sand, concrete, and 70% silica aerogel granulate was presented. The material had an admirable mechanical strength and a K-value of 0.1 W/(m·K). A mix for lightweight concrete was described in [16]. The main challenge was to walk a line between the contradictory properties: low K-value, on the one hand, and workability for the concrete mix and the high strength of the resulting composite, on the other hand.

A side effect of the subcritical production process of aerogel is the complete hydrophobicity of the material. Thus, aerogels cannot absorb water despite their open pores. Hence, solvents (most often alcohols) are used to ensure the homogeneity of the mixture. These reactions are usually performed with ethanol or methanol [17]. Therefore, it is necessary to change the water–cement ratio [18,19].

A scanning electron microscopy investigation of cement hydration revealed that aerogel particles are stable during the process [20], indicating the potential of aerogel admix for cementitious materials. The weaknesses of mixing silica aerogel into a high-strength cement matrix are given in [21]. These include the lack of adhesion at the boundary of the aerogel–cement binder, increased porosity of the composite, high water absorption capacity, and a significant decrease in compressive strength with large volumes of silica aerogel. Moreover, the gaps between the aerogel particles and the cement matrix can serve as capillaries for airflow. The gaps between the aerogel particles and the cement binder appear due to the hydrophobicity of the aerogel. The presence of the capillaries, as well as the fragility of the silica aerogel itself and the increased porosity of the cement matrix, determine the mechanical properties of the composite. Table 1 presents a review of the amount of aerogel mixed into cementitious composites [22].

The review data shows that the presence of granulated silica aerogel in a concrete mix results in a dramatic reduction in the compressive strength of the composite.

Another modification of concrete using aerogel was created in [23]. At first, silica sand was gradually replaced with aerogel. When all the sand was replaced with aerogel, the amount of the cement matrix used was proportionally reduced with increasing aerogel content, thus maintaining a constant cement matrix. The results showed a decrease in the compressive and flexural strength of the composite, as shown in Figure 1.

Table 1. A review of the amount of aerogel used for cementitious composites [22].

Type of Cementitious Composite	Scientific Name of Cementitious Composites with Silica Aerogel	Silica Aerogel Volume (%)
Concretes	Ultra-high-performance aerogel concrete	20–80
	High-performance aerogel concrete	45–70
	Aerogel-incorporated concrete	10–60
	Ultra-lightweight concrete	15–60
Mortars	Silica aerogel-incorporated mortar	20–80
Lightweight composites	Green lightweight composite	1–5
	Lightweight cement-based composite	15–100
	Ultra-lightweight cement composite	-
	Silica aerogel-incorporated concrete	2–8
	Thermal renders with silica aerogel	0–20

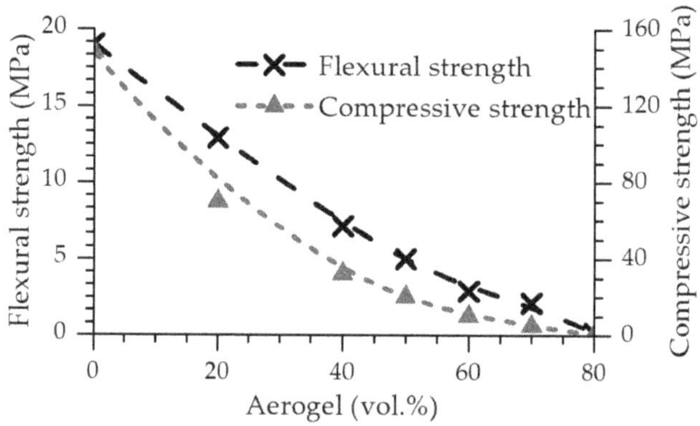

Figure 1. Mechanical properties of concrete depending on the silica aerogel content in the composite. Adapted with permission from Ref. [23]; published by Elsevier, 2015.

The effect of small amounts of aerogel and silica fume was studied in [24]. The samples were exposed to the curing process in water, the wetting–drying effect, and the $MgSO_4$ effect. Silica fume admix enhanced the compressive strength of samples cured with the wetting–drying and $MgSO_4$ effects and decreased the compressive strength of samples cured in water. For different curing conditions, the largest reduction in thermal conductivity of 31.2% was obtained for the water-cured sample containing 0.25% aerogel additive and 10% silica fume. Based on the data collected, it can be concluded that silica fume is not a suitable additive for enhancing insulation properties.

Recent scientific and technological advances allow researchers to assume that the mass application of aerogel as a thermal insulating material in civil engineering will soon be possible. Therefore, it is necessary to investigate the possibilities and prospects of using aerogel as a heat-insulating material in different engineering fields. Thus far, research on the properties of cementitious composites with an aerogel admix has been limited. Attempts have already been made to construct a correlation between the amount of aerogel and the final mechanical and thermal properties of the composite [25–28]; however, there is still insufficient data for a complete review evaluating the feasibility and possibilities of using silica aerogel as a heat-protective additive for dry concrete mixtures.

This study aimed to create a cementitious material that can perform both thermal insulation and bear loads. Two series of composite concrete samples and reference samples were manufactured, and their properties were examined to achieve this goal. The technology for adding the aerogel powder to the mixture was different. In one case, aerogel powder was poured into the dry mix. In the other case, the aerogel was first mixed with isopropyl alcohol and then poured into the dry mix. The thermal conductivity, compressive strength, and bending resistance of the specimens were measured. The surfaces of the samples were examined under a microscope to evaluate the microstructure of the composite. The calculation of heat losses through 1 m^2 of the aerogel-enhanced concrete wall was performed. The calculation of the greenhouse gas emissions from the combustion of fuel required for heating was also considered.

2. Materials and Methods

Figure 2 presents the experimental research program in general.

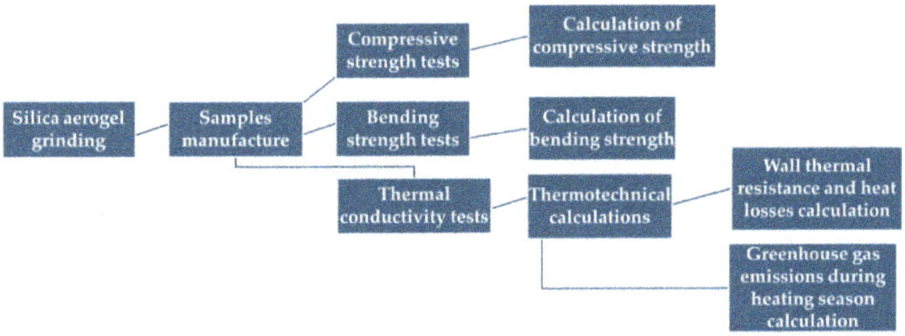

Figure 2. Flowchart for studying the properties of a cementitious composite containing 15 vol.% silica aerogel admix.

2.1. Grinder Choice and Silica Aerogel Admix Grinding

The general advantages of aerogel materials are as follows: environmental safety, fire resistance [29], incombustibility [30], low density (40–150 kg/m^3), hydrophobicity, high reflectivity, extremely low thermal conductivity (0.016 W/(m·K) at 10 °C), ease of installation, durable sheets, wide range of operation temperature [31], and long life [32,33]. However, the material is expensive because of its complex fabrication technique. Silica aerogel is widely used for thermal insulation [29–31]. In this study, silica aerogel granules (manufactured by Guangdong Alison Hi-Tech Co., Ltd., Yingde, China) were crushed to powder. The basic properties of the aerogel granules were provided by the manufacturer as follows: bulk density 40–150 kg/m^3; particle size > 0.5 mm; surface area 500–650 m^2/g; porosity > 90%; pore diameter 20–100 µm. The size of the graded particles was comparable to the concrete powder fraction size, which equals 40 µm or less. The size of the silica aerogel powder was determined using a sieve analyzer. The grinder was set with the minimum possible grinding gap. A JC 6 jaw crusher (Vibrotechnik, LLC, Saint-Petersburg, Russia) was used for grinding. The measured bulk density of the power was 110 kg/m^3. Figure 3 shows the initial and final conditions of the silica aerogel.

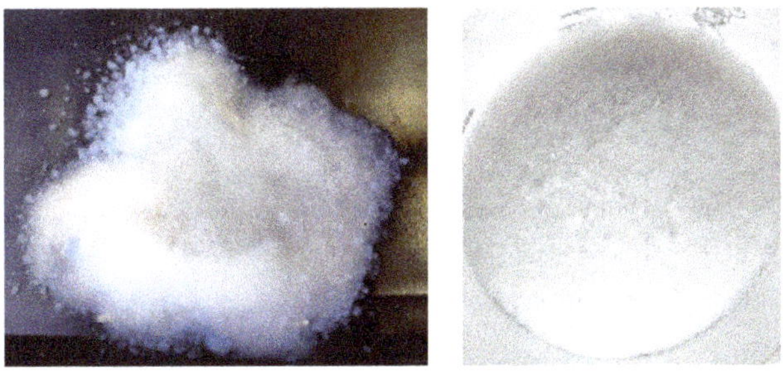

(a) (b)

Figure 3. The initial and final condition of silica aerogel: (a) granules before grinding; (b) grinded powder.

2.2. Concrete Mix Proportion

The composition of the fine-grain concrete mix for 1 L of the mix is presented in Table 2. Grinded aerogel was added to the mix in an amount of 15% by volume.

Table 2. Composition of the fine-grained concrete mix for the reference samples for 1 L.

Portland Cement M400 [g]	Silica Sand [g]	Plasticizer [g]	Water [mL]	Total Volume [L]
687	1253	7	253	1

The properties of the silica sand used for the concrete mix were provided by the manufacturer (REMIX manufacturer LLC, Saint-Petersburg, Russia) as follows: particle size 0–0.63 mm; clay bond amount < 5%; silica amount > 82%. The sand is claimed to be dedusted and thermally treated at 700 °C. The sand complied with Russian state standard GOST 8736-2014 [34] specifications for sand for use in construction works.

The first series of samples that were manufactured and tested are presented in Table 3. In this series, silica aerogel powder was mixed with a dry mix. Then, water was added to the mix. All samples in this study were cured in a humid environment for 28 days after manufacturing [35].

Table 3. The first series of samples, where aerogel was added directly to the concrete mix.

Sample No.	Amount of Mix Used [L]	Amount of Silica Aerogel [%]	Amount of Silica Aerogel [g]
K1–K4	4	0 (reference samples)	0
A1–A3	3	15	66

During the study, it was found that the technology for adding aerogel to the admix should be modified. The causes of technology modification are mentioned in the literature review [17]. In the second series of samples, the aerogel was first mixed with isopropyl alcohol and then added to the dry concrete mix. Information about the samples from the second series is presented in Table 4.

Table 4. The second series of samples, where aerogel was mixed with isopropyl alcohol and then added to the concrete mix.

Sample No.	Amount of Mix Used [L]	Amount of Silica Aerogel [%]	Amount of Silica Aerogel [g]
K2.1–K2.3	3	0 (reference samples)	0
A2.1–A2.3	3	15	66

2.3. Gypsum Putty

The properties of the gypsum putty used in the study were provided by the manufacturer (VOLMA conglomerate, Volgograd, Russia) as follows: drying time (at 20 °C) 24 h; adhesion strength $\geq$ 0.4 MPa; working life: 2 h. The putty complied with Russian standard GOST R 58278-2018 [36]. The mix proportions were chosen in accordance with the manufacturers' recommendations as follows: 0.5 L of water per 1 kg of putty.

2.4. Compressive Strength Tests

The samples for compressive strength tests were manufactured in metal forms. The forms were lubricated with mineral oil, and the concrete mix was poured into the forms. Cubes with side dimensions of 100 mm were made. Figure 4 shows the samples in the wet environment after manufacture.

Figure 4. Curing of manufactured samples in water.

After 28 days of curing in a wet environment, the samples were dried. Every sample was measured and weighed before the tests to quantify the average density of the sample. Compressive strength tests were carried out using a hydraulic press. Figure 5 shows the general view of the machine.

Figure 5. Compressive strength testing.

The analog scale of the press spans 50 tons of load and has 300 divisions. The initial result of the test is the number of divisions on the scale. This number was then converted into MPa to obtain the compression strength of the sample and the average compression strength of the series.

2.5. Three-Point Bending Tests

Three-point bending tests were performed in the Institute of Civil Engineering of Peter the Great St. Petersburg Polytechnic University. An Instron 5965 system (Instron, Norwood, MA, USA) was used. The dual-column model can perform tensile, compression, flexure/bend, peel, tear, shear, and cyclic tests [37]. The operations were controlled from a computer using software that allows automatic measurements and the recording of the measurement results in a graphical form. The beam samples were manufactured for the bending test. A 20 cm-long specimen was placed on two supports. The span between the supports was 15 cm. The moving part of the universal testing machine was placed in the center of the span. The loading of the element continued until the element lost stability, i.e., until the formation of a destructive crack. The stress was calculated as follows:

$$\sigma = (3^* \times F \times l)/(2 \times b \times h^2), \quad (1)$$

where F is the breaking force; l is the length of the span; b is the width of the beam; and h is the height of the beam. Figure 6 shows the general view of the test.

Figure 6. Three-point bending test: the sample in the machine.

2.6. Thermal Conductivity Measurements

The thermal conductivity of the samples was evaluated using a PIT-2.1 thermal conductivity meter (Iztech, Kazan, Russia). The machine creates a stationary heat flow through a flat sample. The heat flow is perpendicular to the front faces of the sample. The power required to create the heat flow and the temperature of the faces was measured. The machine contains refrigerators that are designed to ensure the operation of PIT-2.1 at average sample temperatures below the ambient temperature. The refrigerators are made using Peltier elements [38]. The machine is able to plot the measured parameters during the experiment. The size of the samples was 250 × 250 × 30 mm. The ambient temperature was 21 °C. Figure 7 presents the general view of the machine.

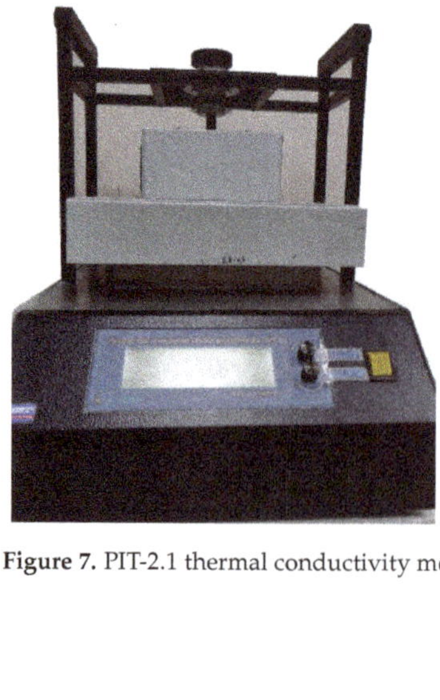

Figure 7. PIT-2.1 thermal conductivity meter.

3. Results

3.1. Compressive Strength Tests

Table 5 presents the calculation of the compression strength of the samples. The results were compared to State Standard (GOST) 26633-2015 [39] to grade the concrete.

Table 5. Compressive strength of manufactured concrete samples.

Sample No.	Weight [kg]	Volume [cm^3]	Average Density [kg/m^3]	Number of Divisions	Load [t]	Compressive Strength [MPa]	Average Compressive Strength [MPa]	Concrete Grade
K1	2.003	1001.6	1983.28	116	19.33	18.93	19.37	B15 (M200)
K2	2.010	1006.7		117	19.50	19.12		
K3	1.994	1004.4		123	20.50	20.10		
K4	2.007	1027.9		118	19.67	19.32		
A1	1.899	1036.4	1861.09	76	12.67	12.42	12.09	B10 (M150)
A2	1.892	1018.3		71	11.83	11.60		
A3	1.923	1015.8		75	12.5	12.26		
K2.1	1.913	970.1	1973.35	128	21.33	20.92	21.08	B15 (M200)
K2.2	1.943	995		128	21.333	20.92		
K2.3	1.951	977.9		131	21.833	21.41		
A2.1	1.907	985.5	1931.64	88	14.67	14.38	14.44	B10 (M150)
A2.2	1.927	996.5		91	15.17	14.87		
A2.3	1.915	994.3		86	14.33	14.05		
A2.4	1.907	985.5		88	14.67	14.38		

Changing the aerogel mixing technology increased the average compressive strength of the samples by 19%. However, the overall reduction in compressive strength relative to the reference samples is 31.5%. Figure 8 illustrates the results of the compression strength tests in graphical form.

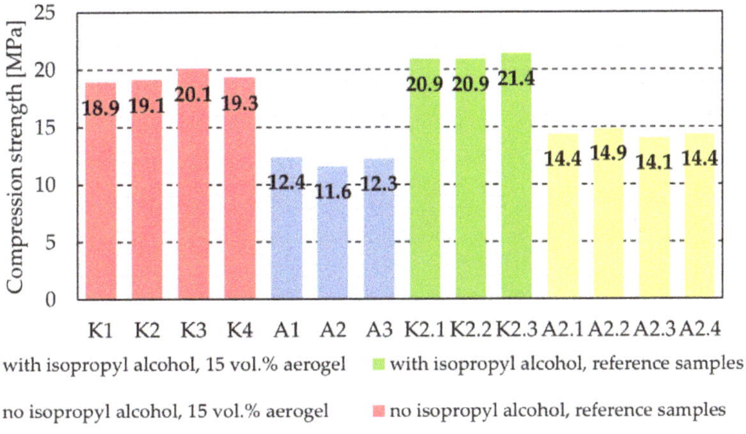

Figure 8. Compression strength of the specimens.

3.2. Three-Point Bending Tests

Bending tests were performed with the second series of samples. The results of the three-point bending tests are presented in Table 6.

Table 6. Bending strength of manufactured concrete samples.

Sample No.	Span [m]	Height of the Sample [m]	Width of the Sample [m]	Breaking Force [N]	Breaking Moment [N·m]	Bending Strength [MPa]	Average Bending Strength [MPa]
K2.1(b)		0.018	0.077	505.24	18.947	4.40	
K2.2(b)	0.15	0.017	0.077	448.36	16.814	4.43	4.61
K2.3(b)		0.020	0.077	681.68	25.563	5.01	
A2.1(b)		0.017	0.057	283.67	10.638	3.76	
A2.2(b)	0.15	0.019	0.057	404.93	15.185	4.23	4.19
A2.3(b)		0.016	0.057	259.2	9.720	3.10	
A2.4(b)		0.014	0.057	225.81	8.468	4.77	

The silica aerogel admix in an amount of 15% by volume decreased the bending strength of the samples by 9.19%. The results of the compression strength tests in graphical form are depicted in Figure 9.

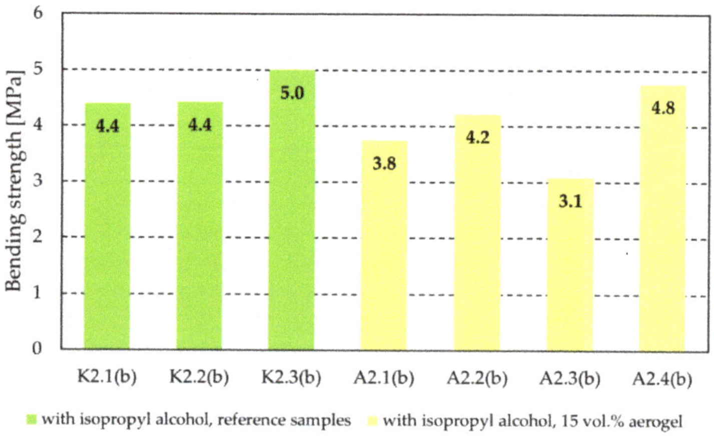

Figure 9. Bending strength of the specimens.

3.3. *Thermal Conductivity Measurements*

The thermal conductivity (K-values) of the aerogel-enhanced samples was measured twice. First, the bare sample was placed into the measuring machine. After the test, the faces of the samples were covered with a fine coat of gypsum putty. The samples were then dried, and the second set of measurements was performed. The second set of measurements was performed to evaluate the influence of the surface porosity on the measurements, as the aerogel admix increases the porosity of the material. The results of the measurements are presented in Table 7.

Table 7. Thermal conductivity of silica aerogel-enhanced concrete samples.

Sample no.	Weight [kg]	Average Density [kg/m^3]	Average K-Value of the Bare Sample [W/(m·K)]	Average K-Value of the Sample with Putty [W/(m·K)]
A1(K)	3709	1902	0.8295	0.642
A2(K)	3708.05	1921	1.1285	0.844

The thermal conductivity of the reference sample is 1.5 W/(m·K). The thermal conductivity of the puttied sample is lower by 45%, and the thermal conductivity of the untreated sample is lower by 25%. The decrease in the thermal conductivity of silica aerogel-enhanced samples compared to the reference sample thermal conductivity is illustrated in Figure 10.

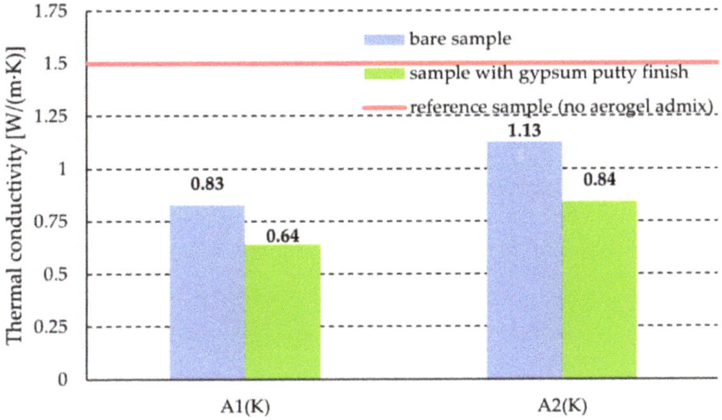

Figure 10. Thermal conductivity of aerogel-enhanced concrete samples.

3.4. Thermotechnical Calculations for Silica Aerogel-Enhanced Concrete

Heat-insulating concrete is used to manufacture structural thermal insulation elements. The high insulating properties of the concrete are achieved because of the low density and high porosity of the concrete mix in accordance with Russian Standard GOST 25820-2021 [40]. The thermal resistance of a regular concrete wall with a width $\delta = 0.3$ m, a density of 2400 kg/m^3, and a K-value $\lambda = 1.5$ W/(m·K) is calculated using the following equation:

$$R = 1/\alpha_{int} + \delta/\lambda + 1/\alpha_{ext} = 1/8.7 + 0.3/1.5 + 1/23 = 0.358\ (m^2 \cdot K)/W. \quad (2)$$

where α_{int} is the heat transfer coefficient of the internal surface of the enclosing structure, $\alpha_{int} = 8.7$; α_{ext} is the heat transfer coefficient on the outer surface of the enclosing structure, $\alpha_{ext} = 23$. To determine the amount of heat required, the heat loss through the enclosing structures should be evaluated. In this particular case, only the heat loss through the external walls of the building is considered. According to Table 3 of the Russian state set of the construction rules, SP 131.13330.2020, "SNiP 23-01-99* Building climatology", Ministry of Construction, Housing and Utilities of the Russian Federation: Moscow, Russia, 2020, the temperature of the coldest five-day period with a probability of 0.92 in St. Petersburg (Russia) and the Leningrad region (Russia) is 249 K (−24 °C), and the heating period in region T is 211 days. In accordance with clause 4 of Russian Standard GOST 30494-2011 [41], the minimum temperature in a room is 291 K (+18 °C), and the maximum relative humidity is 55%. The amount of heat loss Q through an area S of 1 m^2 of a building envelope with thermal resistance R and a temperature difference Δt of 44 K during the heating period T [h] is calculated using the following formula:

$$Q = (S \cdot \times \Delta t \times \cdot T)/R = (1 \times (20 - (-24) \times (211 \times 24)))/0.41 = 543.45\ kWh. \quad (3)$$

The thermal conductivity of the aerogel-enhanced concrete was assumed from the experiments to be 0.844 W/(m·K) (Table 7). Natural gas was chosen as the fuel because it is the most popular type of fuel used for heating in Russia. The parameters of natural gas, such as its calorific value, density, and the amount of emissions per kg of gas used, are considered in the calculation of greenhouse gas emissions. The calculation is presented in Table 8.

Table 8. Calculation of heat loss and greenhouse gas emissions.

Quantity Name	Value
K-value of the material [W/(m·K)]	0.844
R-value of the wall [(m^2·K)/W]	0.514
Façade area [m^2]	1
Temperature difference [K]	42
Heating season [h]	5064
Heat loss through external walls [W·h]	413,893.7
Heat loss through external walls [MJ]	1490
Calorific capacity of natural gas for 1 MJ/m^3	33.5
Natural gas consumption per year [m^3]	44.48
Natural gas density [kg/m^3]	0.68
Mass of consumed natural gas [kg]	30.25
Greenhouse gas emissions [kg CO_2e per 1 kg of consumed gas]	2.64
Total greenhouse gas emissions [kg CO_2e per year]	80
Heat loss through external walls [Gcal]	0.356

The calculations of heat loss through external walls and greenhouse gas emissions during the heating season show that the use of silica aerogel-enhanced concrete would reduce the consumption of natural gas by 30% compared to regular concrete.

3.5. Microscopical Analysis of the Samples

The microscopical analysis was performed using an electronic microscope, Dino-Lite, with a magnification range of 10× to 250×. The observations made with a magnification of 50× are shown in Figure 11. The pores on the section surface of the sample are shown in Figure 11a. The section surface was exposed after the breaking test. The pores on the face surface of the sample are shown in Figure 11b. The major pores are highlighted in red circles.

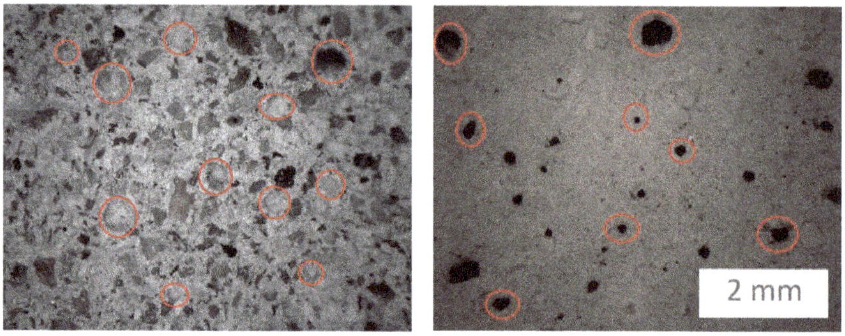

(a) (b)

Figure 11. Sample surfaces at 50× magnification: (**a**) section surface of the sample (after the bending test); (**b**) face surface of the sample. The major pores are highlighted in red circles.

Numerous pores were observed on the sample surfaces. The large number of pores decreases the thermal conductivity of the concrete. However, open pores observed on the surface of a concrete sample can negatively affect the properties of the material.

4. Discussion

The selection of the optimal aerogel-based concrete mix was presented in [42]. The amount of added silica aerogel varied from 50% to 75% by volume. The compressive strength of the samples varied between 3.0 and 23.6 MPa. The most interesting mix contained 60% aerogel granules by volume, and the thermal conductivity of the composite was 0.26 W/(m·K). The compressive strength of the samples was 8.3 MPa. The silica aerogel admix decreased the thermal conductivity of cement paste by 23.7%, 44.8%, and 68.75% as the amount of silica aerogel in the mix was increased from 0% to 20%, from 20% to 40%, and from 40% to 60% by volume, respectively [43]. The data obtained indicate that a large amount of silica aerogel in the mix leads to a decrease in thermal conductivity but has a negative effect on the compressive strength and permeable porosity of the cementitious composite.

Thus, it can be concluded that the nature, type, and quantity of the silica aerogel admix significantly influence the composite properties. A specific trend should be identified for silica aerogel-incorporating concrete composites. As mentioned, pre-treatment can enhance the adhesion between the hydrophobic aerogel and the cement matrix. Still, there are no detailed studies on such reaction mechanisms or the effect of pre-treated aerogel. The impact of pre-treatment on the mechanical and thermal properties of aerogel-based cementitious composites is also not yet clear. The stability of the aerogel particles in the cement matrix may be related to their hydrophobic properties. The particles repel the surrounding water and, therefore, prevent cement hydration [44].

The use of such enhanced concrete would reduce the thickness of the wall and lighten the weight of the structure, increasing cost efficiency and decreasing dead loads. Studying the stability of aerogel particles in concrete is also important because the alkaline environment can affect the aerogel through reactions similar to the well-known alkali-silica reaction. To achieve the indicated results, scientists varied the ratio of the components of the concrete mixture and the technological processes used to manufacture the composite samples.

The importance of passive design is highlighted in [45,46]. Efficient building envelopes, energy-efficient appliances, and renewable energy sources are required to reduce greenhouse gas emissions and the effects of global warming. It is necessary to evaluate the design and increase energy efficiency when choosing materials and additives for building envelopes [47–49]. Many innovations in high-performance construction materials have enabled the design and construction of sustainable and durable infrastructure. For instance, concrete is the most widely used building material in the world. Fibers are used worldwide to develop fiber-reinforced cementitious composites with high ductility to overcome the brittleness of concrete. In recent years, low-heat portland cement has been used to lower the cement hydration heat in an attempt to reduce the risk of thermal cracking. To improve the mechanical properties and microstructure of concrete, nanomaterials, such as silica fume, have been adopted [50]. In this study, a series of concrete composites with silica aerogel admix were manufactured and tested.

The compressive strength of the silica-aerogel-enhanced sample is 37.6% lower than that of the reference sample. Even with the modified technology for adding aerogel, the sample retains a relatively low compressive strength. The compressive strength of the second series of samples is 31.5% lower than that of the reference sample. Hence, silica aerogel admix significantly reduces the compressive strength of the concrete composite.

The bending strength of the silica aerogel-enhanced concrete samples was measured. The average bending strength of the composite was 4.19 MPa. These results are consistent with those presented in [20]. The authors of the abovementioned study used granulated aerogel. Because the results obtained are similar, it can be concluded that the fraction of the silica aerogel admix is not very relevant to the decrease in bending strength.

Aerogel has a low thermal conductivity of about 0.01–0.02 W/(m·K) [51,52]. It is apparent that the incorporation of aerogel particles into concrete will result in a thermally insulating composite since thermal conduction through the solid phase dominates the thermal conductivity of concrete materials [53]. However, the application of aerogel in thermally insulating concrete has not been widely studied, probably due to the high manufacturing cost of aerogel [54]. The grade of hydrophobicity turned out to have a major influence on the compressive and flexural strengths. It was found that the hardened concrete bulk density and the water-cement ratio were correlated with the compressive strength [55]. A high percentage of additive content in the concrete composite leads to an increase in porosity. Open pores form on the surface of the samples as well. The presence of the pores affects the thermal conductivity of the sample. The samples were covered with a fine coat of gypsum putty to close the surface pores. The thermal conductivity of the samples was measured before and after coating them with putty. The thermal conductivity of the reference sample was 1.5 W/(m·K). The thermal conductivity of the sample with putty was 50% lower than the reference value. The thermal conductivity of the sample without putty was 37.4% lower than that of the control sample.

The R-value of the wall made of silica aerogel-enhanced concrete was calculated in the study: R = 0.514 (m^2·K)/W. Annual heat loss through 1 m^2 of such a wall would equal 413.9 kWh. The use of concrete with aerogel admix reduced the weight of the structure and improved the thermal insulation properties. The heat transfer resistance of a silica aerogel-enhanced concrete wall is higher than that of conventional concrete, but it still does not meet the requirements of thermal protection. It is advisable to use this type of concrete in conjunction with a thin layer of insulation.

Further, silica-aerogel-enhanced concrete samples can be tested for weather resistance. Such composite material can be used as an outer layer of enclosing structures, especially during the reconstruction of historical buildings. The weather resistance of the silica aerogel-enhanced concrete was measured using a weatherometer. Based on the observed literature, standards for aerogel-enhanced concrete composites should be developed. In future research, it is planned to evaluate the life cycle of the concrete composite and its economic efficiency.

5. Conclusions

The novelty of this work is the creation of a fine-grain cementitious composite containing 15 vol.% silica aerogel admix. The mechanical and thermal properties of the composite were examined. The dry mix contained portland cement (M400) and silica sand (particle size 0–0.63 mm). Plasticizer was added to the concrete mix to improve the flow rate and rheology of the mix. The detailed properties of the concrete composite, the suggested manufacturing technology improvements, and the assessment of the use of the composite in the enclosing walls have practical significance for further research in aerogel-enhanced cementitious composites. The primary conclusions drawn from the results of the experiments and calculations are as follows:

- Samples of silica-aerogel-enhanced concrete were made. Two series of samples were manufactured. The technology was modified in the second series so that the mechanical properties of the composite were enhanced. For the second batch, the silica aerogel was first mixed with isopropyl alcohol and then added to the dry mix.
- The average compressive strength of the first series' samples was 12 MPa; of the second series' samples, 14 MPa; and for the reference samples, the average compressive strength was 20 MPa.
- The average bending strength of the first series' samples was 4.2 MPa, and the bending strength of the reference samples was 4.6 MPa.
- The average thermal conductivity (K-value) of the samples of the first series was 0.83 W/(m·K); of the samples of the second series, it was 1.13 W/(m·K). The samples were covered with a fine coat of gypsum putty. The average thermal conductivity of the coated samples was 0.64 for the first series and 0.84 for the second series.
- From the calculation of heat loss through external walls and greenhouse gas emissions during the heating season, it can be concluded that the use of silica aerogel-enhanced concrete for external walls would reduce the consumption of natural gas by 30% compared to regular concrete.
- The R-value of the wall made of silica aerogel-enhanced concrete was 0.514 (m^2·K)/W.
- The research can be expanded by manufacturing and testing more samples to specify the composite properties, develop the fabrication technique, and enhance the cost efficiency of the material.

This research will be expanded by investigating the relationship between the properties of the composite and the amount of silica aerogel added to the mixture. To reveal this correlation, a series of samples containing various amounts of silica aerogel will be manufactured and tested. The obtained experimental results will complement the existing data on this problem. Manufacturing plants can use the obtained results to produce aerogel-enhanced cementitious composites for the construction industry. The authors propose the use of such composites in the production of building panels for the construction of panel buildings.

The authors also plan to evaluate the deterioration of the properties of the silica-aerogel-enhanced composite as the material ages. For this purpose, the properties of the samples will be tested after accelerated material aging in a climatic chamber and a xenon weather aging chamber. An increase in the thermal conductivity of the composite during aging will result in an increase in heating expenses. Because not all manufacturers are willing to provide access to the environmental product declaration (EPD), calculations of greenhouse gas emissions have to be made based on available public data on the fuel burned.

Author Contributions: Conceptualization, T.A.K.; Methodology, T.A.K.; Software, D.V.N.; Validation, T.A.K.; Formal analysis, T.A.K. and A.E.D.; Investigation, A.E.D.; Resources, D.V.N.; Data curation, A.E.D.; Writing—original draft, T.A.K.; Writing—review & editing, A.E.D.; Project administration, D.V.N.; Funding acquisition, D.V.N. All authors have read and agreed to the published version of the manuscript.

Funding: This research was funded by the Russian Science Foundation under grant 21-79-10283, date 29 July 2021, https://rscf.ru/en/project/21-79-10283/ (accessed on 2 April 2024).

Data Availability Statement: Data is contained within the article. Further inquiries can be directed to the corresponding author.

Conflicts of Interest: The authors declare no conflicts of interest.

References

1. Gamayunova, O.S. Choice of the Optimal Option of Warming of Residential Houses in Various Climate Zones. *Constr. Ind. Saf.* **2019**, *16*, 89–97.
2. Karaeva, A.P.; Magaril, E.R.; Kiselev, A.V.; Cioca, L.-I. Screening of Factors for Assessing the Environmental and Economic Efficiency of Investment Projects in the Energy Sector. *Int. J. Environ. Res. Public Health* **2022**, *19*, 11716. [CrossRef] [PubMed]
3. Wang, L.; Dong, Y.; Zhou, S.H.; Chen, E.; Tang, S.W. Energy Saving Benefit, Mechanical Performance, Volume Stabilities, Hydration Properties and Products of Low Heat Cement-Based Materials. *Energy Build.* **2018**, *170*, 157–169. [CrossRef]
4. Zubarev, K.; Timofeeva, M. Special Aspects of Calculation of Infiltration in Residential and Public Buildings. *E3S Web Conf.* **2023**, *389*, 06001. [CrossRef]
5. Vasileva, I.L.; Nemova, D.V. Prospects of Using Aerogels in Construction. *AlfaBuild* **2018**, *6*, 135–145.
6. Barnyakov, A.Y.; Barnyakov, M.Y.; Beloborodov, K.I.; Bobrovnikov, V.S.; Buzykaev, A.R.; Golubev, V.B.; Gulevich, B.V.; Danilyuk, A.F.; Kononov, S.A.; Kravchenko, E.A.; et al. Status of Aerogel Production in Novosibirsk. *Nucl. Instrum. Methods Phys. Res. Sect. A Accel. Spectrometers Detect. Assoc. Equip.* **2011**, *639*, 225–226. [CrossRef]
7. Akimov, Y.K.; Zrelov, V.P.; Puzynin, A.I.; Filin, S.V.; Filippov, A.I.; Sheinkman, V.A. An Aerogel Threshold Cerenkov Counter. *Instrum. Exp. Tech.* **2002**, *45*, 634–639. [CrossRef]
8. Ratke, L. Herstellung und Eigenschaften Eines Neuen Leichtbetons: Aerogelbeton. *Beton-und Stahlbetonbau* **2008**, *103*, 236–243. [CrossRef]
9. Peng, F.; Jiang, Y.; Feng, J.; Cai, H.; Feng, J.; Li, L. Thermally Insulating, Fiber-Reinforced Alumina–Silica Aerogel Composites with Ultra-Low Shrinkage up to 1500 °C. *Chem. Eng. J.* **2021**, *411*, 128402. [CrossRef]
10. Liu, P.; Gao, H.; Chen, X.; Chen, D.; Lv, J.; Han, M.; Cheng, P.; Wang, G. In Situ One-Step Construction of Monolithic Silica Aerogel-Based Composite Phase Change Materials for Thermal Protection. *Compos. Part B Eng.* **2020**, *195*, 108072. [CrossRef]
11. Chen, Y.X.; Klima, K.M.; Brouwers, H.J.H.; Yu, Q. Effect of Silica Aerogel on Thermal Insulation and Acoustic Absorption of Geopolymer Foam Composites: The Role of Aerogel Particle Size. *Compos. Part B Eng.* **2022**, *242*, 110048. [CrossRef]
12. Chen, F.; Zhang, Y.; Liu, J.; Wang, X.; Chu, P.K.; Chu, B.; Zhang, N. Fly Ash Based Lightweight Wall Materials Incorporating Expanded Perlite/SiO2 Aerogel Composite: Towards Low Thermal Conductivity. *Constr. Build. Mater.* **2020**, *249*, 118728. [CrossRef]
13. Baetens, R.; Jelle, B.P.; Gustavsen, A. Aerogel Insulation for Building Applications: A State-of-the-Art Review. *Energy Build.* **2011**, *43*, 761–769. [CrossRef]

14. Ganobjak, M.; Brunner, S.; Hofmann, J.; Klar, V.; Ledermann, M.; Herzog, V.; Kämpfen, B.; Kilian, R.; Wehdorn, M.; Wernery, J. Current Trends in Aerogel Use in Heritage Buildings: Case Studies from the Aerogel Architecture Award 2021. *Gels* **2023**, *9*, 814. [CrossRef] [PubMed]
15. Safiuddin, M.; Gonzalez, M.; Cao, J.; Tighe, S.L. State-of-the-Art Report on Use of Nano-Materials in Concrete. *Int. J. Pavement Eng.* **2014**, *15*, 940–949. [CrossRef]
16. Hub, A.; Knippers, J.; Zimmermann, G.; Teichmann, T.; Degner, F. Aerogels as High-Performance Aggregates for Lightweight Concrete—Working Principle and Possible Areas of Use; [Aerogele Als Hochleistungszuschlag Für Leichtbeton—Wirkungsweise Und Einsatzmöglichkeiten]. *Betonw. Fert. Plant Precast Technol.* **2015**, *81*, 60–70.
17. Zhan, W.; Chen, L.; Kong, Q.; Li, L.; Chen, M.; Jiang, J.; Li, W.; Shi, F.; Xu, Z. The Synthesis and Polymer-Reinforced Mechanical Properties of SiO_2 Aerogels: A Review. *Molecules* **2023**, *28*, 5534. [CrossRef]
18. Babashov, V.; Varrik, N.; Karaseva, T. Use of Aerogels to Create Thermal Insulation Materials (a Review). *Proc. VIAM* **2019**, *78*, 32–42. [CrossRef]
19. AlTawaiha, H.; Alhomaidat, F.; Eljufout, T. A Review of the Effect of Nano-Silica on the Mechanical and Durability Properties of Cementitious Composites. *Infrastructures* **2023**, *8*, 132. [CrossRef]
20. Gao, T.; Jelle, B.P.; Gustavsen, A.; Jacobsen, S. Aerogel-Incorporated Concrete: An Experimental Study. *Constr. Build. Mater.* **2014**, *52*, 130–136. [CrossRef]
21. Fickler, S.; Welsch, T.; Schnellenbach-Held, M. Development of High Performance Aerogel Concrete (HPAC) and Statistical Evaluation of Compressive Strength for Practical Use in Construction. *J. Phys. Conf. Ser.* **2021**, *2069*, 012196. [CrossRef]
22. Ślosarczyk, A.; Vashchuk, A.; Klapiszewski, Ł. Research Development in Silica Aerogel Incorporated Cementitious Composites—A Review. *Polymers* **2022**, *14*, 1456. [CrossRef]
23. Ng, S.; Jelle, B.P.; Sandberg, L.I.C.; Gao, T.; Wallevik, Ó.H. Experimental Investigations of Aerogel-Incorporated Ultra-High Performance Concrete. *Constr. Build. Mater.* **2015**, *77*, 307–316. [CrossRef]
24. Ustundag, O.; Sola, O.C. Effect of Aerogel/Silica Fume under Curing Methods on Properties of Cement-Based Mortars. *Rev. Constr.* **2022**, *21*, 368–386. [CrossRef]
25. Hanif, A.; Diao, S.; Lu, Z.; Fan, T.; Li, Z. Green Lightweight Cementitious Composite Incorporating Aerogels and Fly Ash Cenospheres—Mechanical and Thermal Insulating Properties. *Constr. Build. Mater.* **2016**, *116*, 422–430. [CrossRef]
26. Welsch, T.; Schnellenbach-Held, M. High Performance Aerogel Concrete. In *High Tech Concrete: Where Technology and Engineering Meet*; Springer International Publishing: Cham, Switzerland, 2018; pp. 117–124.
27. Ng, S.; Jelle, B.P.; Stæhli, T. Calcined Clays as Binder for Thermal Insulating and Structural Aerogel Incorporated Mortar. *Cem. Concr. Compos.* **2016**, *72*, 213–221. [CrossRef]
28. Ng, S.; Jelle, B.P.; Zhen, Y.; Wallevik, Ó.H. Effect of Storage and Curing Conditions at Elevated Temperatures on Aerogel-Incorporated Mortar Samples Based on UHPC Recipe. *Constr. Build. Mater.* **2016**, *106*, 640–649. [CrossRef]
29. Liu, K.-S.; Zheng, X.-F.; Hsieh, C.-H.; Lee, S.-K. The Application of Silica-Based Aerogel Board on the Fire Resistance and Thermal Insulation Performance Enhancement of Existing External Wall System Retrofit. *Energies* **2021**, *14*, 4518. [CrossRef]
30. Zhang, S.-N.; Pang, H.-Q.; Fan, T.-H.; Ye, Q.; Cai, Q.-L.; Wu, X. Thermal Insulation Performance of SiC-Doped Silica Aerogels under Large Temperature and Air Pressure Differences. *Gels* **2022**, *8*, 320. [CrossRef]
31. Fedyukhin, A.V.; Strogonov, K.V.; Soloveva, O.V.; Solovev, S.A.; Akhmetova, I.G.; Berardi, U.; Zaitsev, M.D.; Grigorev, D.V. Aerogel Product Applications for High-Temperature Thermal Insulation. *Energies* **2022**, *15*, 7792. [CrossRef]
32. Manufacturer's Website Guandong Alison Technology. Available online: https://www.ydalison.com/en/ProductCenters/list.aspx?lcid=13 (accessed on 15 January 2024).
33. Huang, K.-L.; Li, S.-J.; Zhu, P.-H. Effect of Early Curing Temperature on the Tunnel Fire Resistance of Self-Compacting Concrete Coated with Aerogel Cement Paste. *Materials* **2021**, *14*, 5782. [CrossRef]
34. *GOST 8736-2014*; Sand for Construction Works. Specifications; Interstate Council for Standartization, Metrology and Certification (ISC): Moscow, Russia, 2019.
35. Tierney, L.; Safiuddin, M. Insights into Concrete Forming, Reinforcing, and Pouring in Building Construction. *Buildings* **2022**, *12*, 1303. [CrossRef]
36. *GOST R 58278-2018*; Dry Building Putty Mixes Based on Gypsum Binder. Specifications; Federal Agency for Technical Regulation and Metrology (Rosstandart): Moscow, Russia, 2019.
37. Volkova, A.A.; Paykov, A.V.; Stolyarov, O.N.; Semenov, S.G.; Melnikov, B.E. Structure and Properties of Textile Reinforced Concrete. *Mag. Civ. Eng.* **2015**, *59*, 50–56. [CrossRef]
38. Vatin, N.; Sultanov, S.; Krupina, A. Comparison of Thermal Insulation Characteristics of PIR, Mineral Wool, Carbon Fiber, and Aerogel. In *Advances in Intelligent Systems and Computing*; Springer: Cham, Switzerland, 2019; Volume 983, pp. 877–883, ISBN 9783030198671.
39. *GOST 26633-2015*; Heavy-Weight and Sand Concretes. Specifications; Interstate Council for Standartization, Metrology and Certification (ISC): Moscow, Russia, 2019.
40. *GOST 25820-2021*; Lightweight Aggregates Concretes. Specifications; Interstate Council for Standartization, Metrology and Certification (ISC): Moscow, Russia, 2022.
41. *GOST 30494-2011*; Residential and Public Buildings. Microclimate Parameters for Indoor Enclosures; Interstate Council for Standartization, Metrology and Certification (ISC): Moscow, Russia, 2019.

42. Fickler, S.; Milow, B.; Ratke, L.; Schnellenbach-Held, M.; Welsch, T. Development of High Performance Aerogel Concrete. *Energy Procedia* **2015**, *78*, 406–411. [CrossRef]
43. Khamidi, M.F.; Glover, C.; Farhan, S.A.; Puad, N.H.A.; Nuruddin, M.F. Effect of Silica Aerogel on the Thermal Conductivity of Cement Paste for the Construction of Concrete Buildings in Sustainable Cities. In Proceedings of the WIT Transactions on the Built Environment; WIT Press: Southampton, UK, 2014; Volume 137, pp. 665–674.
44. Jia, G.; Li, Z.; Liu, P.; Jing, Q. Applications of Aerogel in Cement-Based Thermal Insulation Materials: An Overview. *Mag. Concr. Res.* **2018**, *70*, 822–837. [CrossRef]
45. Makvandia, G.; Safiuddin, M. Obstacles to Developing Net-Zero Energy (Nze) Homes in Greater Toronto Area. *Buildings* **2021**, *11*, 95. [CrossRef]
46. Kheiri, F. A Review on Optimization Methods Applied in Energy-Efficient Building Geometry and Envelope Design. *Renew. Sustain. Energy Rev.* **2018**, *92*, 897–920. [CrossRef]
47. Zubarev, K.; Zobnina, Y. Methods of Achieving Energy Efficiency in Buildings and Their Evaluation. *E3S Web Conf.* **2023**, *389*, 06025. [CrossRef]
48. Koryakovtseva, T.; Zaborova, D.; Gamayunova, O. Use of Plant and Coal Waste as a Secondary Raw Material in Concrete Composites. *Constr. Ind. Saf.* **2022**, *79*, 27–37.
49. Kamali, M.; Ghahremaninezhad, A. Effect of Glass Powders on the Mechanical and Durability Properties of Cementitious Materials. *Constr. Build. Mater.* **2015**, *98*, 407–416. [CrossRef]
50. Wang, L.; Tang, S. High-Performance Construction Materials: Latest Advances and Prospects. *Buildings* **2022**, *12*, 928. [CrossRef]
51. Pierre, A.C.; Rigacci, A. SiO_2 Aerogels. In *Aerogels Handbook*; Springer: New York, NY, USA, 2011; pp. 21–45.
52. Ebert, H.-P. Thermal Properties of Aerogels. In *Aerogels Handbook*; Springer: New York, NY, USA, 2011; pp. 537–564.
53. Bentz, D.P. Transient Plane Source Measurements of the Thermal Properties of Hydrating Cement Pastes. *Mater. Struct. Constr.* **2007**, *40*, 1073–1080. [CrossRef]
54. Kim, S.; Seo, J.; Cha, J.; Kim, S. Chemical Retreating for Gel-Typed Aerogel and Insulation Performance of Cement Containing Aerogel. *Constr. Build. Mater.* **2013**, *40*, 501–505. [CrossRef]
55. Welsch, T.; Vievers, Y.; Schnellenbach-Held, M.; Bialuschewski, D.; Milow, B. Comparison of Different Aerogel Granules for Use as Aggregate in Concrete. *Gels* **2023**, *9*, 406. [CrossRef]

buildings

MDPI

Article

Combined Effects of Steel and Glass Fibres on the Fracture Performance of Recycled Rubber Concrete

Xiaohui Li [1], Zezhou Pan [1], Hao Zhen [1], Wenhua Luo [1], Zhuangwei Chen [2], Hongming Li [3], Zhichao Wu [3], Feng Liu [1] and Lijuan Li [1,*]

1 School of Civil and Transportation Engineering, Guangdong University of Technology, Guangzhou 510006, China
2 Guangdong Henghui Construction Group Co., Ltd., Guangzhou 510006, China
3 Guangdong Zhongdu Construction Group Co., Ltd., Lianjiang 524499, China
* Correspondence: lilj@gdut.edu.cn

Abstract: As an environmentally friendly construction material, recycled rubber concrete (RRC) is commonly used as a road material owing to its excellent flexural strength and crack resistance. Previous studies have shown that the addition of fibres is an effective method for improving the crack resistance of concrete. The purpose of this study is to investigate the fracture performance of RRC reinforced with steel fibres (SFs) and glass fibres (GFs). A total of 28 RRC mixtures were prepared. The results of the fracture test showed that the addition of SFs and GFs significantly enhanced the RRC fracture performance. The maximum increases or decreases in flexural strength, brittleness coefficient, fracture energy, initial fracture toughness, and unstable fracture toughness were 64.9, −34.6, 775.6, 92.0, and 118.4%, respectively. The ideal GF content is usually in the range of 0.4–0.6% and decreases with increasing SF content. In addition, scanning electron microscope (SEM) tests were conducted to explore the mechanism of the effect of hybrid fibres on RRC at a microscopic level. The results show that SFs were always pulled out, while GFs were pulled apart at the initial defects. At the same time, excessive GFs caused more initial defects. These results are expected to provide theoretical direction and experimental support for the practical application of hybrid fibre-reinforced recycled rubber concrete (HFRRRC).

Keywords: fracture performance; steel fibres; glass fibres; hybrid fibres; recycled rubber concrete

Citation: Li, X.; Pan, Z.; Zhen, H.; Luo, W.; Chen, Z.; Li, H.; Wu, Z.; Liu, F.; Li, L. Combined Effects of Steel and Glass Fibres on the Fracture Performance of Recycled Rubber Concrete. *Buildings* **2024**, *14*, 864. https://doi.org/10.3390/buildings14040864

Academic Editor: Marco Di Ludovico

Received: 17 February 2024
Revised: 16 March 2024
Accepted: 21 March 2024
Published: 22 March 2024

1. Introduction

The widespread use of traditional concrete has led to excessive exploitation of natural aggregates, which exacerbates the depletion of natural resources and environmental damage. Therefore, many researchers propose using recycled resources and abundant sea sand resources as alternatives [1,2]. Meanwhile, with the development of urbanisation, the number of means of transportation is growing rapidly. This not only requires high-quality road materials, but also produces more waste rubber from scrapped car tires. Waste rubber cannot be degraded naturally. Moreover, waste tires are difficult to dissolve at high temperatures; therefore, it is very difficult for waste tires to be recycled into new rubber products. The most common disposal options for waste rubber tires are direct burning and burial, both of which cause irreversible pollution of the natural environment. To overcome the negative impact of waste rubber on the environment, there is an urgent need to develop a harmless method for consuming large quantities of waste rubber [3]. The use of recycled rubber (RR) in the concrete industry is an attempt to move in this direction. The use of RR as a substitute aggregate not only reduces the environmental impact of waste rubber by consuming it in large quantities, but also significantly reduces the consumption of natural aggregates such as sand and gravel [4–6]. The use of RR in the concrete industry has been attempted for a long time. However, it has been found that some of the key mechanical properties of concrete, including workability, compressive strength, tensile strength, stiffness, and modulus of elasticity, are significantly reduced when rubber is added to concrete [7–9]. Because road materials require a lower modulus of elasticity to improve the comfort of vehicle occupants, the decline in these properties makes recycled rubber concrete (RRC) an ideal road material [10,11]. Furthermore, RRC has a higher energy dissipation capacity than ordinary concrete, which makes it more resistant to vehicle impact loads as a road material [12,13].

Strength, stiffness, and stability are three key factors to consider when designing a structure [14,15]. While it is difficult to improve mechanical properties at the structural level, improving them at the material level is an effective approach. In recent years, more and more new types of high-performance concrete have been studied [16–18]. Concrete pavements are typically damaged by cracking owing to excessive bending and tensile stresses, as concrete itself is a material with good compressive properties and poor fracture properties. Therefore, numerous researchers have explored methods to improve the mechanics properties of RRC [19–21]. One effective method to improve the fracture properties of concrete is to add fibres within the concrete [22–25]. From a comprehensive perspective, the incorporation of fibres can significantly enhance the fracture properties of concrete; however, the enhancement effect of different fibres varies. Overall, the improvement in the fracture performance of concrete by fibres mainly depends on the following three aspects: (1) the type and characteristics of fibres, (2) the direction and distribution of fibres, and (3) the bond between the fibres and the cement matrix. Regarding the types and characteristics of fibres, the selection of fibres with a higher tensile strength and elasticity modulus is beneficial for enhancing the fracture properties of concrete [26,27]. Finer and softer fibres mainly limit the development of microcracks into macroscopic cracks, whereas larger and stiffer fibres mainly limit the development of macroscopic cracks and reduce the crack width [28,29]. Therefore, the addition of hybrid fibres is more effective in improving the fracture properties of concrete. With respect to the direction and distribution of fibres, the fracture resistance of each part within the concrete was enhanced when the fibre material was uniformly distributed within the concrete, thus improving the overall fracture resistance of the concrete. When the direction of the fibre was parallel to the direction of the main tensile stress in the concrete, the fibres effectively improved the crack resistance of the concrete. The direction and distribution of the fibre material within the concrete are mainly related to the workability of the fresh concrete, casting process, size of the specimen, and wall effect of the formwork [30]. Wang et al. studied the strengthening effect of four types of fibres on the fracture performance of rubber concrete. Compared with the control group, all fibres significantly enhanced fracture energy and crack propagation after cracking. Among them, steel fibres (SFs) are the best, increasing the fracture energy by about 50 times. The pull-out resistance of different fibres is not affected by the addition of rubber [31]. Gültekin et al. studied the fracture energy, compressive strength, and flexural strength of self compacting concrete, including glass fibres (GFs) or basalt fibres. Although the addition of fibres usually reduces compressive strength, it significantly improves flexural strength and fracture energy. The flexural strength and fracture energy of glass fibre reinforced concrete have been increased by 58.6% and 55.1%, respectively [32]. Muhyaddin studied the hybridisation of GF, micro steel fibre (MSF) and long hooked steel fibre (HSF) to improve the mechanical and fracture properties of ultra-high performance concrete (UHPC). The results show that the performance order is MSF+HSF, MSF+GF, and HSF+GF. The use of a single GF does not improve ductility and may even exhibit brittle failure [33]. In addition, excessive fibre addition not only tends to cause fibre agglomeration, but also weakens the bond interface between the fibres and concrete [34]. Therefore, although the addition of fibres enhances the fracture properties of concrete, the type, characteristics, and content of fibres need to be carefully selected to maximise their role.

At present, there is limited research on the fracture performance of SF and GF reinforced concrete, especially on RRC. Compared to ordinary concrete, RRC often has more internal defects, and the mechanism of the fibre-combined effect in RRC needs to be explored. This is significant for improving the crack resistance performance of rubber concrete pavement, thereby increasing the service life of the pavement. Therefore, this study investigated the fracture performance of RRC containing SFs and GFs, and analysed the influence mechanism of SF and GF on the fracture performance of RRC through microscopic experiments. The results of this research provide important guidelines for producing and using RRC.

2. Experimental Setup

2.1. Raw Materials

In this study, the strength grade and specific gravity of ordinary Portland cement were 42.5R and 3.11, respectively. The properties of cement are shown in Table 1. The experimental water was tapping water from Guangdong University of Technology, with a specific gravity of 1.00. River sand with a maximum particle size of 5 mm and 20 mesh waste rubber were used as fine aggregates. Table 2 shows the river sand parameters obtained according to GB/T14684-2011 [35]. The parameters of the RR provided by the recycled rubber production company are shown in Table 2. The coarse aggregate was made of granite crushed stone with a particle size of 5–16 mm. Table 2 shows the coarse aggregates parameters obtained according to GB/T14685-2011 [36]. The gradation distributions of the sand, RR, and coarse aggregates obtained through the screening test are shown in Figure 1. A straight copper-plated SF with a length of 12 mm was used according to the parameters provided by the supplier. Two lengths of GFs were used in this study, which were 6 and 12 mm. The physical parameters of the fibres are listed in Table 3. Finally, the superplasticiser (SP) was used to enhance the workability of RRC. The solid content and specific gravity of the SP were 9% and 1.02, respectively.

Table 1. The properties of cement.

Contents	CaO	SiO_2	Fe_2O_3	Al_2O_3	SO_3	MgO	Fineness	Loss on Ignition
Composition (%)	63–67	19–23	4–6	3–7	1.9	1	1.1	1.7

Table 2. The basic performance of fine and coarse aggregates.

Aggregate Type	Particle Size (mm)	Apparent Density (kg/m^3)	Bulk Density (kg/m^3)	Water Absorption (%)
Sand	<5	2636	1543	0.5
Recycled rubber	<2.5	750	-	-
Coarse aggregate	5–16	2641	1344	2.1

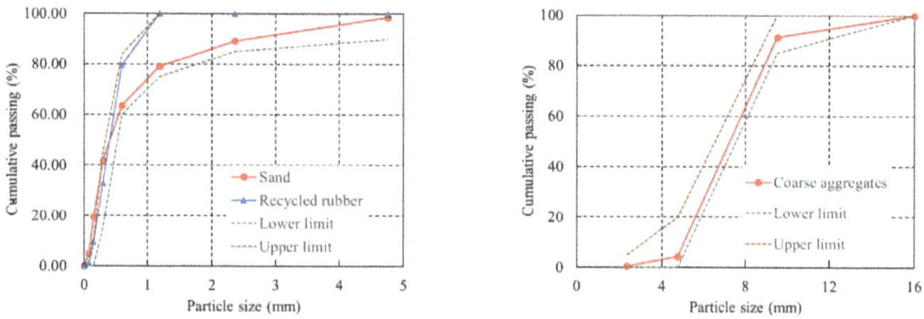

(a) (b)

Figure 1. Passing rate of aggregates: (**a**) fine aggregates; (**b**) coarse aggregates.

Table 3. Properties of SFs and GFs.

Fibre Type	Length (mm)	Specific Gravity	Equivalent Diameter (μm)	Tensile Strength (MPa)	Elastic Modulus (GPa)
Steel fibres (SFs)	12	7.8	200	3000	200
Glass fibres (GFs)	6/12	2.68	14	1700	72

2.2. Concrete Mixtures

To explore the anti-cracking and toughening effects of SF and GF on the fracture performance of RRC, 28 mix proportions were designed in this study, as shown in Table 4. Fibres were added according to the volume percentage of concrete. The content of SFs was designed as 0, 0.4%, 0.8% and 1.2%, and the content of GFs was designed as 0, 0.2%, 0.4% and 0.6%. In this study, rubber content was set to 10%. RR was added according to the volume percentage of the fine aggregate. The SP was added according to the mass percentage of the cement, which was 0.5% in this study. The water-cement ratio was 0.4, and water consumption was adjusted based on the moisture content and water absorption of the aggregates. Each mix proportion is marked in the SxGyLz format. Sx indicates that the SF content is x%. Similarly, Gy indicates that the GF content is y%. Lz is the length of the GFs, which is z mm. If the mix number is without Lz, this implies that no GFs have been added.

Table 4. Test mixture (kg/m^3).

Mix Number	Cement	Water	GFs	SFs	Coarse Aggregate	Sand	Recycled Rubber	SP
S0G0	554.10	245.30	0.00	0.00	966.30	531.20	17.10	2.80
S0.4G0	551.90	244.40	0.00	31.20	962.40	529.10	17.00	2.80
S0.8G0	549.70	243.40	0.00	62.40	958.50	527.00	16.90	2.80
S1.2G0	547.50	242.40	0.00	93.60	954.70	524.80	16.90	2.70
S0G0.2L6	553.00	244.90	5.40	0.00	964.30	530.20	17.00	2.80
S0.4G0.2L6	550.80	243.90	5.40	31.20	960.50	528.00	17.00	2.80
S0.8G0.2L6	548.60	242.90	5.40	62.40	956.60	525.90	16.90	2.70
S1.2G0.2L6	546.40	241.90	5.40	93.60	952.70	523.80	16.80	2.70
S0G0.4L6	551.90	244.40	10.70	0.00	962.40	529.10	17.00	2.80
S0.4G0.4L6	549.70	243.40	10.70	31.20	958.50	527.00	16.90	2.80
S0.8G0.4L6	547.50	242.40	10.70	62.40	954.70	524.80	16.90	2.70
S1.2G0.4L6	545.20	241.40	10.70	93.60	950.80	522.70	16.80	2.70
S0G0.6L6	550.80	243.90	16.10	0.00	960.50	528.00	17.00	2.80
S0.4G0.6L6	548.60	242.90	16.10	31.20	956.60	525.90	16.90	2.70
S0.8G0.6L6	546.40	241.90	16.10	62.40	952.70	523.80	16.80	2.70
S1.2G0.6L6	544.10	240.90	16.10	93.60	948.90	521.70	16.70	2.70
S0G0.2L12	553.00	244.90	5.40	0.00	964.30	530.20	17.00	2.80
S0.4G0.2L12	550.80	243.90	5.40	31.20	960.50	528.00	17.00	2.80
S0.8G0.2L12	548.60	242.90	5.40	62.40	956.60	525.90	16.90	2.70
S1.2G0.2L12	546.40	241.90	5.40	93.60	952.70	523.80	16.80	2.70
S0G0.4L12	551.90	244.40	10.70	0.00	962.40	529.10	17.00	2.80
S0.4G0.4L12	549.70	243.40	10.70	31.20	958.50	527.00	16.90	2.80
S0.8G0.4L12	547.50	242.40	10.70	62.40	954.70	524.80	16.90	2.70
S1.2G0.4L12	545.20	241.40	10.70	93.60	950.80	522.70	16.80	2.70
S0G0.6L12	550.80	243.90	16.10	0.00	960.50	528.00	17.00	2.80
S0.4G0.6L12	548.60	242.90	16.10	31.20	956.60	525.90	16.90	2.70
S0.8G0.6L12	546.40	241.90	16.10	62.40	952.70	523.80	16.80	2.70
S1.2G0.6L12	544.10	240.90	16.10	93.60	948.90	521.70	16.70	2.70

2.3. Preparation of Test Specimens

Beams with lengths (l), widths (t), and heights (h) of 400, 100, and 100 mm, respectively, were used as fracture specimens [37]. The height (a_0) and width of the pre-cracks were 30 and 1 mm, respectively. A schematic diagram of the size of the specimen is shown in Figure 2. Three beams with pre-cracks were prepared for each mix proportion as fracture specimens, for a total of 84. To obtain the compressive strength, three cylinders with diameters of 100 mm and heights of 200 mm were prepared for each mix proportion as axial compression specimens.

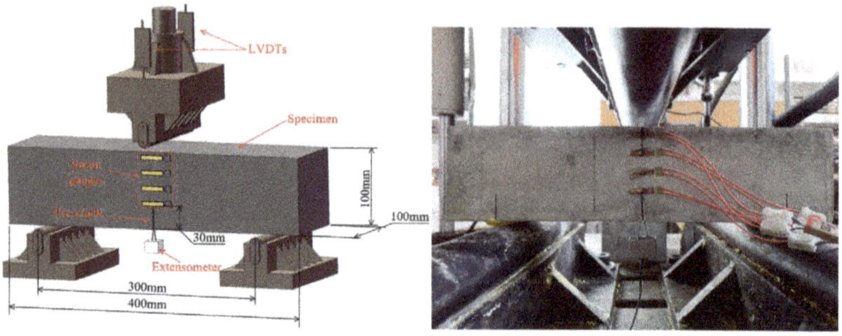

Figure 2. Test setups of the fracture test.

The preparation process of specimens is shown in Figure 3. To evenly distribute the GFs and SFs within the RRC, the preparation process of hybrid fibre-reinforced recycled rubber concrete (HFRRRC) is divided into eight steps. (1) First, cement, sand, and RR were added to a mixer for dry mixing. After starting the mixer, SFs and GFs were uniformly added, which lasted for 60 s. (2) After water and SP were fully mixed, around 70% of the mixture was poured to the mixer and operated for 60 s. (3) The remaining 30% mixture and coarse aggregates were poured into the mixer, and operated the mixer for 180 s to thoroughly mix it. (4) Fresh concrete was poured into the moulds and vibrated for 1 min. The vibration time should not be excessively long to prevent the SF from sinking to the bottom and the rubber particles from floating. (5) Aluminium sheets with a thickness of 1 mm were vertically inserted into the fresh concrete to form a pre-crack. (6) Fresh concrete was smoothened along the surface of the test mould. To prevent the evaporation of internal water and to affect the hydration rate of the cement, the specimens were covered tightly with plastic and wet linen cloths. (7) After the initial setting of the fresh concrete, the aluminium sheets were carefully pulled out with iron tongs. (8) After curing for 24 h, the specimens were demoulded and cured in water for 28 days. Finally, the specimens were removed and wiped dry before mechanical testing.

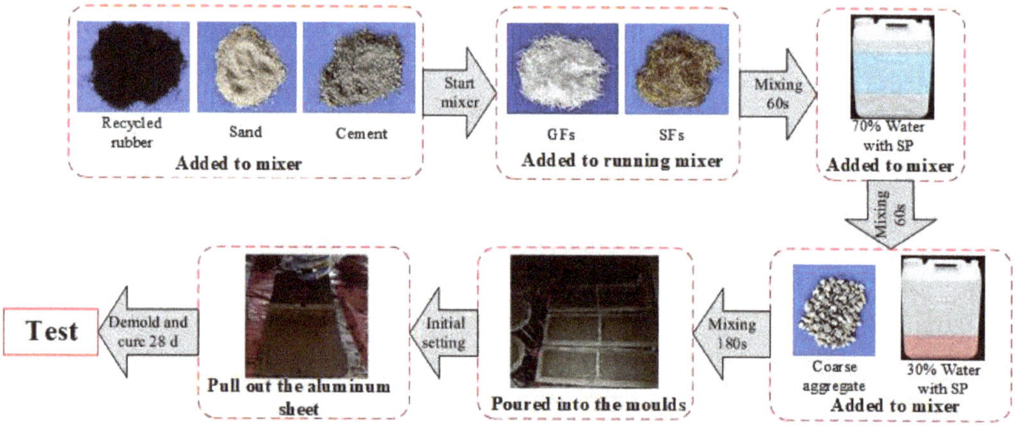

Figure 3. The preparation process of specimens.

2.4. Testing Schemes

Axial deformation in the axial compression test was measured using two linear variable differential transformers (LVDTs). The LVDTs were mounted in the middle of the specimen at a distance of 80 mm. The loading procedure for the compression experiments was displacement controlled, which was 0.18 mm/min. The compressive strength of each specimen was determined according to the standard ASTM C39/C39M [38].

An electro-hydraulic servo static and dynamic universal testing machine with a maximum capacity of 500 kN was used for the fracture tests. The fracture test setup is illustrated in Figure 2. The support span, s, was 300 mm. Two LVDTs were symmetrically placed in the mid span of the test specimens to measure the deflection. Four strain gauges were used to test the crack development in each specimen. Crack mouth opening displacement (CMOD) was measured using an extensometer. To fix the extensometer, two aluminium sheets with a thickness (h_0) of 1 mm were symmetrically bonded to both sides of the pre-crack using an adhesive. To eliminate load eccentricity, the top and bottom of the specimens were levelled with high-strength gypsum before the test. A load was applied with a displacement control of 0.2 mm/min. During the test, strain gauges, LVDTs, extensometer, and load data were recorded at a frequency of 1 Hz.

The analysis indices of the fracture test included the flexural strength f_t, fracture energy G_f, initial fracture toughness K^{ini}, and unstable fracture toughness K^{un}. The flexural strength of the specimens was calculated as follows:

$$f_t = \frac{3P_{max}s}{2t(h-a_0)^2} \tag{1}$$

where P_{max} is the maximum load during the testing. The equation for calculating the fracture energy G_f is as follows:

$$G_f = \frac{\int_0^{\delta_0} P d\delta + mg\frac{s}{l}\delta_0}{t(h-a_0)} \tag{2}$$

where P is the load recorded during the test; δ_0 is the deflection required for calculating the fracture energy and is determined as 2 mm; m represents the mass of the specimen; g is the gravitational acceleration, i.e., 9.8 N/kg.

In this study, fracture toughness was calculated using the double-K fracture criterion proposed by Xu and Reinhardt [39]. According to this criterion, the initial fracture toughness and unstable fracture toughness are the key indicators for determining the fracture performance of concrete. When the stress intensity factor of concrete was less than the initial fracture toughness, cracks did not develop. When the stress intensity factor was greater than the initial fracture toughness and less than the unstable fracture toughness, cracks developed stably. Crack instability develops when the stress intensity factor exceeds the unstable fracture toughness. Therefore, the crack toughness can be calculated as follows:

$$K^{ini} = \frac{3P_{ini}s}{2h^2t}\sqrt{a_0}F\left(\frac{a_0}{h}\right) \tag{3}$$

$$F\left(\frac{a_0}{h}\right) = \frac{1.99 - \frac{a_0}{h}\left(1-\frac{a_0}{h}\right)\left(2.15 - \frac{3.93a_0}{h} + 2.7\left(\frac{a_0}{h}\right)^2\right)}{\left(1+2\frac{a_0}{h}\right)\left(1-\frac{a_0}{h}\right)^{\frac{3}{2}}} \tag{4}$$

where $F(a_0/h)$ is calculated using Equation (4). P_{ini} denotes cracking load. Determining the initial load is key to processing the test data, and there are generally two methods. The first is the strain gauge method [40,41]. The strain gauges are pasted on both sides of the pre-crack end, and the retraction point of the strain value of the load–strain curve is considered as the cracking point. This method is simple, easy to implement, and widely used. The second method is the test curve method [37,42]. As stated in the standard and other studies, the load corresponding to the turning point of the rising section of the load–CMOD curve from a straight section to a curved section is the cracking load. This method can overcome some of the disadvantages of the strain gauge method (e.g., the appearance times of surface cracks and internal cracks are inconsistent). In this study, the second method is used to determine the cracking point.

To calculate the unstable fracture toughness, the initiation load P_{ini} and pre-crack height a_0 in Equations (3) and (4) were replaced with the maximum load P_{max} and critical crack length a_c, respectively. The critical crack length was calculated as follows:

$$a_c = \frac{2}{\pi}(h + h_0)\arctan\sqrt{\frac{tE_f}{32.6P_{max}}\text{CMOD}_c - 0.1135} - h_0 \tag{5}$$

where h_0 is the thickness of the aluminium sheet; CMOD_c is the CMOD corresponding to the maximum load P_{max}; E_f is the rupture modulus, which was calculated as:

$$E_f = \frac{1}{tc_i}\left[3.70 + 32.6\tan^2\left(\frac{\pi}{2}\frac{a_0 + h_0}{h + h_0}\right)\right] \tag{6}$$

where c_i denotes the initial compliance. This was calculated as the ratio of P to CMOD at any point on the straight segment of the P–CMOD curve. P_{ini} and its corresponding CMOD were selected uniformly for this study.

Finally, samples were obtained from the fracture surface of the specimen for the scanning electron microscope (SEM) test.

3. Result and Discussion

3.1. Failure Mode Analysis

The final morphologies of the fractured specimens are shown in Figure 4. For the specimens without SFs, as shown in the left column of Figure 4, the cracks mostly developed vertically upward after they appeared, and the crack widths were not large. When the GF content was 0.6%, as shown in Figure 4m,y, the development path of the cracks became more tortuous. This indicates that when SFs were not added, the macroscopic cracks were not effectively confined and always developed along the high-stress position. Further, a small crack width indicates that the deformation of the specimen was small when its failure and the deformation ability was poor. Specimens with different SF contents are shown in the rows in Figure 4. With the increase in SF content, the crack above pre-crack changes from vertical to tortuous. This indicates that the presence of SFs can enhance the crack resistance of RRC so that crack development avoids these enhanced areas. In addition, the crack width increased with increasing SF content. This indicates that the deformation during the RRC fracture failure was larger, and the deformation capacity was better with the increase in the SF content.

According to the failure mode analysis, the effects of the two fibres on the fracture performance of the RRC are as follows: When only the GFs were added, the cracks in the specimens were relatively vertical. This shows that GF had little effect on the restriction of macroscopic cracks, and the cracks appeared at the position of the greatest stress. The GFs that appeared in the macroscopic cracks were either pulled out or broken. Therefore, the GFs mainly acted before the appearance of macroscopic cracks. Whether it was a 6 mm or 12 mm GF, it had no significant effect on the final failure mode of the fracture specimen. In contrast, SFs had a greater influence on the final failure mode. The main reason for this was that the bridging effect of SFs can continue after the appearance of macroscopic cracks. Steel fibres reduce the maximum stress at the crack tip by altering the strain field at the crack tip. Studies have shown that as long as the crack width does not exceed 1/4 of the length of the SF, SF-reinforced concrete still has a load-bearing capacity [43]. Therefore, SFs can improve the deformation ability and toughness of RRC.

Figure 5 shows the crack development in fracture tests. At the beginning, as shown in Figure 5a, the fibres are uniformly distributed in the RRC and there is a number of initial defects. As the load increases, microcracks first appear at initial defects and the interface between the aggregates and matrix, as shown in Figure 5b. At this point, GFs can suppress the development of microcracks. When the load continues to increase to the cracking load, microcracks connect to form macroscopic cracks. Cracks usually avoid fibres and appear in weak areas of RRC. Therefore, the presence of SFs makes macroscopic cracks more tortuous, as shown in Figure 5c. Finally, as shown in Figure 5d, the load continued to increase until it approached the ultimate load of the specimen. At this point, most GFs break, some of SFs are pulled out and fail. Only a portion of SFs play a bridging role. The development time from Figure 5c to 5d is relatively long. In other words, the SFs gradually improve the fracture toughness of RRC during the process of being pulled out.

3.2. Load–Deflection Curves

Figure 6 illustrates the load–deflection curves of different series. The load–deflection curves for each series were divided into ascending and descending sections. The influence of SFs and GFs on the load–deflection curves of each series was mainly reflected in the peak load and deformation capacity.

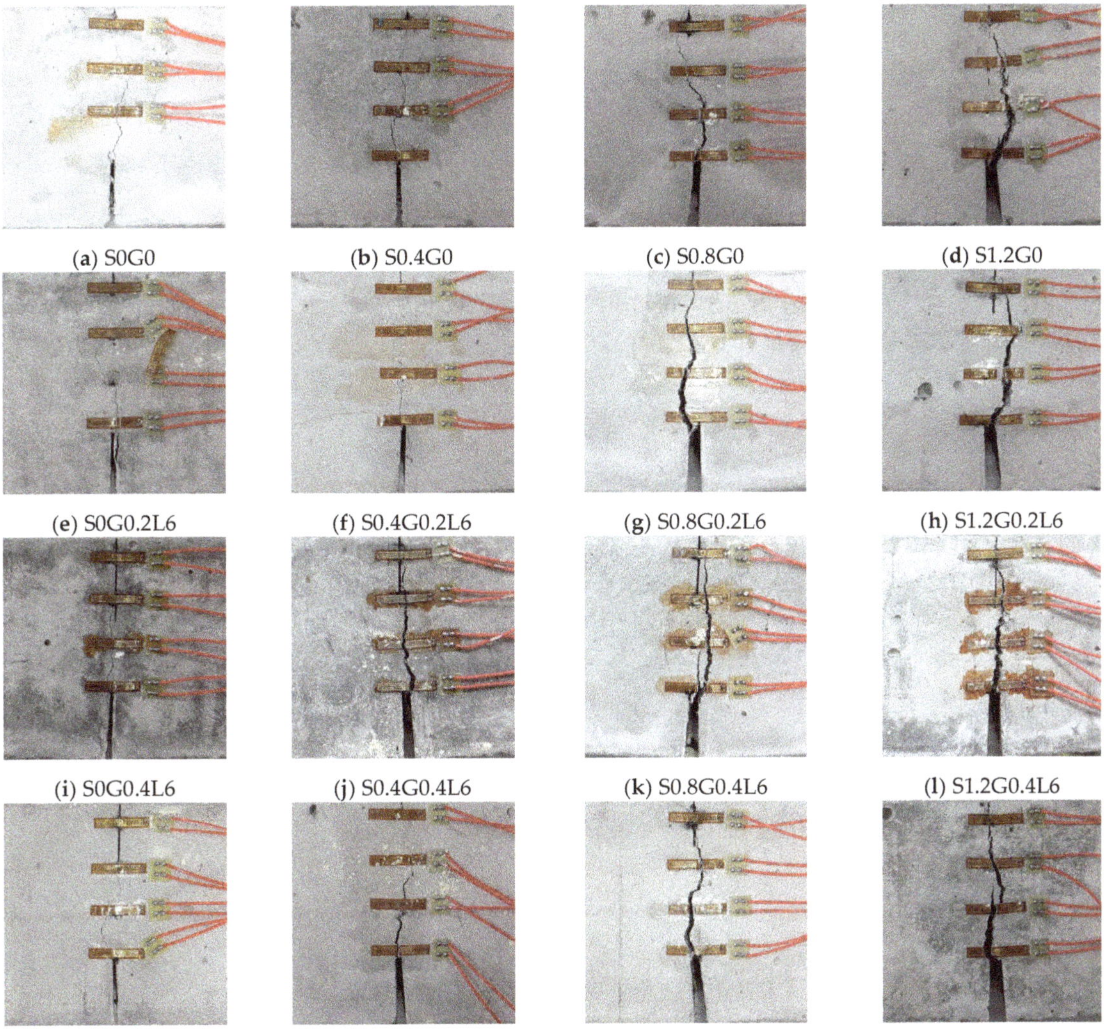

(**a**) S0G0 (**b**) S0.4G0 (**c**) S0.8G0 (**d**) S1.2G0

(**e**) S0G0.2L6 (**f**) S0.4G0.2L6 (**g**) S0.8G0.2L6 (**h**) S1.2G0.2L6

(**i**) S0G0.4L6 (**j**) S0.4G0.4L6 (**k**) S0.8G0.4L6 (**l**) S1.2G0.4L6

Figure 4. *Cont.*

(**m**) S0G0.6L6 (**n**) S0.4G0.6L6 (**o**) S0.8G0.6L6 (**p**) S1.2G0.6L6

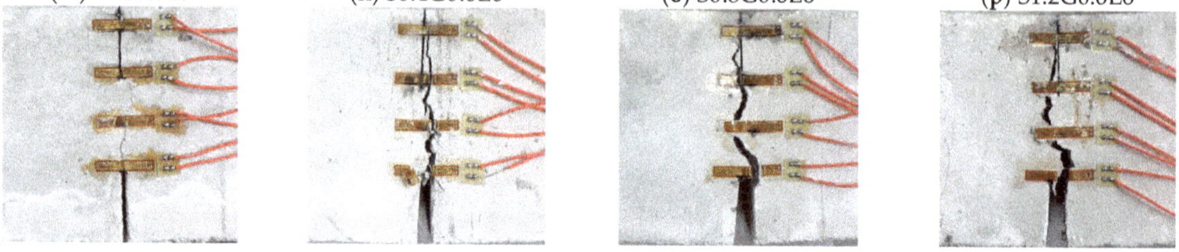

(**q**) S0G0.2L12 (**r**) S0.4G0.2L12 (**s**) S0.8G0.2L12 (**t**) S1.2G0.2L12

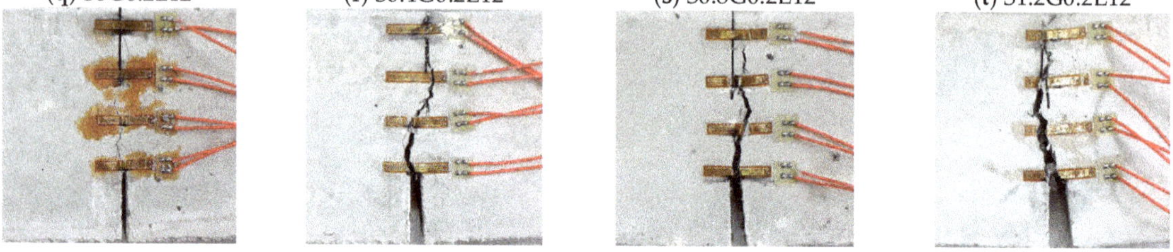

(**u**) S0G0.4L12 (**v**) S0.4G0.4L12 (**w**) S0.8G0.4L12 (**x**) S1.2G0.4L12

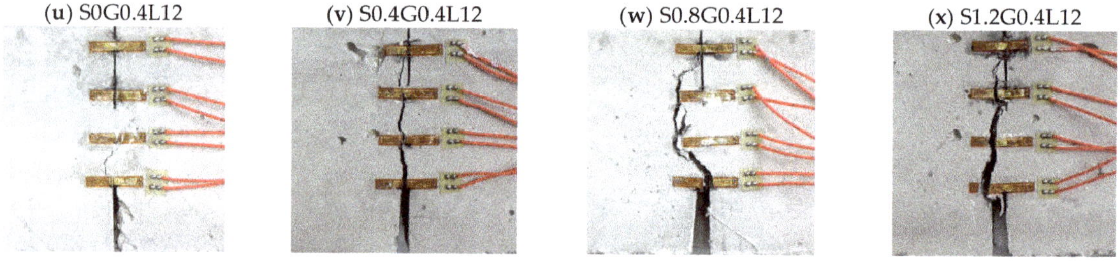

(**y**) S0G0.6L12 (**z**) S0.4G0.6L12 (**aa**) S0.8G0.6L12 (**ab**) S1.2G0.6L12

Figure 4. Failure modes of HFRRRC.

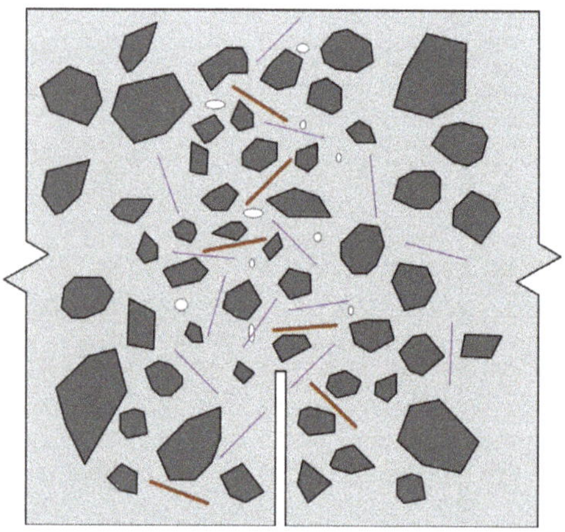

(**a**)

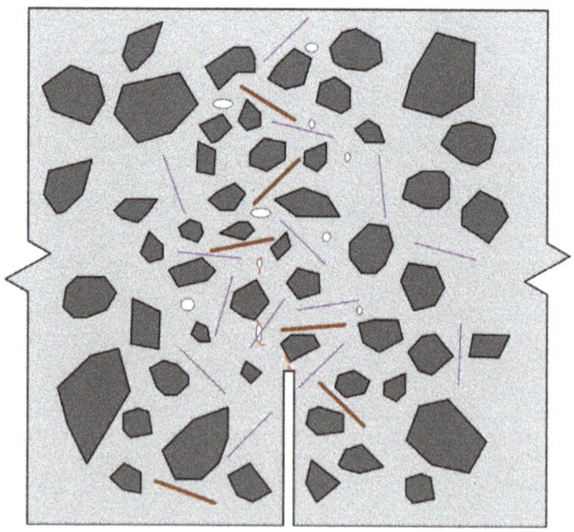

(**b**)

Figure 5. *Cont.*

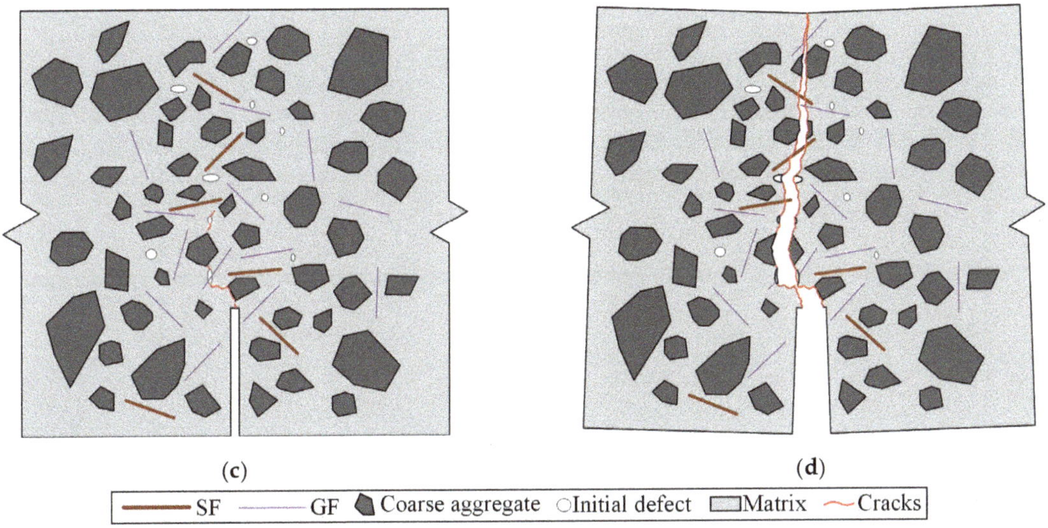

Figure 5. Schematic diagram of cracks development in HFRRRC: (**a**) initial state; (**b**) the microcracks appeared; (**c**) the microcracks developed into macrocracks; (**d**) final state.

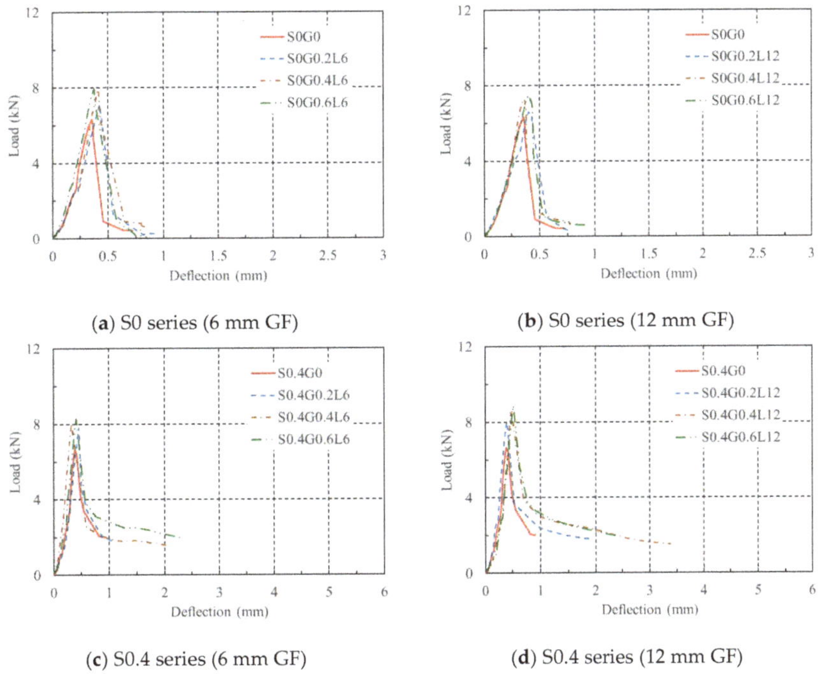

Figure 6. *Cont.*

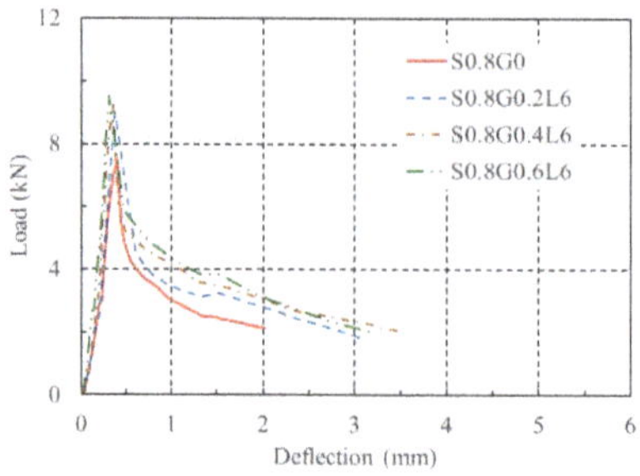

(**e**) S0.8 series (6 mm GF)

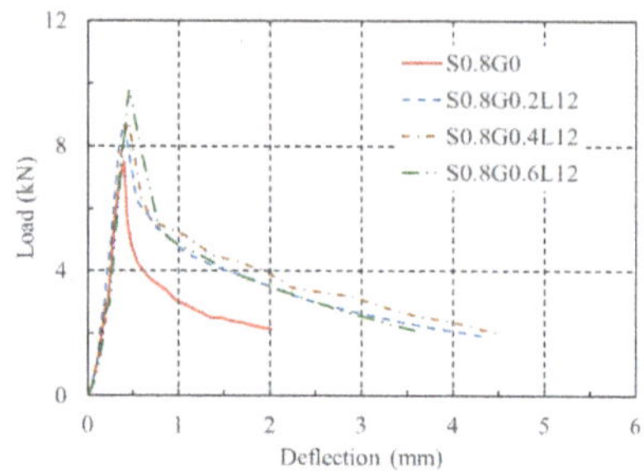

(**f**) S0.8 series (12 mm GF)

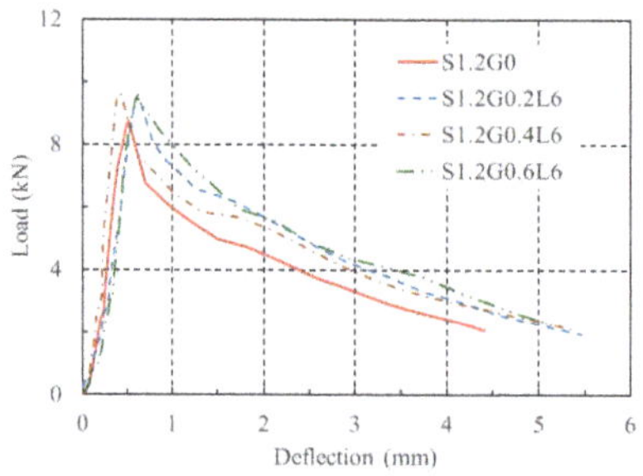

(**g**) S1.2 series (6 mm GF)

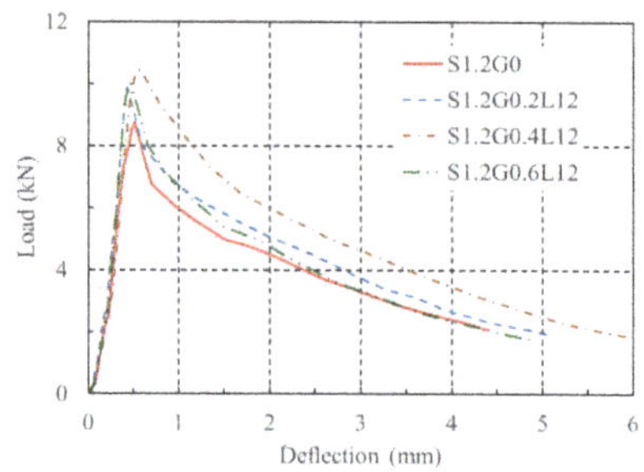

(**h**) S1.2 series (12 mm GF)

Figure 6. Load–deflection curves of fracture tests.

From the perspective of the peak load, when the SF content was less than 1.2%, the peak load of the curves increased with the increase in GF content. In this case, when the GF contents of 6 and 12 mm was 0.6%, the peak load was maximised. When the SF content was 1.2%, the peak load of the curves first increased and then decreased as the GF content increased. When the GF content of 6 and 12 mm was 0.4%, the peak load reached its maximum. Therefore, an appropriate amount of GF can increase the peak load. This was because the GFs had a bridging effect and improved the mechanical properties of concrete. However, excessive GFs reduce the workability of fresh concrete and increase the void ratio inside the concrete, which was harmful to the mechanical properties of concrete. When comparing the load–deflection curves for different SF contents, the peak load increases with the increase in SF content. This was because the SFs can play a bridging role after the concrete had cracked so that the concrete on both sides of the crack can continue to carry loads.

The inclusion of fibres can effectively improve the deformation ability of RRC. The deflection corresponding to the peak load was defined as the peak deflection. As shown in Figure 5a,b, when no SF was added, the peak deflection first increases and then decreases with the increase in GF content. The peak deflections of the specimens with GFs were greater than those of the S0G0 group. Both GF lengths had the highest peak deflection when the content was 0.2%. After the peak deflection, the load value decreased rapidly, indicating brittle failure, and the slope of the descending curve was similar to that of the S0G0 group. This indicates that an appropriate number of GFs can improve the deformation capacity of the RRC before the peak load, but there was not much improvement in the deformation capacity after the peak load. This was due to the fact that GFs can limit the formation and development of microcracks in the ascending section of the load–deflection curve. However, excessive GFs can increase the initial defects of the RRC. During the descending section of the load–deflection curve, there were evident macroscopic cracks in the RRC, and the GFs broke or were pulled out to fail. Therefore, the deformation ability of the RRC after the peak load was not significantly improved when only GFs were added. As shown in Figure 5c–f, for the specimens with SFs, the descending section of the load–deflection curve can be divided into two sub-phases: a rapid descent phase and a stabilisation phase. As the SF content increases, the transition zone between these two subphases becomes smoother. The load corresponding to the transition zone was greater. This indicates that SFs can effectively limit the speed of the development of macroscopic cracks. The transition from the rapid descent phase to the stabilisation phase at higher load levels increased the energy consumed and improved the toughness of the RRC. By comparing the 6 mm GFs series with the 12 mm GFs series, it can be observed that the existence of SFs better improves the deformation capacity of the 12 mm GFs series after peak loading.

3.3. Index Analysis

Table 5 lists the values of each index. The values of each index in Table 5 represent the average values of the three specimens. If the following two situations occur, they were considered separately: (1) if both the maximum and minimum values differ from the median value by more than 15% of the median value, the median value was considered as the representative value; and (2) if only one of the maximum and minimum values differs from the median value by more than 15% of the median value, the value with a large difference was discarded, and the average of the other two values was considered as the representative value. In addition, to better reflect the impact of SFs and GFs on the indexes, Table 5 also lists the change rates of each index. The rates of change were relative to the S0G0 group. Positive values indicate an increase and negative values indicate a decrease. In addition, because we had previously studied the compressive strength and elastic modulus, this study focuses on the effects of SFs and GFs on the fracture index of RRC.

Table 5. Test results.

Mix Number	Compressive Strength		Flexural Strength		Brittleness Coefficient		Fracture Energy		Initial Fracture Toughness		Unstable Fracture Toughness	
	RV * (MPa)	PC * (%)	RV (MPa)	PC (%)	RV	PC (%)	RV (J/m^2)	PC (%)	RV (MPa·m$^{1/2}$)	PC (%)	RV (MPa·m$^{1/2}$)	PC (%)
S0G0	34.36	-	5.81	-	5.91	-	228	-	0.60	-	1.24	-
S0.4G0	36.93	7.5	6.13	5.5	6.03	1.92	405	77.3	0.79	30.2	1.74	40.3
S0.8G0	35.06	2.0	6.87	18.3	5.10	−13.7	917	301.7	0.88	46.1	2.29	84.8
S1.2G0	38.09	10.9	8.03	38.1	4.74	−19.8	1531	570.8	0.93	54.2	2.32	87.8
S0G0.2L6	34.24	−0.3	6.53	12.3	5.25	−11.3	297	30.3	0.72	19.0	1.79	44.5
S0.4G0.2L6	36.57	6.4	7.18	23.6	5.09	−13.9	480	110.4	0.91	50.0	1.94	57.1
S0.8G0.2L6	38.57	12.3	8.26	42.2	4.67	−21.0	1063	365.7	0.90	49.4	2.44	97.6
S1.2G0.2L6	42.55	23.8	8.78	51.0	4.85	−18.0	1755	669.2	0.96	58.3	2.63	112.8
S0G0.4L6	34.77	1.2	7.27	25.1	4.78	−19.1	333	46.0	0.80	32.2	1.71	38.5
S0.4G0.4L6	34.77	1.2	7.40	27.4	4.70	−20.6	736	222.4	0.91	51.4	1.88	51.8
S0.8G0.4L6	40.05	16.6	8.48	45.9	4.72	−20.1	1175	415.0	0.89	47.7	2.31	87.0
S1.2G0.4L6	36.87	7.3	8.84	52.0	4.17	−29.4	1737	661.2	0.98	61.6	2.70	118.4
S0G0.6L6	35.26	2.6	7.31	25.7	4.83	−18.4	351	53.6	0.90	49.5	1.95	57.8
S0.4G0.6L6	39.53	15.0	7.59	30.5	5.21	−11.9	903	295.6	1.08	79.5	2.07	67.5
S0.8G0.6L6	37.58	9.4	8.75	50.5	4.30	−27.3	1283	462.1	1.09	80.3	2.39	93.4
S1.2G0.6L6	33.73	−1.8	8.73	50.1	3.87	−34.6	1798	687.9	1.07	77.8	2.69	117.7
S0G0.2L12	35.68	3.8	6.09	4.8	5.86	−0.9	298	30.6	0.91	51.3	1.31	5.6
S0.4G0.2L12	37.95	10.4	7.37	26.7	5.15	−12.8	755	230.9	1.01	67.4	1.63	31.6
S0.8G0.2L12	39.81	15.9	7.99	37.5	4.98	−15.8	1331	483.3	0.94	56.3	2.25	82.2
S1.2G0.2L12	41.59	21.0	9.06	55.9	4.59	−22.4	1727	656.8	1.02	68.6	2.24	80.8
S0G0.4L12	36.36	5.8	6.71	15.4	5.42	−8.3	267	16.9	0.90	49.3	1.33	7.2
S0.4G0.4L12	38.49	12.0	7.88	35.5	4.89	−17.4	901	294.7	0.99	63.2	1.71	38.6
S0.8G0.4L12	38.99	13.5	8.07	38.8	4.83	−18.3	1403	514.6	0.95	56.7	2.26	82.3
S1.2G0.4L12	43.60	26.9	9.59	64.9	4.55	−23.1	1998	775.6	1.16	92.0	2.57	107.6
S0G0.6L12	37.18	8.2	6.83	17.4	5.45	−7.9	325	42.5	0.95	58.0	1.43	15.6
S0.4G0.6L12	39.90	16.1	8.16	40.4	4.89	−17.3	942	312.9	0.97	61.1	1.73	39.9
S0.8G0.6L12	38.02	10.7	9.01	55.0	4.22	−28.6	1357	494.7	1.09	80.1	2.05	65.5
S1.2G0.6L12	40.28	17.2	9.19	58.1	4.38	−25.9	1702	645.9	1.13	86.7	2.29	85.5

* RV: representative value; PC: percentage changes.

3.3.1. Compressive Strength

As shown in Table 5, when the SF contents were 0 and 0.4%, the compressive strength increased with the GF content. When the SF content was greater than 0.4%, the compressive strength first increased and then decreased with increasing GF content. Specifically, when only GFs were added at 0.6%, the 6 mm and 12 mm GFs resulted in the largest increase in the compressive strength, which was 2.6 and 8.2%, respectively. When only 1.2% SFs were added, the greatest gain in compressive strength was 10.9%. However, when the 12 mm GF content was 0.4% and SF content was 1.2%, the largest gain in compressive strength was 26.9%. Therefore, although the separate addition of the two fibres can improve the compressive strength of RRC, the combination of GFs and SFs resulted in the largest compressive strength.

3.3.2. Flexural Strength

The flexural strength of the fracture test was calculated using Equation (1) and is shown in Table 5 and Figure 7. When the SF content was less than 1.2%, the flexural strength of each series increased with the GF content; however, the growth rate decreased. In this case, the flexural strength of the RRC reached its maximum at a GF content of 0.6%. When the SF content was 1.2%, the flexural strength first increased and then decreased with an increase in the GF content. The flexural strength of the RRC reached a maximum at a GF content of 0.4%, which was similar to 0.3–0.6% in the literature [44]. Specifically, when only GFs were added at 0.6%, the 6 mm and 12 mm GFs resulted in the largest increase in the flexural strength, which was 25.7 and 17.4%, respectively. For the specimens with SFs, 6 mm and 12 mm GFs make the highest improvements in flexural strength, which were 52.0 and 64.9%, respectively, when the SF content was 1.2% and the GF content was 0.4%.

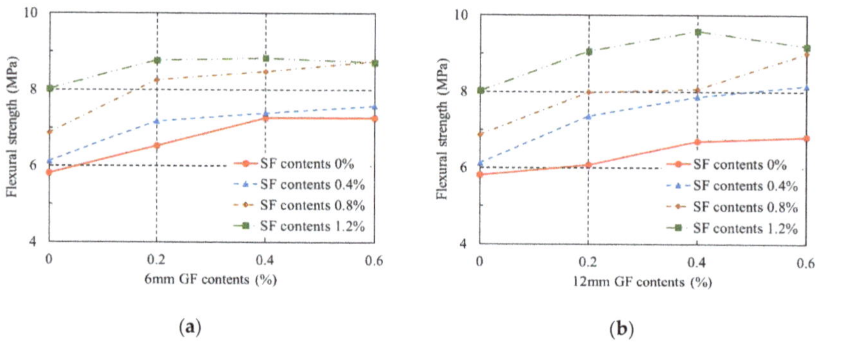

Figure 7. Flexural strength: (**a**) 6 mm GF; (**b**) 12 mm GF.

In addition, when only 1.2% SF was added, the greatest increase in flexural strength was 38.1%. For the specimen with only 12 mm GF, the growth rate of the flexural strength reached a maximum of 17.4% when the GF content was 0.6%. However, in the case of hybrid fibres, the growth rate of the flexural strength reached a maximum of 64.9% when the SF content was 1.2% and the 12 mm GF content was 0.4%. Therefore, the separate addition of SFs and GFs can improve the flexural strength of the RRC. The combination of GFs and SFs resulted in the largest flexural strength. The positive mixing effect was caused by the different properties of the SFs and GFs. At the initial stage, a larger quantity of GFs effectively reduces the stress concentration at the microcrack tip. Therefore, the stress can be redistributed and more matrix materials can be used, improving the strength of the RRC specimen. Due to the significant impact of GFs on the workability of RRC, it is not advisable to exceed 0.4% of GF content when the total volume fraction of hybrid fibres is large. Due to the high strength and elastic modulus of SFs, a skeleton was formed inside the concrete. Simultaneously, the SFs played a bridging role when a macroscopic crack appeared until the SFs were pulled out and failed.

3.3.3. Brittleness Coefficient

The brittleness coefficient, defined as the ratio of the compressive strength to the flexural strength, is an important index for RRC subjected to tensile failure. RRC with larger brittleness coefficients exhibited greater brittleness. The brittleness coefficients of the RRC are listed in Table 5 and Figure 8. For the specimens without adding SFs, when the GF content was 0.4%, 6 and 12 mm GFs make the maximum decrease in brittleness coefficient, which were −19.1 and −8.3%, respectively. For the specimens with SFs, when the 6 mm GF content and SF content was 0.6% and 1.2%, respectively, the maximum decrease in the brittleness coefficient was −34.6%. When the 12 mm GF content and SF content was 0.6% and 0.8%, respectively, the maximum decrease in brittleness coefficient was −28.6%. Overall, the addition of fibres can reduce the brittleness coefficient of the RRC. It can be noted that when 12 mm GFs and SFs were added together, the brittleness coefficient only slightly decreased with the increase in GF content, when the content of GF was larger than 0.4%. Therefore, the relationship between the brittleness coefficient and the GF content is quite complex. For 6 mm GF, the optimal content was 0.4–0.6%. For 12 mm GF, the optimal dosage was 0.4%.

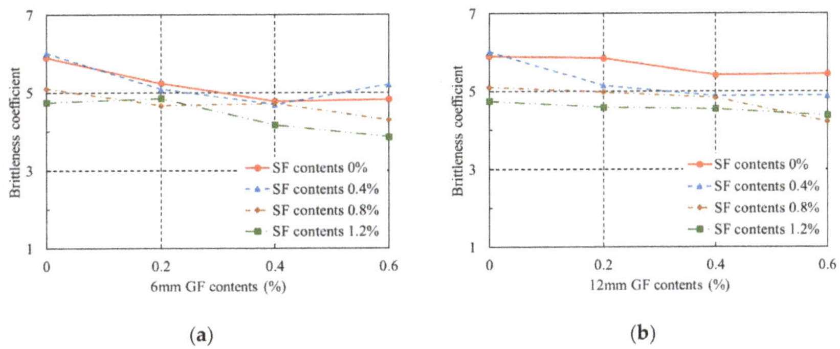

Figure 8. Brittleness coefficient: (**a**) 6 mm GF; (**b**) 12 mm GF.

3.3.4. Fracture Energy

As the most useful material parameter in the analysis of cracked concrete structures, the fracture energy was calculated using Equation (2) [45]. It is shown in Table 5 and Figure 9. The fracture energy reflects the amount of energy dissipated by the RRC fracture. When only GFs were added, the fracture energy increased slightly with increasing GF content. When only 0.6% GFs were added, the 6 and 12 mm GFs resulted in the highest improvements in the fracture energy, which were 53.6% and 42.5%, respectively. When only SFs were added, the fracture energy increased significantly with increasing SF content, with the greatest improvement being 570.8%. When the 6 mm GF content and SF content was 0.6% and 1.2%, respectively, the maximum increase in the fracture energy was 687.9%. When the 12 mm GF content and SF content was 0.4% and 1.2%, respectively, the maximum increase in fracture energy was 775.6%. Overall, the fracture energy increased with the increase in the content of two types of fibres, especially SFs. The reason was that even at lower fibre content, the softening part of the load–deflection curves of RRC containing SFs were longer [33]. The 12 mm GFs caused a greater increase in fracture energy than the 6 mm GFs. This was because the propagation of cracks typically avoids the area where the fibres were located, and longer GFs make it more difficult for cracks to avoid them.

In addition, when only 1.2% SFs were added, the greatest gain in fracture energy was 570.8%. For the specimen with only 12 mm GF, the growth rate of the fracture energy reached a maximum of 42.5% when the GF content was 0.6%. However, in the case of hybrid fibres, the growth rate of the fracture energy reached a maximum of 775.6% when the content of SF was 1.2% and the content of 12 mm GF was 0.4%. Therefore, although both SFs and GFs can increase the fracture energy of RRC, SFs play a major role. According to

Figure 9, without adding SFs, the lines became gentler with increasing GF content. However, when SFs were added, the slope of the lines increased, indicating that the two fibres had a positive mixing effect. This was because brittle GFs fail quickly after the appearance of macroscopic cracks. GFs alone cannot effectively limit the crack development after peak loads. When SFs were added, SFs prevented the rapid development of macroscopic cracks. The specimens containing SFs had more tortuous cracks, indicating that more GFs and RRC played a role in increasing the energy consumed for crack development. The failure modes of the RRC confirmed this, as described in Section 3.1.

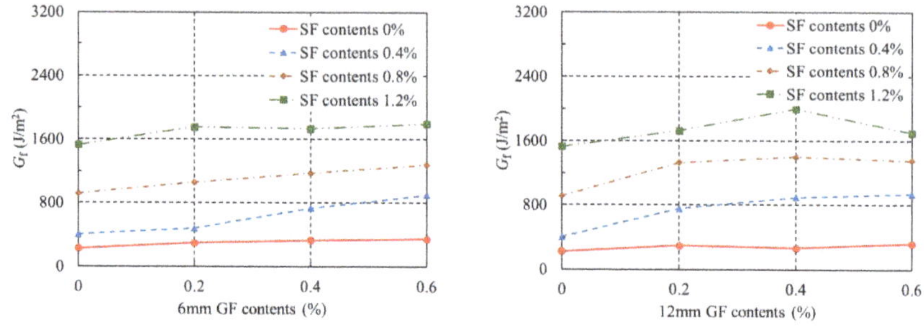

(a) (b)

Figure 9. Fracture energy: (**a**) 6 mm GF; (**b**) 12 mm GF.

3.3.5. Initial Fracture Toughness

The initial fracture toughness results are presented in Table 5 and Figure 10. Equation (3) shows that the initial fracture toughness is proportional to the cracking load. Therefore, the effect of SFs and GFs on the cracking load was equal to their effect on the initial fracture toughness. In general, the initial fracture toughness of the RRC was increased by the inclusion of fibres. When only 1.2% SFs was added, the highest gain in the initial fracture toughness was 54.2%. When only 0.6% GFs were added, the 6 mm and 12 mm GFs resulted in the highest improvement in the initial fracture toughness, which was 49.5 and 58.0%, respectively. When the 12 mm GF content and SF content was 0.4% and 1.2%, respectively, the initial fracture toughness gained a maximum increase of 92.0%. Before the RRC cracked, the RRC and fibres were synergistically deformed by the adhesive force. Compared with RRC without fibres, the high elastic modulus of SFs allows RRC to withstand greater loads under the same deformation, thereby increasing the cracking load. The increase in initial fracture toughness was not significant with the increase in SF content. Literature [41] introduced 1% SF into concrete, which only increased the initial fracture toughness by 2.6%. Simultaneously, because of the higher quantity of GFs in the same content, the formation and development of microscopic cracks were limited by the GFs better. Therefore, GF seems to contribute more significantly to the initial fracture toughness than SF. Compared with single fibres, hybrid fibres were more effective in the enhancement of the initial fracture toughness of RRC.

3.3.6. Unstable Fracture Toughness

The unstable fracture toughness results are listed in Table 5 and shown in Figure 11. In general, the unstable fracture toughness of RRC increased with an increase in the SF content but only slightly increased through increasing the content of GF. When only 1.2% SFs were added, the greatest gain in the unstable fracture toughness was 87.8%. When only 0.6% GFs were added, both lengths of GFs resulted in the highest improvements in the unstable fracture toughness, which were 57.8% and 15.6%, respectively. When the 6 mm GF content and SF content was 0.4% and 1.2%, respectively, the maximum increase in the unstable fracture toughness was 118.4%. Compared to the initial fracture toughness, SFs had a more significant improvement in unstable fracture toughness. This conclusion can also be drawn from literature [41]. For the RRC specimens reinforced with the two types of fibres, the increase in the unstable fracture toughness was not significant with the increase in the GF content. This was because unstable fracture toughness is the critical point at which cracks develop from stability to instability. The bridging effect of the GFs mainly occurs in the microcrack stage. After the appearance of macroscopic cracks, most of GFs quickly fail. Therefore, the amount of GF content has little effect on the unstable fracture toughness. At this time, the SFs play a major bridging role, slowing down the development of macroscopic cracks, and thereby improving the unstable fracture toughness of the RRC.

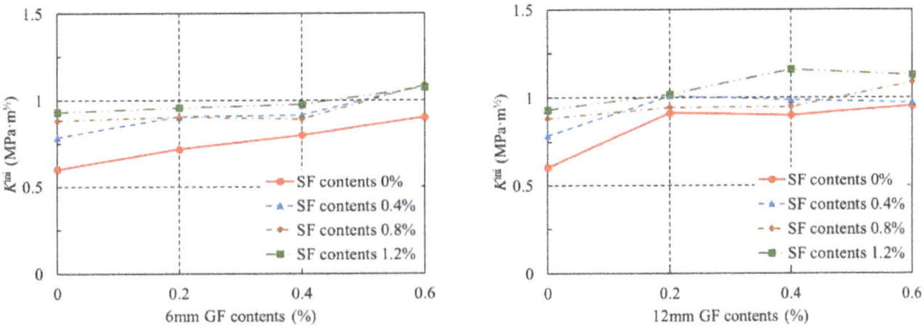

(**a**) (**b**)

Figure 10. Initial fracture toughness: (**a**) 6 mm GF; (**b**) 12 mm GF.

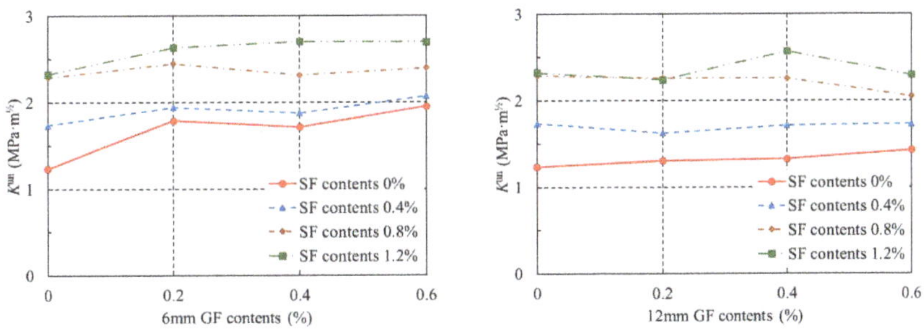

(**a**) (**b**)

Figure 11. Unstable fracture toughness: (**a**) 6 mm GF; (**b**) 12 mm GF.

4. Mechanism Analysis

Through the analysis of the failure modes and various indices, it can be inferred that the incorporation of SFs is useful for improving the fracture performance of RRC. Before the cracking load, there was a strong bond between the SFs and RRC. The high elastic modulus and tensile strength of SFs can improve the bearing capacity of RRC, thereby improving the peak and cracking loads. As illustrated in Figure 12a, after the peak load, because the tensile strength of the SFs was greater than the bonding strength with the RRC, the SFs were always pulled out rather than broken. Previous literature has indicated that, until the tensile deformation of SF-reinforced concrete reaches 1/4 of the fibre length, the tensile stress decreases to 0 [37]. Therefore, the SFs located at the tip of the crack can still play a bridging role so that the concrete on both sides of the crack can continue to bear the load. As the SFs were pulled out, the neutral axis and the tip of the crack slowly developed upward. The energy consumed by this process is characterised by an increase in fracture energy.

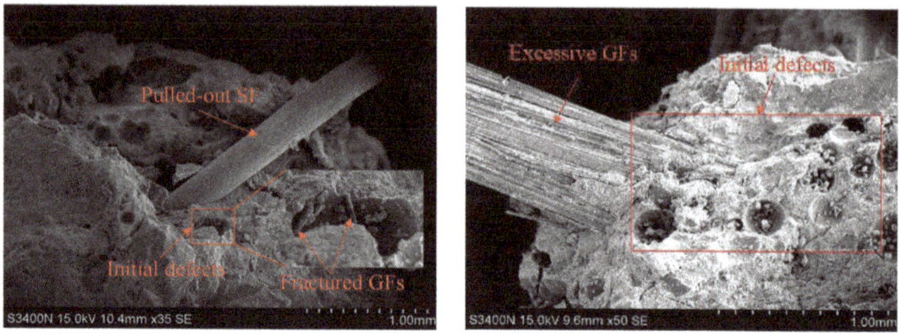

(a) (b)

Figure 12. SEM test results: (**a**) the failure modes of fibres at the crack; (**b**) initial defects caused by excessive GFs.

The influence of the GFs on the fracture performance of RRC mainly occurred before the cracking load. As RRC is a heterogeneous material, there may be certain initial internal defects. When RRC is subjected to external loads, microcracks usually develop from the initial defect and eventually interconnect into macroscopic cracks. In particular, when excessive GFs were added, the poor workability increased the initial defects of the RRC, as shown in Figure 12b. Therefore, the GF content should not be excessively high. When an appropriate GF content is used, the fracture performance of the RRC can be effectively improved, and the workability of the RRC is less affected. Because there were significantly more GFs than SFs at a given fibre content, the formation and growth of microcracks may be efficiently suppressed. Because GFs are brittle materials and microcracks are present on the glass surface, their strength is generally lower than the theoretical strength. Therefore, most of the GFs located in the macroscopic cracks broke after the cracking load. Therefore, the GFs had no significant bridging effect on the descending segment after the peak load.

The fracture properties of the RRC were maximised by the combined addition of SFs and GFs. This indicates that the SFs and GFs had a positive mixing effect on the fracture properties of RRC. In the preliminary loading period, the GFs was useful for inhibiting the formation and growth of microcracks. As the load continued to increase, the microcracks interconnected to form macroscopic cracks. At this point, the bridging effect of the SFs slowed the development of cracks. The two parts bridged by the SFs could continue to bear the load, allowing more RRC and GFs to work. This was verified using fracture energies. As shown in Figure 9, after the addition of SFs, the fracture energy increased more significantly with an increase in GF content. However, the variation in the experimental results is complex because the fibre content has an important impact on the workability of fresh RRC. Further studies are required to elucidate the mechanisms underlying HFRRRC.

5. Conclusions

Fracture experiments were conducted on 28 different concrete mixes in order to examine the effects of SFs and GFs on the fracture characteristics of RRC. Analysis was performed on the failure mechanism, load–deflection curves, flexural strength, fracture energy, brittleness coefficient, initial fracture toughness, and unstable fracture toughness. The test findings supported the following deductions:

(1) The fractured specimens without SFs were damaged by vertical cracks and their crack widths were smaller. The two GF lengths had no significant effect on the final morphology of the cracks. With the addition of the SFs, the cracks became more curved as the crack width increased. The SFs primarily played a bridging role after RRC cracking. At this point, the GFs located in the cracks had failed.

(2) In general, the peak load increased with increasing dosages of both fibres. However, with 1.2% SFs, the excessive addition of GFs decreased the peak load. For specimens

with only GFs, the peak deflection increased and then decreased, indicating brittle damage. The addition of SFs significantly improved the post-peak deformation capacity and smoothed the descending section of the load–deflection curves. The specimens of the 12 mm GFs series exhibited a greater enhancement of the deformation capacity when SFs were added.

(3) When only SFs were added, the maximum increases or decreases in flexural strength, brittleness coefficient, fracture energy, initial fracture toughness, and unstable fracture toughness with increasing SF content were 38.1, −19.8, 570.8, 54.2, and 87.8%, respectively. When only GFs were added, the maximum increases or decreases in the above-mentioned indices with increasing GF content were 25.7, −19.1, 53.6, 58.0, and 57.8%, respectively. When SFs and GFs were added together, the maximum increases and decreases in the above-mentioned indices with increasing GF content were 64.9, −34.6, 775.6, 92.0, and 118.4%, respectively. Thus, although the addition of SFs or GFs alone also improved the mechanical performance indices of the RRC, the combination of SFs and GFs produced the most significant improvements.

(4) In this study, the optimum GF content was found to be 0.4–0.6%. It decreases with increasing SF content ensuring the workability of the RRC.

(5) The mechanism for the positive mixing effect of SFs and GFs was that they acted at different stages during the fracture test. A large number of GFs restricted the development of microcracks in the early stages of the test. After the RRC cracked, which caused the GFs to fail, the SFs still acted as bridges.

(6) To prevent the cracking of rubber concrete pavement caused by external factors, adding SFs and GFs together is an effective solution. Before practical application, more performance indicators need to be quantitatively analysed. The durability and impact resistance of HFRRRC still needs to be studied.

Author Contributions: Conceptualisation, L.L. and X.L.; methodology, X.L. and F.L.; software, Z.P.; validation, Z.C., H.L. and Z.W.; formal analysis, W.L.; investigation, H.Z.; resources, H.L. and Z.W.; data curation, Z.P.; writing—original draft preparation, X.L.; writing—review and editing, W.L.; visualisation, H.Z.; supervision, L.L.; project administration, Z.C.; funding acquisition, L.L. and F.L. All authors have read and agreed to the published version of the manuscript.

Funding: This research was funded by the National Natural Science Foundation of China, grant numbers 12072080 and 12032009.

Data Availability Statement: The original contributions presented in the study are included in the article, further inquiries can be directed to the corresponding authors.

Conflicts of Interest: Author Zhuangwei Chen was employed by the company Guangdong Henghui Construction Group Co., Ltd. Authors Hongming Li and Zhichao Wu were employed by the company Guangdong Zhongdu Construction Group Co., Ltd. The remaining authors declare that the research was conducted in the absence of any commercial or financial relationships that could be construed as a potential conflict of interest.

References

1. Neupane, R.; Imjai, T.; Makul, N.; Garcia, R.; Kim, B.; Chaudhary, S. Use of recycled aggregate concrete in structural members: A review focused on Southeast Asia. *J. Asian Archit. Build.* **2023**, 1–24. [CrossRef]
2. Xiong, Z.; Mai, G.; Pan, Z.; Chen, Z.; Jian, J.; Wang, D.; Ling, Z.; Li, L. Synergistic effect of expansive agents and glass fibres on fatigue bending performance of seawater sea sand concrete. *Constr. Build. Mater.* **2024**, *421*, 135665. [CrossRef]
3. Asaro, L.; Gratton, M.; Seghar, S.; Hocine, N.A. Recycling of Rubber Wastes by Devulcanization. *Resour. Conserv. Recy.* **2018**, *133*, 250–262. [CrossRef]
4. Li, Y.; Chai, J.; Wang, R.; Zhou, Y.; Tong, X. A Review of the Durability-Related Features of Waste Tyre Rubber as a Partial Substitute for Natural Aggregate in Concrete. *Buildings* **2022**, *12*, 1975. [CrossRef]
5. Tang, Y.; Feng, W.; Chen, Z.; Nong, Y.; Guan, S.; Sun, J. Fracture behavior of a sustainable material: Recycled concrete with waste crumb rubber subjected to elevated temperatures. *J. Clean. Prod.* **2021**, *318*, 128553. [CrossRef]
6. Gupta, T.; Chaudhary, S.; Sharma, R.K. Assessment of Mechanical and Durability Properties of Concrete Containing Waste Rubber Tire as Fine Aggregate. *Constr. Build. Mater.* **2014**, *73*, 562–574. [CrossRef]

7. Pham, T.M.; Renaud, N.; Pang, V.; Shi, F.; Hao, H.; Chen, W. Effect of Rubber Aggregate Size on Static and Dynamic Compressive Properties of Rubberized Concrete. *Struct. Concr.* **2021**, *23*, 2510–2522. [CrossRef]
8. Guo, S.; Dai, Q.; Si, R.; Sun, X.; Lu, C. Evaluation of Properties and Performance of Rubber-modified Concrete for Recycling of Waste Scrap Tire. *J. Clean. Prod.* **2017**, *148*, 681–689. [CrossRef]
9. Xie, J.; Guo, Y.; Liu, L.; Xie, Z. Compressive and flexural behaviours of a new steel-fibre-reinforced recycled aggregate concrete with crumb rubber. *Constr. Build. Mater.* **2015**, *79*, 263–272. [CrossRef]
10. Pacheco-Torres, R.; Cerro-Prada, E.; Escolano, F.; Varela, F. Fatigue Performance of Waste Rubber Concrete for Rigid Road Pavements. *Constr. Build. Mater.* **2018**, *176*, 539–548. [CrossRef]
11. Msallam, M.; Asi, I. Improvement of Local Asphalt Concrete Binders Using Crumb Rubber. *J. Mater. Civil Eng.* **2018**, *30*, 04018048. [CrossRef]
12. Chen, J.; Zhuang, J.; Shen, S.; Dong, S. Experimental Investigation on the Impact Resistance of Rubber Self-compacting Concrete. *Structures* **2022**, *39*, 691–704. [CrossRef]
13. Gupta, T.; Tiwari, A.; Siddique, S.; Sharma, R.K.; Chaudhary, S. Response Assessment under Dynamic Loading and Microstructural Investigations of Rubberized Concrete. *J. Mater. Civil Eng.* **2017**, *29*, 04017062. [CrossRef]
14. Guo, X.; Xiong, Z.; Luo, Y.; Qiu, L.; Liu, J. Experimental Investigation on the Semi-rigid Behaviour of Aluminium Alloy Gusset Joints. *Thin Wall. Struct.* **2015**, *87*, 30–40. [CrossRef]
15. Xiong, Z.; Guo, X.; Luo, Y.; Zhu, S. Elasto-plastic Stability of Single-layer Reticulated Shells with Aluminium Alloy Gusset Joints. *Thin Wall. Struct.* **2017**, *115*, 163–175. [CrossRef]
16. Fang, Z.; Wu, J.; Xian, B.; Zhao, G.; Fang, S.; Ma, Y.; Jiang, H. Shear Performance and Design Recommendations of Single Embedded Nut Bolted Shear Connectors in Prefabricated Steel–UHPC Composite Beams. *Steel Compos. Struct.* **2024**, *50*, 319–336. [CrossRef]
17. Lin, J.; Luo, R.; Su, J.; Guo, Y.; Chen, W. Coarse synthetic fibers (PP and POM) as a replacement to steel fibers in UHPC: Tensile behavior, environmental and economic assessment. *Constr. Build. Mater.* **2024**, *412*, 134654. [CrossRef]
18. Fang, Z.; Wu, J.; Xu, X.; Ma, Y.; Fang, S.; Zhao, G.; Jiang, H. Grouped rubber-sleeved studs–UHPC pocket connections in prefabricated steel–UHPC composite beams: Shear performance under monotonic and cyclic loadings. *Eng. Struct.* **2024**, *305*, 117781. [CrossRef]
19. He, L.; Cai, H.; Huang, Y.; Ma, Y.; Bergh, W.V.D.; Gaspar, L.; Valentin, J.; Vasiliev, Y.E.; Kowalski, K.J.; Zhang, J. Research on the Properties of Rubber Concrete Containing Surface-modified Rubber Powders. *J. Build. Eng.* **2021**, *35*, 101991. [CrossRef]
20. Xu, Y.; Wang, J.; Zhang, P.; Guo, J.; Hu, S. Enhanced Effect and Mechanism of Colloidal Nano-SiO_2 Modified Rubber Concrete. *Constr. Build. Mater.* **2023**, *378*, 131203. [CrossRef]
21. Kang, J.; Liu, Y.; Yuan, J.; Chen, C.; Wang, L.; Yu, Z. Effectiveness of Surface Treatment on Rubber Particles towards Compressive Strength of Rubber Concrete: A Numerical Study on Rubber-cement Interface. *Constr. Build. Mater.* **2022**, *350*, 128820. [CrossRef]
22. Zhu, H.; Xiong, Z.; Song, Y.; Zhou, K.; Su, Y. Effect of Expansion Agent and Glass Fiber on the Dynamic Splitting Tensile Properties of Seawater–Sea-Sand Concrete. *Buildings* **2024**, *14*, 217. [CrossRef]
23. Pan, Z.; Liu, F.; Li, H.; Li, X.; Wang, D.; Ling, Z.; Zhu, H.; Zhu, Y. Performance Evaluation of Thermal Insulation Rubberized Mortar Modified by Fly Ash and Glass Fiber. *Buildings* **2024**, *14*, 221. [CrossRef]
24. Wang, X.; Shao, J.; Wang, J.; Ma, M.; Zhang, B. Influence of Basalt Fiber on Mechanical Properties and Microstructure of Rubber Concrete. *Sustainability* **2022**, *14*, 12517. [CrossRef]
25. Zhen, H.; Xiong, Z.; Song, Y.; Li, L.; Qiu, Y.; Zou, X.; Chen, B.; Chen, D.; Liu, F.; Ji, Y. Early Mechanical Performance of Glass Fibre-reinforced Manufactured Sand Concrete. *J. Build. Eng.* **2024**, *83*, 108440. [CrossRef]
26. Merli, R.; Preziosi, M.; Acampora, A.; Lucchetti, M.C.; Petrucci, E. Recycled Fibers in Reinforced Concrete: A Systematic Literature Review. *J. Clean. Prod.* **2020**, *248*, 119207. [CrossRef]
27. Gong, C.; Kang, L.; Zhou, W.; Liu, L.; Lei, M. Tensile Performance Test Research of Hybrid Steel Fiber—Reinforced Self-Compacting Concrete. *Materials* **2023**, *16*, 1114. [CrossRef]
28. Bhosale, A.; Rasheed, M.A.; Prakash, S.S.; Raju, G. A study on the efficiency of steel vs. synthetic vs. hybrid fibers on fracture behavior of concrete in flexure using acoustic emission. *Constr. Build. Mater.* **2019**, *199*, 256–268. [CrossRef]
29. Gong, Y.; Song, J.; Zhang, Y.; Jiang, T. Sustainable Utilization of Foundry Waste: Cracking Property Analysis of Basalt Fiber-Reinforced Rubber Concrete. *J. Mater. Civil Eng.* **2023**, *35*, 04023181. [CrossRef]
30. Zhao, Y.; Bi, J.; Wang, Z.; Huo, L.; Guan, J.; Zhao, Y.; Sun, Y. Numerical Simulation of the Casting Process of Steel Fiber Reinforced Self-compacting Concrete: Influence of Material and Casting Parameters on Fiber Orientation and Distribution. *Constr. Build. Mater.* **2021**, *312*, 125337. [CrossRef]
31. Wang, J.; Dai, Q.; Guo, S.; Si, R. Study on Rubberized Concrete Reinforced with Different Fibers. *ACI Mater. J.* **2019**, *116*, 21–31. [CrossRef]
32. Gültekin, A.; Beycioğlu, A.; Arslan, M.; Serdar, A.; Dobiszewska, M.; Ramyar, K. Fresh Properties and Fracture Energy of Basalt and Glass Fiber–Reinforced Self-Compacting Concrete. *J Mater. Civil. Eng.* **2021**, *34*, 04021406. [CrossRef]
33. Muhyaddin, G. Mechanical and fracture characteristics of ultra-high performance concretes reinforced with hybridization of steel and glass fibers. *Heliyon* **2023**, *9*, e17926. [CrossRef] [PubMed]
34. Akça, K.R.; Çakır, Ö.; Ipek, M. Properties of Polypropylene Fiber Reinforced Concrete using Recycled Aggregates. *Constr. Build. Mater.* **2015**, *98*, 620–630. [CrossRef]

35. *GB/T 14684-2011;* Sand for Construction. General Administration of Quality Supervision. Inspection and Quarantine: Beijing, China, 2011.
36. *GB/T 14685-2011;* Pebble and Crushed Stone for Construction. General Administration of Quality Supervision. Inspection and Quarantine: Beijing, China, 2011.
37. *DL/T 5332-2005;* Norm for Fracture Test of Hydraulic Concrete. Inspection and Quarantine: Beijing, China, 2005.
38. *ASTM C39/C39M-21;* Standard Test Method for Compressive Strength of Cylindrical Concrete Specimens. ASTM International: West Conshohocken, PA, USA, 2021.
39. Xu, S.; Reinhardt, H.W. A Simplified Method for Determining Double-K Fracture Parameters for Three-point Bending Tests. *Int. J. Fract.* **2000**, *104*, 181–209. [CrossRef]
40. Li, X.; Dong, W.; Wu, Z.; Chang, Q. Experimental Investigation on Double-K Fracture Parameters for Small Specimens of Concrete. *Eng. Mech.* **2010**, *27*, 166–171. (In Chinese)
41. Xie, J.; Li, J.; Lu, Z.; Li, Z.; Fang, C.; Huang, L.; Li, L. Combination Effects of Rubber and Silica Fume on the Fracture Behaviour of Steel-fibre Recycled Aggregate Concrete. *Constr. Build. Mater.* **2019**, *203*, 164–173. [CrossRef]
42. Abdallah, M.A.; Elakhras, A.A.; Reda, R.M.; Sallam, H.E.-D.M.; Moawad, M. Applicability of CMOD to Obtain the Actual Fracture Toughness of Rightly-Cracked Fibrous Concrete Beams. *Buildings* **2023**, *13*, 2010. [CrossRef]
43. Soetens, T.; Matthys, S. Different Methods to Model the Post-cracking Behaviour of Hooked-end Steel Fibre Reinforced Concrete. *Constr. Build. Mater.* **2014**, *73*, 458–471. [CrossRef]
44. Akbari, J.; Abed, A. Experimental evaluation of effects of steel and glass fibers on engineering properties of concrete. *Frat. Integrita Strut.* **2020**, *54*, 116–127. [CrossRef]
45. Karihaloo, B.; Abdalla, H.; Imjai, T. A simple method for determining the true specific fracture energy of concrete. *Mag. Concr. Res.* **2003**, *55*, 471–481. [CrossRef]

buildings

Article

Geopolymerization of Coal Gangue via Alkali-Activation: Dependence of Mechanical Properties on Alkali Activators

Xiaoping Wang [1,2], Feng Liu [1], Zezhou Pan [1], Weizhi Chen [1], Faheem Muhammad [3], Baifa Zhang [3,*] and Lijuan Li [1,*]

1 School of Civil and Transportation Engineering, Guangdong University of Technology, Guangzhou 510006, China
2 School of Architecture and Civil Engineering, Huangshan University, Huangshan 245041, China
3 School of Environmental Science and Engineering, Guangdong University of Technology, Guangzhou 510006, China
* Correspondence: zhangbaifa@gdut.edu.cn (B.Z.); lilj@gdut.edu.cn (L.L.)

Abstract: Coal gangue (CG) is a residual product from coal mining and washing processes. The reutilization of CG to produce geopolymers is a low-carbon disposal strategy for this material. In this study, the calcined CG ($CG_{700^\circ C}$) was used as aluminosilicate precursors, and the effects of alkali activators (i.e., $Na_2SiO_3/NaOH$, NaOH concentration, and liquid–solid) on the mechanical characteristics and microstructure of $CG_{700^\circ C}$-based geopolymers were investigated. The findings indicated that the specimens with a liquid–solid ratio of 0.50 ($G_{2.0\text{-}10\text{-}0.50}$) exhibited a compact microstructure and attained a compressive strength of 24.75 MPa. Moreover, increasing the $Na_2SiO_3/NaOH$ mass ratio has shortened the setting times and facilitated geopolymer gel formation, resulting in a denser microstructure and improved compressive strength. The higher NaOH concentrations of alkali activators facilitated the dissolution of $CG_{700^\circ C}$ particles, and the geopolymerization process was more dependent on the condensation of SiO_4 and AlO_4 ions, which promoted the formation of geopolymer networks. Conversely, an increase in the liquid–solid ratio from 0.50 to 0.65 had a negative impact on compressive strength enhancement, impeding the polycondensation rate. Examination through scanning electron microscopy and mercury intrusion porosimetry revealed that employing a lower $Na_2SiO_3/NaOH$ mass ratio ($G_{1.2\text{-}10\text{-}0.55}$), smaller NaOH concentrations ($G_{2.0\text{-}8\text{-}0.55}$), and a higher liquid–solid ratio ($G_{2.0\text{-}10\text{-}0.65}$) led to the presence of larger pores, resulting in decreased 28 days compressive strength values (15.87 MPa, 13.25 MPa, and 14.92 MPa, respectively), and a less compact structure. The results suggest that the performance of $CG_{700^\circ C}$-based geopolymers is significantly influenced by alkali activators.

Keywords: coal gangue; geopolymers; alkali activators; mechanical properties; microstructure

Citation: Wang, X.; Liu, F.; Pan, Z.; Chen, W.; Muhammad, F.; Zhang, B.; Li, L. Geopolymerization of Coal Gangue via Alkali-Activation: Dependence of Mechanical Properties on Alkali Activators. *Buildings* **2024**, *14*, 787. https://doi.org/10.3390/buildings14030787

Academic Editor: Antonio Caggiano

Received: 30 January 2024
Revised: 8 March 2024
Accepted: 12 March 2024
Published: 14 March 2024

1. Introduction

Geopolymers, proposed by Joseph Davidovits [1], are inorganic polymeric materials that exhibit excellent performance in several areas, including fire resistance, acid resistance, and the solidification of heavy metal wastes [2–4]. Over the last few decades, geopolymers have received considerable attention owing to their superior performance and smaller environmental footprint [5–8]. Statistically, the CO_2 emission of the geopolymer products with a typical feedstock is approximately 44–64% lower than that of ordinary Portland cement [9]. Geopolymers are normally synthesized through aluminosilicate precursors and alkali activators [10]. The precursors primarily originate from byproducts (e.g., fly ash [11–13], ground granulated blast-furnace slag [14], and red mud [15]) or natural minerals (e.g., halloysite [16,17] and kaolinite [18,19]), while the alkali activators are typically composed of potassium (or sodium) hydroxides and silicates.

Coal gangue (CG) is a byproduct that is separated during the coal mining and washing processes. The production of CG is approximately 10–20% of the total coal output [20–23].

As of now, enormous amounts of CG have accumulated around the world. CG waste in China has reached 7 billion tons and continues to grow [10]. In the past, CG was normally used for stacking or landfilling; these discharge methods not only occupied land resources but also caused significant soil, water, and air pollution [24,25]. In addition, researchers have demonstrated the use of CG in building materials such as coarse aggregates [26,27], sintered bricks [28], and paved roads [29]. However, the mechanical properties of the products are negatively influenced by the CG waste, which has more porosity and higher water absorption. Karimipour et al. [27] reported that when the natural aggregates were replaced by CG aggregates at 25%, the compressive strength and elastic modulus decreased by 29% and 20%, respectively. Guan et al. [30] showed that CG concrete exhibited lower frost resistance in a freeze-thaw environment compared to normal concrete. Therefore, there is an urgent need to explore alternative uses for CG.

CG waste contains a large amount of silicate and aluminate materials, the main components of CG waste are kaolinite, quartz, muscovite, etc. [20–22]. Studies have attempted to use CG to prepare geopolymers. Zhang et al. [31] found a thermal temperature of 800 °C was suitable for the pozzolanic reactivity of CG, and the calcined CG prepared geopolymers have good performance. As the temperature reached 900 °C, the mechanical properties of CG-based geopolymers decreased, as reported by Zhou et al. [32]. Li et al. [20] discussed the effects of CG particle size on the compressive strength of geopolymers and the results showed that the 200 mesh of CG could obtain the desired compressive strength for geopolymers. Han et al. [21] investigated the geopolymers incorporating CG and slag precursors, and their findings showed that the compressive strength of geopolymers was more than 40 MPa. Furthermore, Geng et al. [22] mixed CG with red mud to produce geopolymers, and the results showed that CG with red mud at a ratio of 2:8 was appropriate for preparing geopolymers. However, the present research on CG-based geopolymers is mainly focused on calcination temperature, particle size, or the mixing materials for the precursors. In addition, alkali activators, which are essential in determining the composition and mechanical properties, significantly influence the performance of geopolymers [33–37].

A study reported that the silicate modulus (SiO_2/Na_2O) and the alkalinity of the solution significantly influenced the autogenous shrinkage of alkali-activated slag cement (AASC) [34]. The findings revealed that the silicate modulus of 1.0 enhanced geopolymerization in the AASC paste. Moreover, the NaOH concentration on the mechanical characteristics of geopolymers derived from fly ash was investigated [35,36,38], Phoongernkham et al. [35] reported that the NaOH concentration of 14 mol/L was optimum for preparing the geopolymers, which improved the mechanical strength, which is consistent with the conclusions conducted by Somna et al. [38]. However, Guo et al. [36] reported that a modulus of 1.5 was favored for fly ash-based geopolymers, this was in contrast to the result (1.0) of AASC. In addition, Guo et al. [36,37] reported that water significantly influenced kaolin-based geopolymers by facilitating the dissolution of aluminosilicates or ion diversion during the geopolymerization process. When more water was introduced into the geopolymers, the geopolymerization process would be suppressed, leading to a reduction in the compressive strength. The previous study on various precursor materials such as slag cement, fly ash, and kaolin indicated different effects of alkali activators on mechanical properties. However, the application of alkali activators in CG-based geopolymers has rarely been investigated, resulting in an inadequate understanding of the effects of how alkali activators impact the mechanical properties of CG-based geopolymers.

In this study, the CG-based geopolymers were prepared using CG as a precursor and a variety of alkali activators consisting of NaOH and Na_2SiO_3 solutions. The effects of the Na_2SiO_3/NaOH mass ratio, NaOH concentration, and liquid–solid ratio on the mechanical properties of CG-based geopolymers were investigated. The mineral composition, microstructure, and mechanism properties of the CG-based geopolymers have been analyzed via X-ray diffraction (XRD), Fourier-transform infrared (FTIR) spectroscopy, scanning electron microscopy (SEM), and energy-dispersive X-ray (EDX) spectroscopy. Furthermore, the pore characteristics and parameters of the geopolymers were detected by mercury intrusion porosimetry (MIP). This study aimed to provide a better understanding of alkali activators on the geopolymerization process of CG-based geopolymers, thereby contributing to the reutilization of CG waste.

2. Materials and Methods

2.1. Materials

The CG was acquired from Hongqi Coal Industry Co., Ltd., located in Zhengzhou, China. The chemical compositions were measured using X-ray fluorescence spectroscopy (Table 1). The main chemical components of CG were SiO_2 and Al_2O_3, which together accounted for 86.12 wt% of the total content. CG powers were subjected to calcination in a muffle furnace at 700 °C for a duration of 3 h, with a heating rate of 10 °C/min. The calcined CG was denoted as $CG_{700°C}$.

Table 1. Main chemical composition of CG (wt%).

Chemical Composition	SiO_2	Al_2O_3	Fe_2O_3	CaO	Na_2O	MgO	K_2O	TiO_2	Others
CG	62.84	23.28	4.44	1.73	1.04	1.99	3.22	0.98	0.48

The CG was gray before being calcined (Figure 1a), and the $CG_{700°C}$ became red as it contained iron oxides during the calcination (Figure 1b). The particle size distributions of $CG_{700°C}$ were shown in Figure 1c, with a d_{50} particle size of 5.59 μm.

Figure 1. Photos of CG of CG (**a**) and $CG_{700°C}$ (**b**); Particle distributions of $CG_{700°C}$ (**c**).

The XRD patterns of the samples denoted as CG and $CG_{700°C}$ were displayed in Figure 2a. The peak at approximately 26.3° (2θ) signified the existence of quartz, and the peak at approximately 12.3° (2θ) corresponded to kaolinite. Moreover, traces of muscovite and albite were contained in CG. After calcination, the reflections corresponding to kaolinite disappeared, and the spectrum of $CG_{700°C}$ featured a broad diffraction band centered at ~23.7° (2θ), indicating that the kaolinite had been transformed into an amorphous phase. However, the quartz existed in the $CG_{700°C}$ owing to its thermal stability. The SEM image of $CG_{700°C}$ is shown in Figure 2b,c. The SEM images at low magnification showed that $CG_{700°C}$ possessed irregular particles and a loose surface morphology (Figure 2b), and the SEM images at high magnification showed that the microstructure of $CG_{700°C}$ displayed a lamella shape (Figure 2c).

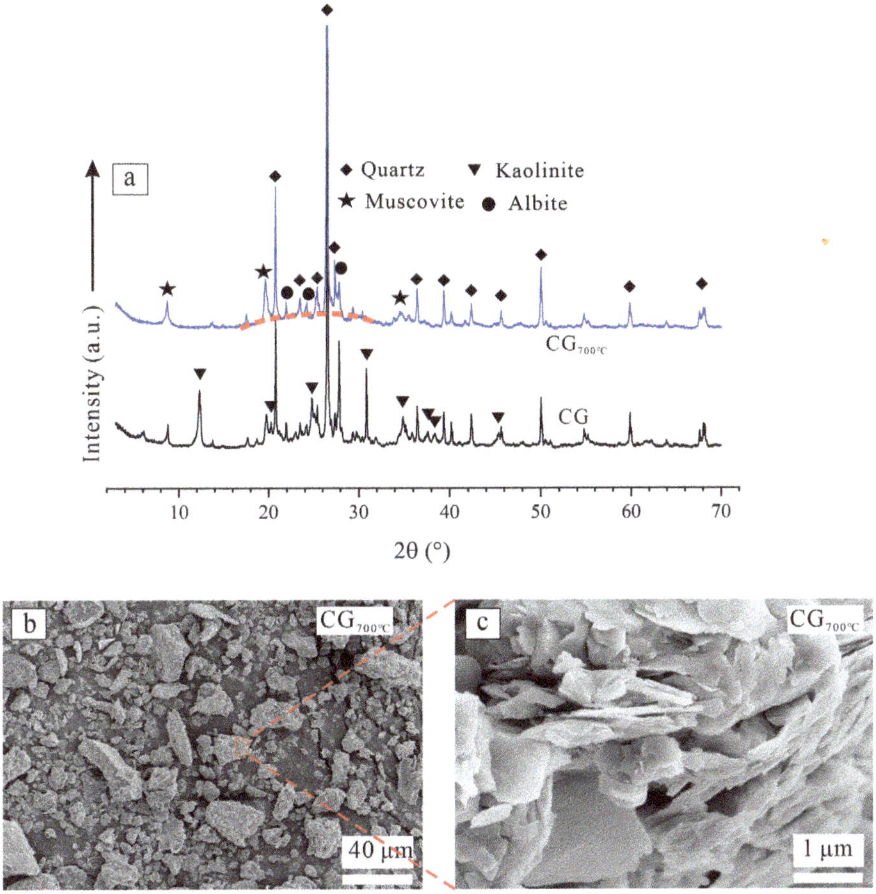

Figure 2. (**a**) XRD patterns of CG and $CG_{700^{\circ}C}$; SEM images of $CG_{700^{\circ}C}$ (**b**) and (**c**).

The alkali activators used in the study included a sodium silicate solution (Na_2O 8.5 wt%, SiO_2 27.3 wt%, H_2O 64.2 wt%, and a modulus of 3.3) and sodium hydroxide (NaOH) particles with a purity exceeding 96%. The chemical reagents were procured from commercial suppliers, while the distilled water used was supplied by the laboratory.

2.2. Geopolymers Preparation

The preparation process was depicted in Figure 3, and the experimental conditions were listed in Table 2.

Preparation process. First, NaOH was added to distilled water to prepare a NaOH solution, which was then mixed with a Na_2SiO_3 solution to form an alkali activator. Second, $CG_{700^{\circ}C}$ was blended with the alkali activators to produce a homogeneous gel slurry, which was forged into silica molds measuring 20 mm × 20 mm × 20 mm. Subsequently, a thin polyethylene film was applied to shield the slurry from water evaporation. The resulting specimens were then subjected to curing in a controlled environment at 60 °C for 24 h, followed by demolding and further curing at ambient temperature (~25 °C) for varying periods (3, 7, and 28 days) before testing.

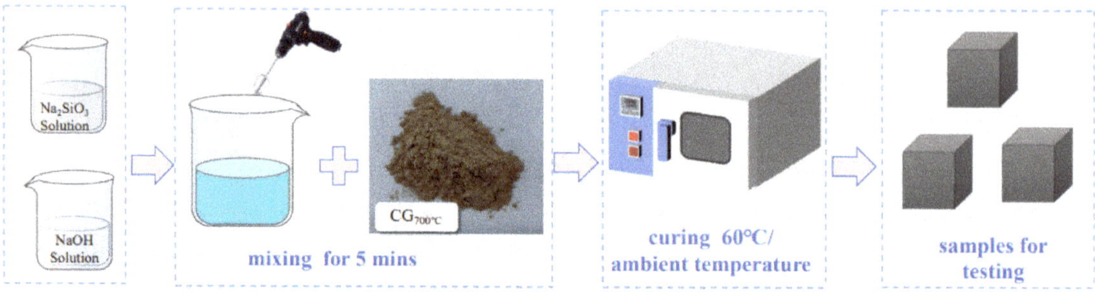

Figure 3. Preparation of $CG_{700^\circ C}$-based geopolymers.

Table 2. Mixture proportions of $CG_{700^\circ C}$-based geopolymers.

Specimens	$CG_{700^\circ C}$ (g)	Na_2SiO_3 (g)	NaOH (g)	Fluidity (mm)
$G_{1.2\text{-}10\text{-}0.55}$	180	54	45	113.9
$G_{1.6\text{-}10\text{-}0.55}$	180	60.92	38.08	113.3
$G_{2.0\text{-}10\text{-}0.55}$	180	66	33	112.5
$G_{2.4\text{-}10\text{-}0.55}$	180	69.88	29.12	111.6
$G_{2.0\text{-}8\text{-}0.55}$	180	66	33	112.7
$G_{2.0\text{-}12\text{-}0.55}$	180	66	33	112.2
$G_{2.0\text{-}14\text{-}0.55}$	180	66	33	111.8
$G_{2.0\text{-}10\text{-}0.50}$	180	60	30	100.1
$G_{2.0\text{-}10\text{-}0.60}$	180	72	36	124.8
$G_{2.0\text{-}10\text{-}0.65}$	180	78	39	136.2

The prepared geopolymers were marked as '$G_{X\text{-}Y\text{-}Z}$', where 'X' represents the Na_2SiO_3/NaOH mass ratio, 'Y' represents the NaOH concentration, and 'Z' represents the liquid–solid mass ratio. For example, '$G_{2.0\text{-}10\text{-}0.55}$' indicated the geopolymer specimens with a Na_2SiO_3/NaOH mass ratio of 2.0, a NaOH concentration of 10 mol/L, and a liquid–solid mass ratio of 0.55.

2.3. *Characterisation Methods*

The fluidity of the geopolymer slurry was tested using a truncated conical device from Hebei, China. The setting times of the geopolymer slurry were tested using a Vicat device from Shanghai, China. Compressive strength tests on the geopolymer specimens were conducted using a STS100K machine from Fujian, China, with a loading rate of 0.5 mm/min. Each group consisted of three specimens, and the compressive strength value was determined by averaging the results from the testing process.

The XRD patterns of the raw materials and the geopolymers were obtained using a Bruker D8 Advance diffractometer (Mannheim, Germany) within the range of 3–70° (2θ), the scanning rate was 3°/min. The FTIR spectra of the geopolymers were recorded using a Nicolet IS50 spectrometer (ThermoFisher, Waltham, MA, USA) across the range of 4000–400 cm^{-1}, the resolution was 4 cm^{-1}. The SEM images and EDX spectra of the geopolymers were captured using an S-3400N-II instrument (Hitachi, Tokyo, Japan) at a voltage of 35 kV and a current of 10 mA. The pore size distribution and the total pore volume were determined using an AutoPore IV 9510 porosimeter (Micromeritics, Norcross, GA, USA) with intrusion pressures ranging from 0.5 to 33,000 psi.

3. Results and Discussion

3.1. *Setting Time and Compressive Strength*

3.1.1. Fluidity and Setting Time

Figure 4 illustrates the initial and final setting times of geopolymers, while Table 2 indicates the fluidity of geopolymers. The fluidity of the specimens changed from 100.1 to

136.2 mm as the mixture proportions of alkali activators altered. However, the samples of $G_{X\text{-}10\text{-}0.55}$ and $G_{2.0\text{-}Y\text{-}0.55}$ seemed to be unaffected (around 112.5 mm) by the Na_2SiO_3/NaOH mass ratio or NaOH concentration change, but the paste fluidity increased more quickly when the liquid–solid mass ratio rose from 0.50 to 0.65, suggesting that the liquid–solid mass ratio significantly impacted paste fluidity. As the Na_2SiO_3/NaOH mass ratio increased, the setting times of the $CG_{700°C}$-based geopolymer gels decreased. The final setting times of $G_{1.2\text{-}10\text{-}0.55}$ and $G_{2.4\text{-}10\text{-}0.55}$ were 127 and 87 min, respectively, representing a reduction of 23.7%. Similarly, studies have reported that the precursor binder activated by silicates showed faster hardening than hydroxides [39]. The high soluble silicate contents in the geopolymerization process favored the occurrence of geopolymer gels, which were beneficial for the formation of the pastes, and shortened the setting time of the hardened geopolymers.

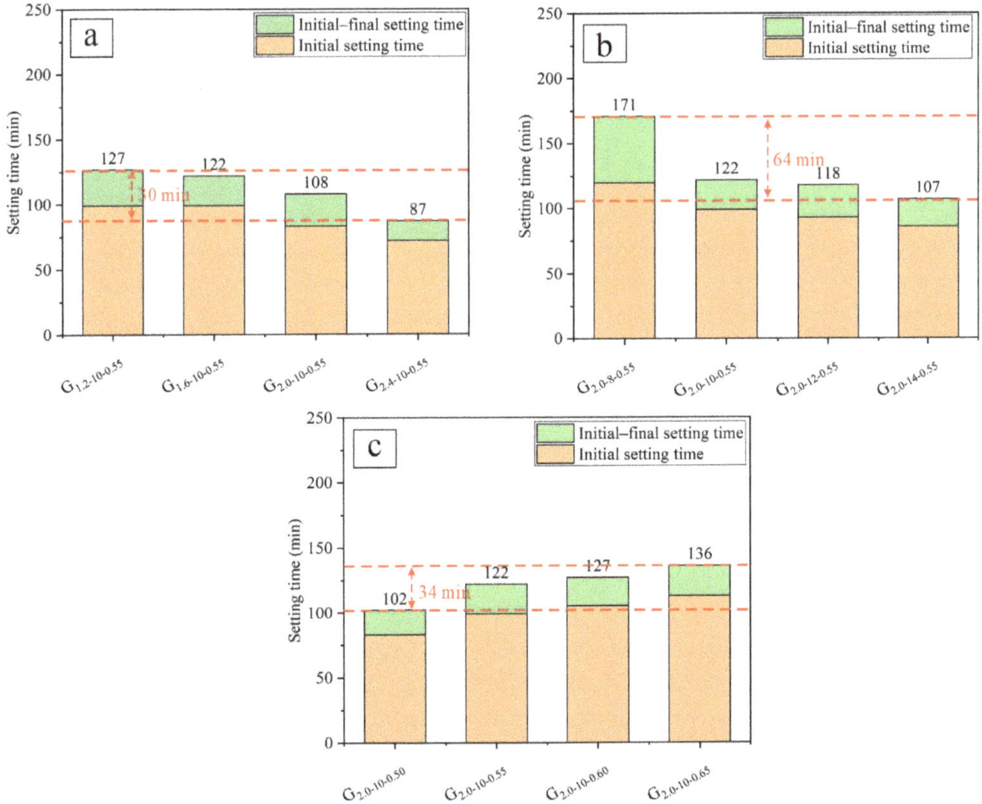

Figure 4. Setting times of geopolymers: (**a**) $G_{X\text{-}10\text{-}0.55}$, (**b**) $G_{2.0\text{-}Y\text{-}0.55}$, and (**c**) $G_{2.0\text{-}10\text{-}Z}$.

In addition, as the NaOH concentration increased from 8 to 14 mol/L, it was observed that the final setting time of the $CG_{700°C}$-based geopolymer gels gradually decreased, decreasing from 171 to 107 min, indicating that increasing alkalinity (i.e., the OH^- concentration) promoted the diffusion of Si and Al from the $CG_{700°C}$. The ions that were diffused through the hydrate reaction products then rapidly reacted with alkali solutions, which accelerated the reaction process of geopolymerization. These results were coherent with those observation by Bernal et al. [40]. The higher alkaline conditions were more strongly influenced by the dissolution of the slag, resulting in a reduction in setting times.

However, increasing the liquid–solid mass ratio prolonged both the initial and final setting times of the geopolymer gels. For example, the final setting time of $G_{2.0\text{-}10\text{-}0.65}$ was

136 min, while $G_{2.0\text{-}10\text{-}0.50}$ was 102 min. These results indicate that too much water inhibits the reaction rate of geopolymer gels. In other words, the presence of excess water will decrease the alkalinity of activators that adversely influence the dissolution of the $CG_{700°C}$, thereby extending the setting times.

3.1.2. Compressive Strength

Figure 5 shows the compressive strengths and damages of different geopolymer samples. The strength of the geopolymers gradually increased over the curing duration, which indicated that geopolymerization was going, and the failure of geopolymer samples exhibited brittle damage (Figure 5d). Notably, the compressive strength of the geopolymers increased rapidly within 3 days of curing, reaching a level that was ~65% of the strength achieved after 28 days of curing. Furthermore, the strength between 3 and 7 days was slightly higher than that within the first 3 days. This trend was due to two factors: (1) more soluble silicates reacted during the early stages, and (2) the ambient temperature slowed down the geopolymerization process.

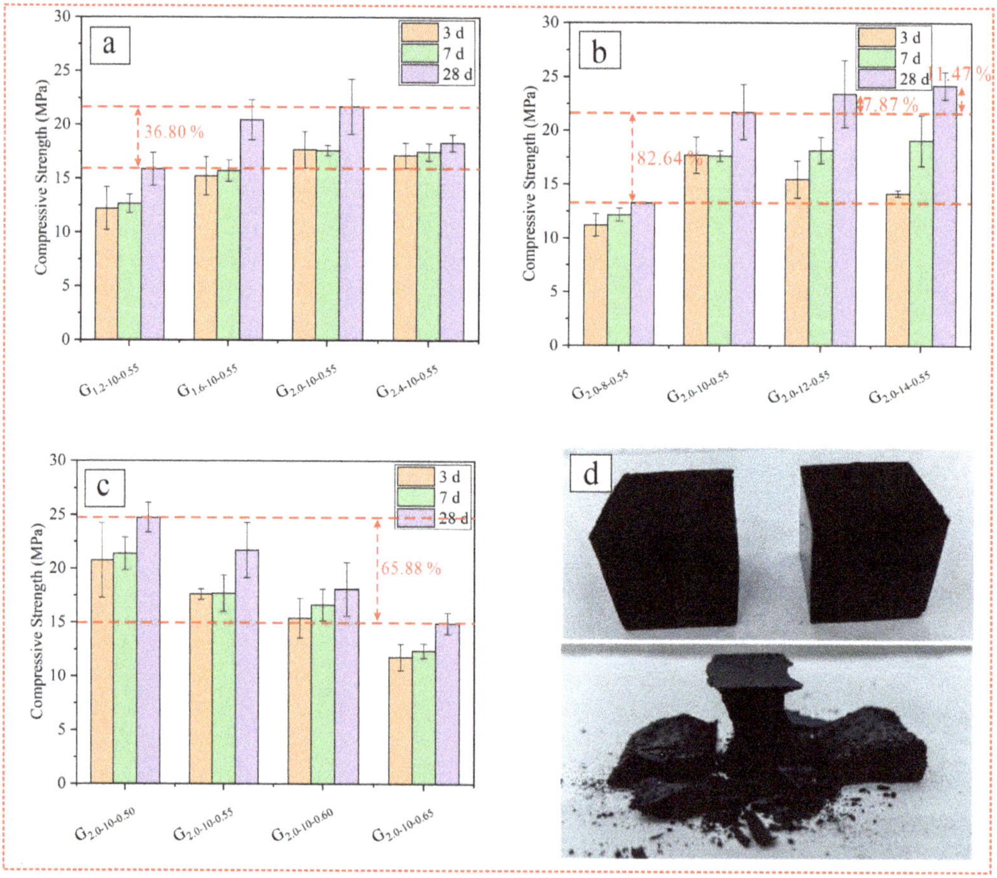

Figure 5. Compressive strength of geopolymers (**a**) $G_{X\text{-}10\text{-}0.55}$, (**b**) $G_{2.0\text{-}Y\text{-}0.55}$, (**c**) $G_{2.0\text{-}10\text{-}Z}$, and (**d**) damage photos of samples.

In addition, with an increasing $Na_2SiO_3/NaOH$ mass ratio, the compressive strength of the geopolymers initially rose before subsequently declining. The 28 days compressive strength of $G_{2.0\text{-}10\text{-}0.55}$ (21.71 MPa) was 36.80% higher than that of $G_{1.2\text{-}10\text{-}0.55}$ (15.87 MPa).

However, $G_{2.4\text{-}10\text{-}0.55}$ exhibited a lower compressive strength (18.34 MPa) than $G_{2.0\text{-}10\text{-}0.55}$ (21.71 MPa). This decrease can be attributed to the presence of surplus soluble silicate, leading to more geopolymer gels precipitating at an early stage, and those gels adhered to the raw materials would inhibit the condensation of SiO_4 and AlO_4 ions. These findings align with previous research findings [41,42]. Guo et al. [36] found the 28 days compressive strength of FA-based geopolymers exhibited an increase from 17.1 MPa to 59.3 MPa as the modulus (SiO_2/Na_2O) changed from 1.0 to 1.5.

Moreover, with an increase in NaOH concentration from 8 to 14 mol/L, there was an improvement in the compressive strength of the geopolymer. Specifically, the compressive strength of $G_{2.0\text{-}14\text{-}0.55}$ increased by 94.11% compared to that of $G_{2.0\text{-}8\text{-}0.55}$. Notably, the compressive strength saw an 82.64% increase as the NaOH concentration increased from 8 to 10 mol/L, followed by incremental increases of 7.87% and 3.60% as the NaOH concentration further increased from 10 to 12 mol/L and 12 to 14 mol/L, respectively. These results suggested that the low NaOH concentration inhibited the dissolution of $CG_{700°C}$ particles, thereby leading to the lower compressive strength of the geopolymers [43]. Moreover, the literature has shown the geopolymerization process was not only dependent on the dissolution rate of the precursor but also more relied on the condensation of SiO_4 and AlO_4 ions, which resulted in the formation of geopolymer networks. Therefore, the compressive strength experienced a rapid increase with the elevation of the NaOH concentration from 8 to 10 mol/L, followed by a slower rate of increase as the NaOH concentration continued to rise. Somna et al. [38] suggested the NaOH concentrations of ground fly ash-based geopolymers ranged from 9.5 mol/L to 14.0 mol/L.

With the rise in the liquid–solid ratio, the 28 days compressive strength gradually decreased. The compressive strength decreased from 24.75 MPa ($G_{2.0\text{-}10\text{-}0.50}$) to 14.92 MPa ($G_{2.0\text{-}10\text{-}0.65}$) as the liquid–solid ratio increased from 0.50 to 0.65. These results highlighted that the excess water in the geopolymerization process was unfavorable for the mechanical properties of $CG_{700°C}$-based geopolymers. Zhang et al. [44] discovered that additional voids were introduced when the batched water evaporated in the geopolymer slurry, which was detrimental to enhancing compressive strength. Zhang et al. [37] pointed out that the lower liquid–solid ratio was beneficial for the polycondensation rate.

The relations between the compressive strength and the microstructure of the geopolymers will be discussed in the next sections.

3.2. *Geopolymer Structure*

3.2.1. XRD Analysis

Figure 6 displays the XRD patterns of the $CG_{700°C}$-based geopolymers. The XRD spectra highlighted a wide reflection in the range of 18–34° (2θ), with a peak focused at ~28.3° (2θ). In comparison to the raw material of $CG_{700°C}$ (Figure 1a), the XRD patterns of the geopolymers shifted towards a higher angle (~28.3° vs. ~23.7° (2θ)), indicating that the connection between Si-O bonds and Al-O bonds increased, and more geopolymer gels were formed [45]. The geopolymer gels were probably N-A-S-H owing to the lower calcium content in the precursor [16,46], which contributed to the compressive strength of the samples. Additionally, minerals such as quartz, muscovite, and albite were detected in the geopolymers owing to their stable chemical reactivity, indicating that a certain amount of $CG_{700°C}$ did not participate in the geopolymerization reaction [47].

Furthermore, with an increase in the $Na_2SiO_3/NaOH$ mass ratio, the samples featured no new diffraction peaks (Figure 6a), indicating that an increase in the soluble silicate content had minimal impact on the compound synthesis of the geopolymers. However, compared with the XRD patterns of $G_{1.2\text{-}10\text{-}0.55}$, those of $G_{2.0\text{-}10\text{-}0.55}$ and $G_{2.4\text{-}10\text{-}0.55}$ featured lower intensities of quartz diffraction peaks, suggesting a greater consumption of quartz crystals in $G_{2.0\text{-}10\text{-}0.55}$ and $G_{2.4\text{-}10\text{-}0.55}$, as the increased $Na_2SiO_3/NaOH$ mass ratio favored geopolymerization.

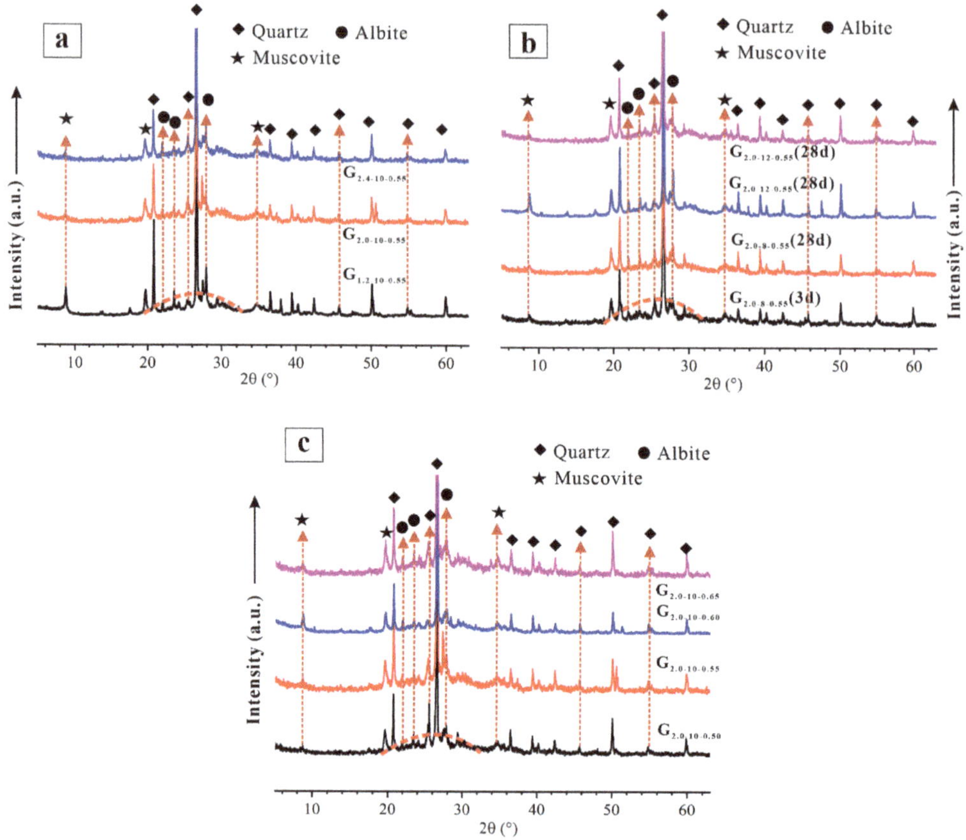

Figure 6. XRD patterns of geopolymers: (**a**) $G_{X\text{-}10\text{-}0.55}$, (**b**) $G_{2.0\text{-}Y\text{-}0.55}$, and (**c**) $G_{2.0\text{-}10\text{-}Z}$.

The elevation of the NaOH concentration seemed to reduce the intensity of quartz diffraction peaks in the geopolymers. Compared with $G_{2.0\text{-}14\text{-}0.55}$, $G_{2.0\text{-}8\text{-}0.55}$ exhibited a more pronounced intensity of quartz diffraction peaks, indicating that fewer $CG_{700°C}$ participated in geopolymerization at lower alkalinity conditions. Thus, a smaller amount of geopolymer gels was formed, resulting in lower compressive strength in $G_{2.0\text{-}8\text{-}0.55}$. Additionally, with an increase in the curing duration (Figure 6b, 3 vs. 28 d), $G_{2.0\text{-}8\text{-}0.55}$ (28 d) exhibited a more significant decrease in the intensity of the quartz diffraction peaks compared to $G_{2.0\text{-}8\text{-}0.55}$ (3 d).

In addition, with an increase in the liquid–solid ratio, the XRD spectra featured no new diffraction peaks (Figure 6c), and the intensity of quartz diffraction peaks was not significantly influenced, suggesting that more water content did not influence the chemical composition of the geopolymers, and no new geopolymer gels formed with an increasing liquid–solid ratio. These findings align with those of Cui et al. [48]. Cui et al. [48] reported that adding more water could contribute to the dissolution of fly ash; however, this led to an increase in the porosity and pore sizes of fly ash-based geopolymers.

3.2.2. FTIR Spectroscopy

Figure 7 shows the FTIR spectra of $CG_{700°C}$ and geopolymer samples within the spectral range of 400–1800 cm^{-1}. The spectra of the selected geopolymers exhibited a broad band centered at ~1020 cm^{-1}, indicative of the asymmetrical stretching vibration linked to Si-O and Al-O bonds [49,50]. The bands observed at 801 and 780 cm^{-1} corresponded to the bending and stretching vibrations of Al-O bonds, consistent with findings from previous research [51]. Additionally, the bands at 480 cm^{-1} corresponded to the stretching vibration of Si-O-T bands, where T denotes either Si or Al, as supported by existing literature [52,53]. Moreover, the bands identified at 697 cm^{-1} corresponded to the internal extension of Si–O bonds, as referenced in a previous study [54]. Furthermore, the bands detected at 1646 cm^{-1} were related to the bending vibration of O-H bonds, in alignment with findings from various other studies [55–57].

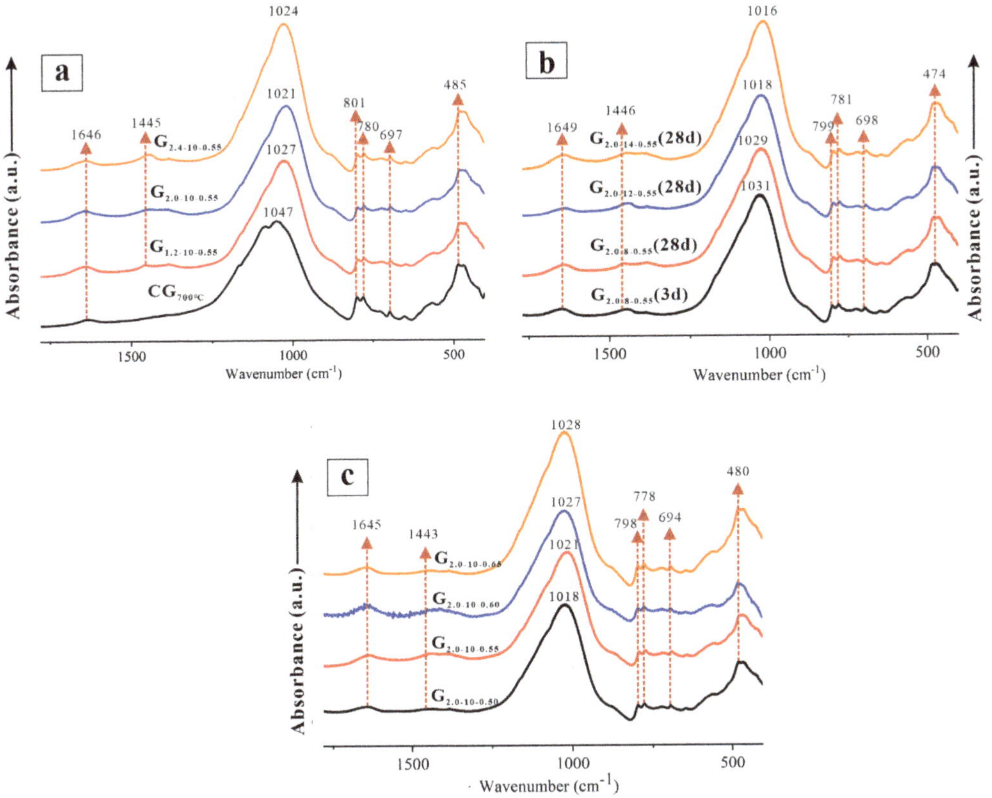

Figure 7. FTIR spectra of geopolymers: (**a**) $G_{X\text{-}10\text{-}0.55}$, (**b**) $G_{2.0\text{-}Y\text{-}0.55}$, and (**c**) $G_{2.0\text{-}10\text{-}Z}$.

In addition, the main band (wavenumbers: 900–1200 cm^{-1}) of the $CG_{700°C}$-based geopolymers shifted to lower wavenumbers after alkali activation, indicating that more amorphous products were produced [58–60]. The main bands shifted to lower wavenumbers in the spectra of $G_{2.0\text{-}10\text{-}0.55}$ (1021 cm^{-1}) and $G_{2.4\text{-}10\text{-}0.55}$ (1024 cm^{-1}), which compared to the specimen of $G_{1.2\text{-}10\text{-}0.55}$ (1027 cm^{-1}) (Figure 7a). As the curing time of $G_{2.0\text{-}8\text{-}0.55}$ increased (Figure 7b), the wavenumbers of the main band in the spectra decreased from 1031 cm^{-1} (3 d) to 1029 cm^{-1} (28 d); Moreover, as the NaOH concentration increased from 8 to 14 mol/L or the liquid–solid ratio decreased from 0.65 to 0.50, the main bands shifted to lower wavenumbers. The literature [58–61] demonstrated that the adsorption band in the range around ~1030 cm^{-1} was associated with the asymmetric stretching vibration of Si-O-T bonds (where T = Si or Al), and the lower wavenumber specimens were induced by more Si atoms being replaced by Al atoms during geopolymerization. Moreover, the absorption band at 1445 cm^{-1} corresponded to the stretching vibration of C-O bonds [61], which was related to the carbonation of geopolymer gels. The raw material ($CG_{700°C}$) did not show the absorption band due to no alkali activators applied.

3.2.3. SEM/EDX Results

Figures 8–10 displays the SEM/EDX results of selected geopolymers. The microstructures of geopolymers with different Na_2SiO_3/NaOH mass ratios are shown in Figure 8a–g. The low-magnification SEM images (Figure 8a,c,e) showed that many pores existed in the $CG_{700°C}$-based geopolymers. The high-magnification SEM images of the geopolymers revealed that the matrix featured some isolated particles and microcracks within the matrix of the geopolymers (Figure 8b,d,f), which contributed to the compressive strength.

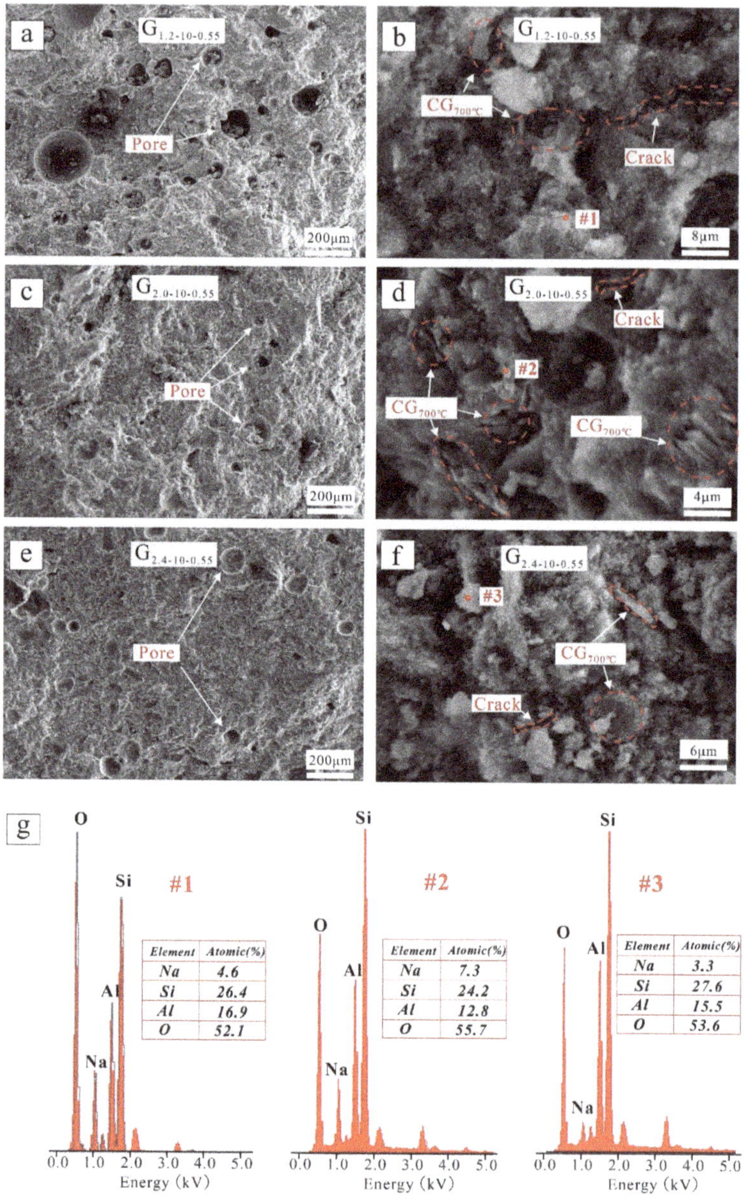

Figure 8. SEM images of geopolymers: (**a**,**b**) $G_{1.2\text{-}10\text{-}0.55}$, (**c**,**d**) $G_{2.0\text{-}10\text{-}0.55}$, and (**e**,**f**) $G_{2.4\text{-}10\text{-}0.55}$; (**g**) EDX results of spots highlighted in images (**b**,**d**,**f**).

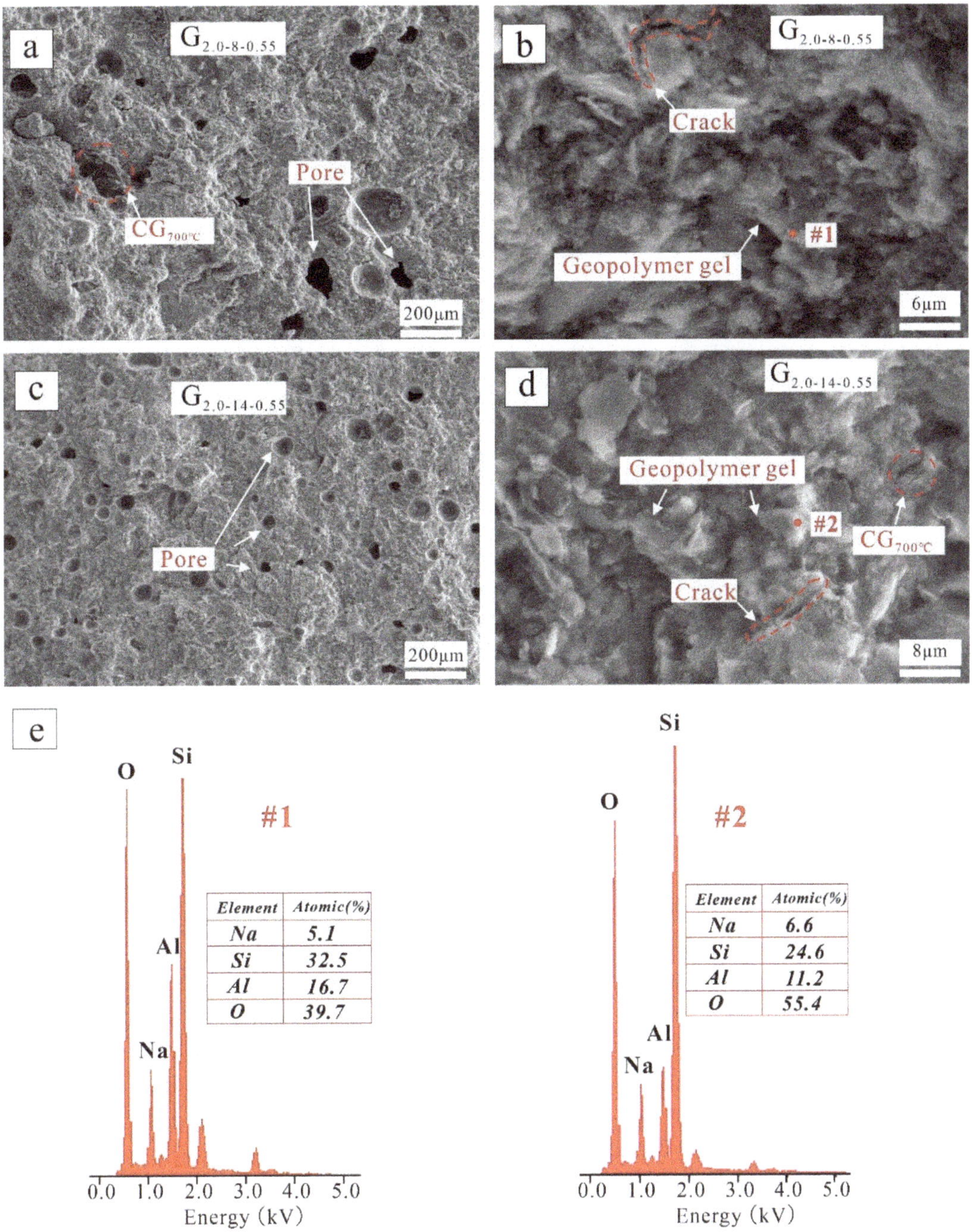

Figure 9. SEM images of geopolymers: (**a**,**b**) $G_{2.0\text{-}8\text{-}0.55}$, (**c**,**d**) $G_{2.0\text{-}14\text{-}0.55}$, and (**e**) EDX results of spots highlighted in images (**b**,**d**).

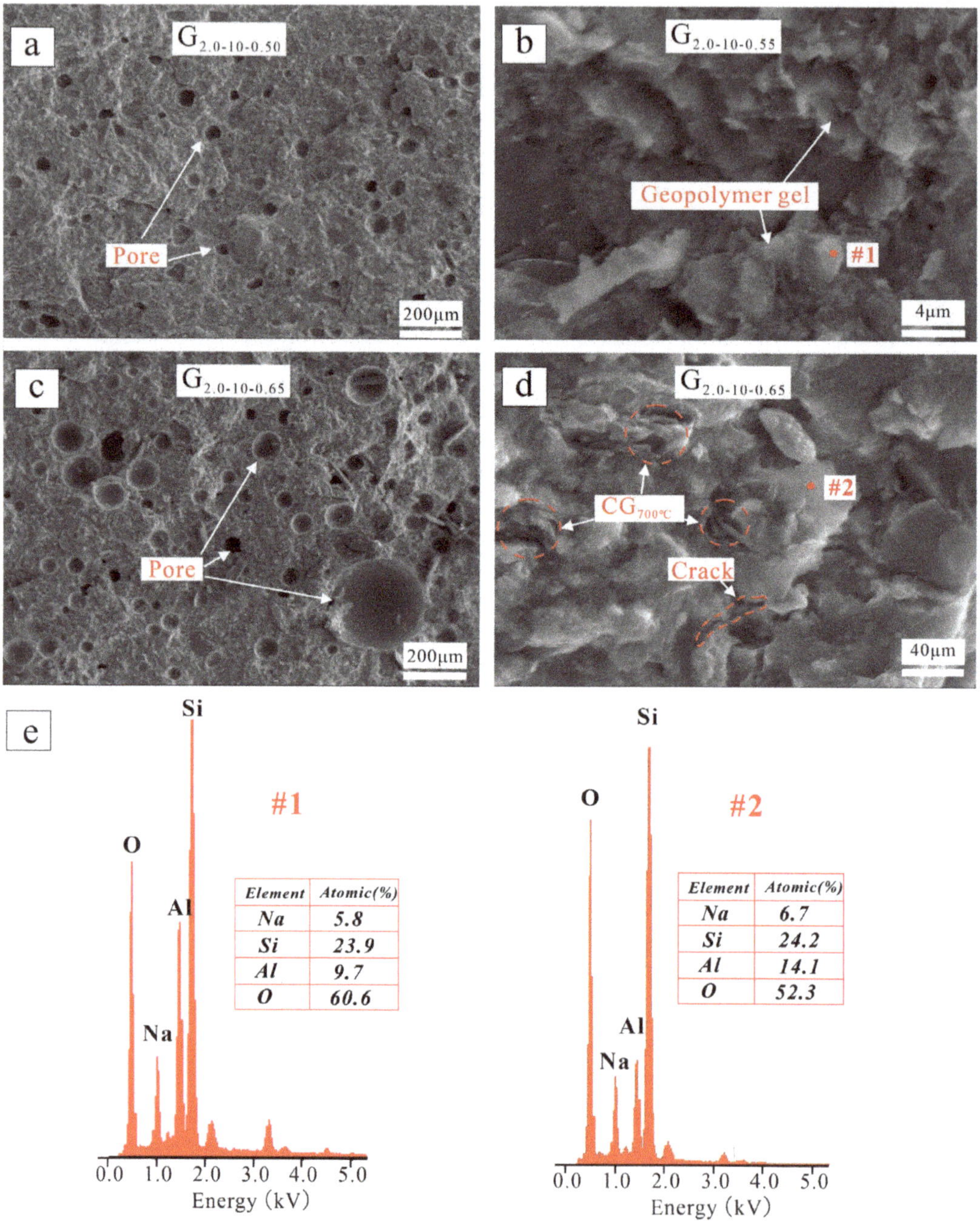

Figure 10. SEM images of geopolymers: (**a**,**b**) $G_{2.0\text{-}10\text{-}0.50}$, (**c**,**d**) $G_{2.0\text{-}10\text{-}0.65}$, and (**e**) EDX results of spots highlighted in images (**b**,**d**).

Compared with $G_{1.2\text{-}10\text{-}0.55}$, both $G_{2.0\text{-}10\text{-}0.55}$ and $G_{2.4\text{-}10\text{-}0.55}$ (Figure 8c,e) exhibited homogeneous and compact microstructures and smaller pore sizes. These findings demonstrated that increasing the Na_2SiO_3/NaOH mass ratios favored the geopolymerization of the

$CG_{700°C}$-based geopolymers, providing more soluble silicates to participate in geopolymerization reactions. The EDX results (Figure 8g, spots 1, 2, and 3) indicated that the geopolymer gels were primarily composed of Si, Al, Na, and O, forming N-A-S-H compounds. Moreover, the Si/Al ratios varied among the selected geopolymers. For instance, the Si/Al ratio of $G_{1.2\text{-}10\text{-}0.55}$ was 1.56, considerably lower than those of $G_{2.0\text{-}10\text{-}0.55}$ (1.89) and $G_{2.4\text{-}10\text{-}0.55}$ (1.78). This discrepancy suggests that fewer Si atoms participated in geopolymerization at lower soluble silicate concentrations, consistent with the compressive strength findings in Figure 5.

However, as the concentration of NaOH increased from 8 to 14 mol/L, $G_{2.0\text{-}8\text{-}0.55}$ exhibited a relatively loose structure (Figure 9a), and $G_{2.0\text{-}14\text{-}0.55}$ exhibited a denser matrix (Figure 9c). In addition, isolated particles (Figure 9a,d) and microcracks (Figure 9b,d) were observed in the geopolymer matrix. The EDX results (Figure 9e, spot 1, 2) showed that $G_{2.0\text{-}8\text{-}0.55}$ exhibited smaller Si/Al ratios than that of $G_{2.0\text{-}14\text{-}0.55}$ (1.94 vs. 2.19), which resulted in lower compressive strength for $G_{2.0\text{-}8\text{-}0.55}$.

Figure 10 illustrates the SEM images and EDX results of $G_{2.0\text{-}10\text{-}0.50}$ and $G_{2.0\text{-}10\text{-}0.65}$. The low-magnification SEM images (Figure 10a,c) showed that $G_{2.0\text{-}10\text{-}0.50}$ and $G_{2.0\text{-}10\text{-}0.65}$ exhibited no significant differences in morphology, except for the smaller pore sizes observed in $G_{2.0\text{-}10\text{-}0.50}$. Additionally, more isolated particles occurred in $G_{2.0\text{-}10\text{-}0.65}$ (Figure 10d). The smaller pore sizes of $G_{2.0\text{-}10\text{-}0.50}$ were related to the higher compressive strength; the correlations would be discussed in the MIP.

3.2.4. MIP Results

Figure 11 displays the pore sizes and pore volume distribution of the selected geopolymers, while Table 3 provides the corresponding pore parameters obtained by MIP. According to Zhang et al. [62], pores were categorized into three categories: gel pores (1–10 nm), medium capillary pores (10–100 nm), and large capillary pores (100–1000 nm). The specimens $G_{2.0\text{-}8\text{-}0.55}$ exhibited a larger critical pore size of 67.51 nm, and specimens $G_{2.0\text{-}10\text{-}0.50}$ exhibited a smaller pore size of 25.01 nm. All of the samples had medium capillary pores (Figure 11a).

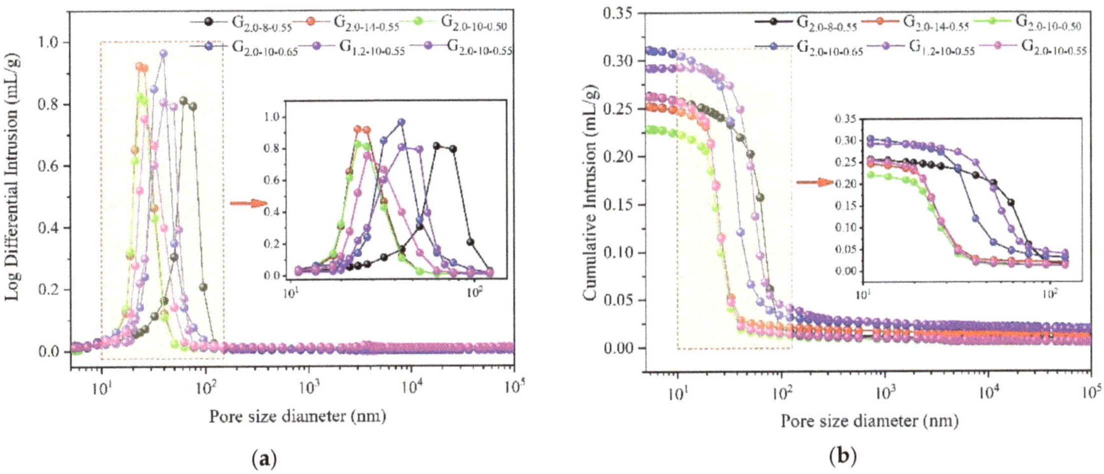

Figure 11. MIP results of geopolymers: (**a**) log differential intrusion; (**b**) cumulative intrusion.

As the Na_2SiO_3/NaOH mass ratio elevated from 1.2 to 2.0, the average pore sizes of $G_{1.2\text{-}10\text{-}0.55}$ and $G_{2.0\text{-}10\text{-}0.55}$ were 41.03 and 27.24 nm, respectively, and the porosity of the corresponding samples decreased from 39.44% to 37.61%, indicating that more soluble silicates were beneficial for reducing both the pore size and pore volume. The cumulative intrusion of the geopolymers is depicted in Figure 10b. $G_{2.0\text{-}10\text{-}0.55}$ exhibited a lower cumulative intrusion (0.251 mL/g) than that of $G_{1.2\text{-}10\text{-}0.55}$ (0.287 mL/g), and those also demonstrated that the increase in the Na_2SiO_3/NaOH mass ratio resulted in the formation of a more compact microstructure.

Table 3. The pore parameters of geopolymers determined via MIP.

Samples	Total Pore Area (m^2/g)	Average Pore Diameter (nm)	Porosity (%)
$G_{1.2\text{-}10\text{-}0.55}$	28.373	41.03	39.44
$G_{2.0\text{-}10\text{-}0.55}$	36.193	27.24	37.61
$G_{2.0\text{-}8\text{-}0.55}$	21.777	48.39	38.22
$G_{2.0\text{-}14\text{-}0.55}$	38.755	26.05	37.06
$G_{2.0\text{-}10\text{-}0.50}$	36.818	24.84	36.79
$G_{2.0\text{-}10\text{-}0.65}$	34.918	35.66	41.93

Moreover, $G_{2.0\text{-}14\text{-}0.55}$ exhibited a critical pore size of 26.31 mm and a porosity of 37.06%, which was lower than those of $G_{2.0\text{-}8\text{-}0.55}$. These indicated that increasing the concentration of NaOH favored the pore microstructure. A previous study [63] reported that pore size significantly affected the mechanical properties of cement materials, particularly at larger pore sizes (>50 nm). Thus, the compressive strength of specimens with 14 mol/L ($G_{2.0\text{-}14\text{-}0.55}$) was improved.

With the increase in the liquid–solid ratio from 0.50 to 0.65, the porosity of the geopolymers increased from 36.79% to 41.93%, and the average pore size increased from 24.84 to 35.66 mm (Table 3). In a study by Cui et al. [48], it was observed that with an increase in the H_2O/Na_2O ratio from 13 to 17, the pore size increased from 7.1 to 18.1 nm. Additionally, the smaller pore area confirmed that the pore diameter of $G_{2.0\text{-}10\text{-}0.50}$ was smaller than that of $G_{2.0\text{-}10\text{-}0.65}$, contributing to its increased compressive strength.

3.3. Summary and Discussion

This research revealed that alkali activators significantly influenced the compressive strength and microstructure of geopolymers, although the chemical composition of $CG_{700^\circ C}$-based geopolymers remained relatively unaffected. For example, an increase in the Na_2SiO_3/NaOH mass ratio from 1.2 to 2.4 resulted in a 36.80% enhancement in compressive strength after 28 days. Similarly, by increasing the NaOH concentration from 8 to 14 mol/L or reducing the liquid–solid ratio from 0.65 to 0.50, the 28 days compressive strength increased by 94.11% and 65.88%, respectively. Notably, the XRD patterns of the geopolymers did not feature new diffraction peaks (Figure 6), aligning with the results of previous research. Guo et al. [36] investigated the impact of modulus (SiO_2/Na_2O) on the compressive strength of FA-based geopolymers. The 28 days compressive strength increased from 17.1 MPa to 59.3 MPa as the modulus changed from 1.0 to 1.5. Zhang et al. [37] found the geopolymerization rate of calcined kaolin-based geopolymer accelerated and strengthened with an increase in the NaOH solution concentration from 3 mol/L to 12 mol/L. Somna et al. [38] suggested the NaOH concentrations of ground fly ash-based geopolymers ranged from 9.5 mol/L to 14.0 mol/L. Moreover, Zhang et al. [37] pointed out that the lower liquid–solid ratio promoted the polycondensation rate of the geopolymers.

Moreover, in the process of using $CG_{700^\circ C}$-based geopolymers, more attention should be paid to the alkali activators. In the present study, the NaOH concentration should be controlled at an appropriate dosage, and the Na_2SiO_3/NaOH mass ratio and the liquid–solid ratio should be controlled at 2.0 and 0.5, respectively, which can result in a dense microstructure and a higher compressive strength for the geopolymers. In the next research, some limitations are needed to be considered, such as the sample damage, the stress-strain curves, and the toughness improvement of $CG_{700^\circ C}$-based geopolymer concrete.

4. Conclusions

Alkali activators significantly influence the mechanical properties and microstructure of geopolymers. This research delved into the impact of varying factors such as the $Na_2SiO_3/NaOH$ mass ratio, NaOH concentration, and liquid–solid ratio on the mechanical properties of geopolymers derived from coal gangue calcined at 700 °C ($CG_{700°C}$). By conducting a series of spectroscopic and microscopic tests, the following conclusions were obtained.

As the $Na_2SiO_3/NaOH$ mass ratio increased from 1.2 to 2.4, the 28 days compressive strength of $G_{2.0\text{-}10\text{-}0.55}$ (21.71 MPa) was 36.80% higher than that of $G_{1.2\text{-}10\text{-}0.55}$ (15.87 MPa). The surplus soluble silicate ($G_{2.4\text{-}10\text{-}0.55}$) would lead to more geopolymer gels precipitating at an early stage, which inhibited the condensation of SiO_4 and AlO_4 ions. Moreover, a lower NaOH concentration inhibited the dissolution rate of $CG_{700°C}$ particles; the specimens ($G_{2.0\text{-}8\text{-}0.55}$) exhibited a loose microstructure, which was inferior to the development of the compressive strength (13.25 MPa). Additionally, with an increase in the liquid–solid ratio from 0.50 to 0.65, the specimens with a liquid–solid ratio of 0.50 ($G_{2.0\text{-}10\text{-}0.50}$) showed higher compressive strength (24.75 MPa), and the specimens of $G_{2.0\text{-}10\text{-}0.65}$ exhibited larger pore sizes and lower compressive strength (14.92 MPa).

The $CG_{700°C}$-based geopolymer gels were composed of N-A-S-H compounds, and the chemical composition was not considerably influenced by the alkali activators. With a lower $Na_2SiO_3/NaOH$ mass ratio ($G_{1.2\text{-}10\text{-}0.55}$), fewer Si atoms participated in geopolymerization, the microstructure of the geopolymers became loose, and more pores were produced, thus compressive strength decreased. The porosity of the geopolymers for the smaller NaOH concentrations ($G_{2.0\text{-}8\text{-}0.55}$) and higher liquid–solid ratio ($G_{2.0\text{-}10\text{-}0.65}$) were 38.22% and 41.93%, respectively.

These findings indicated that alkali activators significantly influenced the mechanical properties and microstructure of $CG_{700°C}$-based geopolymers. The results can provide theoretical support for the reutilization of CG waste.

Author Contributions: Conceptualization, X.W., F.L. and B.Z.; data curation, X.W. and B.Z.; formal analysis, X.W., Z.P. and W.C.; funding acquisition, F.L., B.Z. and L.L.; investigation, Z.P. and W.C.; methodology, X.W. and B.Z.; resources, F.L. and L.L.; project administration, F.L. and B.Z.; software Z.P.; supervision, F.L.; writing—original draft preparation, X.W. and W.C.; writing—review and editing, X.W., F.M. and B.Z. All authors have read and agreed to the published version of the manuscript.

Funding: This research was funded by the National Natural Science Foundation of China, (Grant No. 12072080, 12032009, and 42302032); and the Basic and Applied Basic Research Foundation of Guangdong Province, (Grant No. 2023A1515012180).

Data Availability Statement: The original data will be available upon requirement. The data are not publicly available due to the project privacy.

Acknowledgments: The authors gratefully thanks all technical personnel from the Structural Laboratory of Guangdong University of Technology for their assistance during the experiment.

Conflicts of Interest: The authors declare that there have been no known competing financial interests or personal relationships that could have appeared to influence the work reported in this paper.

References

1. Davidovits, J. Geopolymers: Inorganic Polymeric New Materials. *J. Therm. Anal. Calorim.* **1991**, *37*, 1633–1656. [CrossRef]
2. Duxson, P.; Provis, J.L.; Lukey, G.C.; Deventer, J. The Role of Inorganic Polymer Technology in the Development of "Green Concrete". *Cem. Concr. Res.* **2007**, *37*, 1590–1597. [CrossRef]
3. Rickard, W.D.A.; Vickers, L.; van Riessen, A. Performance of Fibre Reinforced, Low Density Metakaolin Geopolymers under Simulated Fire Conditions. *Appl. Clay Sci.* **2013**, *73*, 71–77. [CrossRef]
4. Zhang, J.; Shi, C.; Zhang, Z.; Ou, Z. Durability of Alkali-Activated Materials in Aggressive Environments: A Review on Recent Studies. *Constr. Build. Mater.* **2017**, *152*, 598–613. [CrossRef]
5. McLellan, B.C.; Williams, R.P.; Lay, J.; van Riessen, A.; Corder, G.D. Costs and Carbon Emissions for Geopolymer Pastes in Comparison to Ordinary Portland Cement. *J. Clean. Prod.* **2011**, *19*, 1080–1090. [CrossRef]

6. Shi, C.; Jiménez, A.F.; Palomo, A. New Cements for the 21st Century: The Pursuit of an Alternative to Portland Cement. *Cem. Concr. Res.* **2011**, *41*, 750–763. [CrossRef]
7. Farooq, M.; Bhutta, A.; Banthia, N. Tensile Performance of Eco-Friendly Ductile Geopolymer Composites (EDGC) Incorporating Different Micro-Fibers. *Cem. Concr. Compos.* **2019**, *103*, 183–192. [CrossRef]
8. Zhang, B.; Feng, Y.; Xie, J.; Dai, J.; Chen, W.; Xue, Z.; Li, L.; Li, Y.; Li, J. Effects of Pretreated Recycled Powder Substitution on Mechanical Properties and Microstructures of Alkali-Activated Cement. *Constr. Build. Mater.* **2023**, *406*, 133360. [CrossRef]
9. Cloete, S.; Giuffrida, A.; Romano, M.C.; Zaabout, A. The Swing Adsorption Reactor Cluster for Post-Combustion CO_2 Capture from Cement Plants. *J. Clean. Prod.* **2019**, *223*, 692–703. [CrossRef]
10. Zhang, X.; Feng, X.; Wang, Z.; Jian, J.; Chen, S.; Luo, W.; Zhang, C. Experimental Study on the Physico-Mechanical Properties and Microstructure of Foam Concrete Mixed with Coal Gangue. *Constr. Build. Mater.* **2022**, *359*, 129428. [CrossRef]
11. Zhang, B.; Yu, T.; Deng, L.; Li, Y.; Guo, H.; Zhou, J.; Li, L.; Peng, Y. Ion-Adsorption Type Rare Earth Tailings for Preparation of Alkali-Based Geopolymer with Capacity for Heavy Metals Immobilization. *Cem. Concr. Compos.* **2022**, *134*, 104768. [CrossRef]
12. Pavithra, P.; Reddy, M.S.; Dinakar, P.; Rao, B.H.; Satpathy, B.K.; Mohanty, A.N. A Mix Design Procedure for Geopolymer Concrete with Fly Ash. *J. Clean. Prod.* **2016**, *133*, 117–125. [CrossRef]
13. Wang, Q.; Guo, H.; Yu, T.; Yuan, P.; Deng, L.; Zhang, B. Utilization of Calcium Carbide Residue as Solid Alkali for Preparing Fly Ash-Based Geopolymers: Dependence of Compressive Strength and Microstructure on Calcium Carbide Residue, Water Content and Curing Temperature. *Materials* **2022**, *15*, 973. [CrossRef]
14. El-Hassan, H.; Ismail, N. Effect of Process Parameters on the Performance of Fly Ash/GGBS Blended Geopolymer Composites. *J. Sustain. Cem.-Based Mater.* **2018**, *7*, 122–140. [CrossRef]
15. Singh, S.; Aswath, M.U.; Ranganath, R.V. Effect of Mechanical Activation of Red Mud on the Strength of Geopolymer Binder. *Constr. Build. Mater.* **2018**, *177*, 91–101. [CrossRef]
16. Zhang, B.; Guo, H.; Yuan, P.; Li, Y.; Wang, Q.; Deng, L.; Liu, D. Geopolymerization of Halloysite via Alkali-Activation: Dependence of Microstructures on Precalcination. *Appl. Clay Sci.* **2020**, *185*, 105375. [CrossRef]
17. Zhang, B.; Yuan, P.; Guo, H.; Deng, L.; Li, Y.; Li, L.; Wang, Q.; Liu, D. Effect of Curing Conditions on the Microstructure and Mechanical Performance of Geopolymers Derived from Nanosized Tubular Halloysite. *Constr. Build. Mater.* **2021**, *268*, 121186. [CrossRef]
18. Zhang, B.; Guo, H.; Deng, L.; Fan, W.; Yu, T.; Wang, Q. Undehydrated Kaolinite as Materials for the Preparation of Geopolymer through Phosphoric Acid-Activation. *Appl. Clay Sci.* **2020**, *199*, 105887. [CrossRef]
19. Zhang, B.; Yu, T.; Guo, H.; Chen, J.; Liu, Y.; Yuan, P. Effect of the SiO_2/Al_2O_3 Molar Ratio on the Microstructure and Properties of Clay-Based Geopolymers: A Comparative Study of Kaolinite-Based and Halloysite-Based Geopolymers. *Clays Clay Miner.* **2022**, *70*, 882–902. [CrossRef]
20. Li, Z.; Gao, Y.; Zhang, J.; Zhang, C.; Chen, J.; Liu, C. Effect of Particle Size and Thermal Activation on the Coal Gangue Based Geopolymer. *Mater. Chem. Phys.* **2021**, *267*, 124657. [CrossRef]
21. Han, J.Y.; Song, X.Y.; Gao, Z.H. Excitation Effect of Soluble Glass on Composite System with Calcined Coal Gangue and Slag. *Appl. Mech. Mater.* **2012**, *174–177*, 30–34. [CrossRef]
22. Geng, J.; Zhou, M.; Li, Y.; Chen, Y.; Han, Y.; Wan, S.; Zhou, X.; Hou, H. Comparison of Red Mud and Coal Gangue Blended Geopolymers Synthesized through Thermal Activation and Mechanical Grinding Preactivation. *Constr. Build. Mater.* **2017**, *153*, 185–192. [CrossRef]
23. Zhou, X. Study on the Reaction Degree of Calcined Coal Gangue Powder in Blended Cement by Selective Solution Method. *Procedia Earth Planet. Sci.* **2009**, *1*, 634–639. [CrossRef]
24. Ju, F.; Li, B.; Guo, S.; Xiao, M. Dynamic Characteristics of Gangues during Vertical Feeding in Solid Backfill Mining: A Case Study of the Wugou Coal Mine in China. *Environ. Earth Sci.* **2016**, *75*, 1389. [CrossRef]
25. Liu, H.; Liu, Z. Recycling Utilization Patterns of Coal Mining Waste in China. *Resour. Conserv. Recycl.* **2010**, *54*, 1331–1340. [CrossRef]
26. Liu, X. Low-Carbon Utilization of Coal Gangue under the Carbon Neutralization Strategy: A Short Review. *J. Mater. Cycles Waste Manag.* **2023**, *25*, 1978–1987. [CrossRef]
27. Karimipour, A. Effect of Untreated Coal Waste as Fine and Coarse Aggregates Replacement on the Properties of Steel and Polypropylene Fibres Reinforced Concrete. *Mech. Mater.* **2020**, *150*, 103592. [CrossRef]
28. Wu, J.; Bai, G.; Wang, P.; Liu, Y. Mechanical Properties of a New Type of Block Made from Shale and Coal Gangue. *Constr. Build. Mater.* **2018**, *190*, 796–804. [CrossRef]
29. Zhao, H.; Wang, S.; Wang, R.; Shen, L.; Wang, Q. Utilization of Raw Coal Gangue as Coarse Aggregates in Pavement Concrete. *Constr. Build. Mater.* **2023**, *378*, 131062. [CrossRef]
30. Guan, X.; Qiu, J.; Song, H.; Qin, Q.; Zhang, C. Stress–Strain Behaviour and Acoustic Emission Characteristic of Gangue Concrete under Axial Compression in Frost Environment. *Constr. Build. Mater.* **2019**, *220*, 476–488. [CrossRef]
31. Zhang, W.; Dong, C.; Huang, P.; Sun, Q.; Li, M.; Chai, J. Experimental Study on the Characteristics of Activated Coal Gangue and Coal Gangue-Based Geopolymer. *Energies* **2020**, *13*, 2504. [CrossRef]
32. Zhou, S.; Dong, J.; Yu, L.; Xu, C.; Jiao, X.; Wang, M. Effect of Activated Coal Gangue in North China on the Compressive Strength and Hydration Process of Cement. *J. Mater. Civ. Eng.* **2019**, *31*, 04019022. [CrossRef]

33. Dassekpo, J.-B.M.; Feng, W.; Miao, L.; Dong, Z. Effect of Alkali Activators on Loess Geopolymer: Potential Waterproof Repair Material. *J. Mater. Civ. Eng.* **2021**, *33*, 04021247. [CrossRef]
34. Chen, W.; Li, B.; Wang, J.; Thom, N. Effects of Alkali Dosage and Silicate Modulus on Autogenous Shrinkage of Alkali-Activated Slag Cement Paste. *Cem. Concr. Res.* **2021**, *141*, 106322. [CrossRef]
35. Phoo-ngernkham, T.; Sata, V.; Hanjitsuwan, S.; Ridtirud, C.; Hatanaka, S.; Chindaprasirt, P. High Calcium Fly Ash Geopolymer Mortar Containing Portland Cement for Use as Repair Material. *Constr. Build. Mater.* **2015**, *98*, 482–488. [CrossRef]
36. Guo, X.; Shi, H.; Dick, W.A. Compressive Strength and Microstructural Characteristics of Class C Fly Ash Geopolymer. *Cem. Concr. Compos.* **2010**, *32*, 142–147. [CrossRef]
37. Zhang, Z.; Xiao, Y.; Huajun, Z.; Yue, C. Role of Water in the Synthesis of Calcined Kaolin-Based Geopolymer. *Appl. Clay Sci.* **2009**, *43*, 218–223. [CrossRef]
38. Somna, K.; Jaturapitakkul, C.; Kajitvichyanukul, P.; Chindaprasirt, P. NaOH-Activated Ground Fly Ash Geopolymer Cured at Ambient Temperature. *Fuel* **2011**, *90*, 2118–2124. [CrossRef]
39. Puligilla, S.; Chen, X.; Mondal, P. Understanding the Role of Silicate Concentration on the Early-Age Reaction Kinetics of a Calcium Containing Geopolymeric Binder. *Constr. Build. Mater.* **2018**, *191*, 206–215. [CrossRef]
40. Bernal, S.A.; Provis, J.L.; Rose, V.; Mejía de Gutierrez, R. Evolution of Binder Structure in Sodium Silicate-Activated Slag-Metakaolin Blends. *Cem. Concr. Compos.* **2011**, *33*, 46–54. [CrossRef]
41. Zhang, B.; Huang, D.; Li, L.; Lin, M.; Liu, Y.; Fang, W.; Lu, J.; Liu, F.; Li, Y.; Liu, Y.; et al. Effect of Magnesium Salt Contamination on the Microstructures and Properties of Metakaolinite-Based Geopolymer: The Role of $MgCl_2$ and $MgSO_4$. *J. Mater. Res. Technol.* **2022**, *20*, 4500–4514. [CrossRef]
42. Zhang, M.; Islam, J. Use of Nano-Silica to Reduce Setting Time and Increase Early Strength of Concretes with High Volumes of Fly Ash or Slag. *Constr. Build. Mater.* **2012**, *29*, 573–580. [CrossRef]
43. Longhi, M.A.; Walkley, B.; Rodríguez, E.D.; Kirchheim, A.P.; Zhang, Z.; Wang, H. New Selective Dissolution Process to Quantify Reaction Extent and Product Stability in Metakaolin-Based Geopolymers. *Compos. Part B Eng.* **2019**, *176*, 107172. [CrossRef]
44. Zhang, Z.; Yao, X.; Zhu, H. Potential Application of Geopolymers as Protection Coatings for Marine Concrete: I. Basic Properties. *Appl. Clay Sci.* **2010**, *49*, 1–6. [CrossRef]
45. White, C.E.; Provis, J.L.; Llobet, A.; Proffen, T.; van Deventer, J.S.J. Evolution of Local Structure in Geopolymer Gels: An *In Situ* Neutron Pair Distribution Function Analysis. *J. Am. Ceram. Soc.* **2011**, *94*, 3532–3539. [CrossRef]
46. Chen, Z.; Li, J.-S.; Zhan, B.-J.; Sharma, U.; Poon, C.S. Compressive Strength and Microstructural Properties of Dry-Mixed Geopolymer Pastes Synthesized from GGBS and Sewage Sludge Ash. *Constr. Build. Mater.* **2018**, *182*, 597–607. [CrossRef]
47. He, J.; Jie, Y.; Zhang, J.; Yu, Y.; Zhang, G. Synthesis and Characterization of Red Mud and Rice Husk Ash-Based Geopolymer Composites. *Cem. Concr. Compos.* **2013**, *37*, 108–118. [CrossRef]
48. Cui, Y.; Wang, D.; Wang, Y.; Sun, R.; Rui, Y. Effects of the n(H_2O: Na_2O_{eq}) Ratio on the Geopolymerization Process and Microstructures of Fly Ash-Based Geopolymers. *J. Non-Cryst. Solids* **2019**, *511*, 19–28. [CrossRef]
49. Lu, C.; Wang, Q.; Liu, Y.; Xue, T.; Yu, Q.; Chen, S. Influence of New Organic Alkali Activators on Microstructure and Strength of Fly Ash Geopolymer. *Ceram. Int.* **2022**, *48*, 12442–12449. [CrossRef]
50. Dehghani, A.; Aslani, F.; Ghaebi Panah, N. Effects of Initial SiO_2/$Al2O_3$ Molar Ratio and Slag on Fly Ash-Based Ambient Cured Geopolymer Properties. *Constr. Build. Mater.* **2021**, *293*, 123527. [CrossRef]
51. Yu, Q.; Li, S.; Li, H.; Chai, X.; Bi, X.; Liu, J.; Ohnuki, T. Synthesis and Characterization of Mn-Slag Based Geopolymer for Immobilization of Co. *J. Clean. Prod.* **2019**, *234*, 97–104. [CrossRef]
52. Sun, Z.; Vollpracht, A. One Year Geopolymerisation of Sodium Silicate Activated Fly Ash and Metakaolin Geopolymers. *Cem. Concr. Compos.* **2019**, *95*, 98–110. [CrossRef]
53. Koshy, N.; Dondrob, K.; Hu, L.; Wen, Q.; Meegoda, J.N. Synthesis and Characterization of Geopolymers Derived from Coal Gangue, Fly Ash and Red Mud. *Constr. Build. Mater.* **2019**, *206*, 287–296. [CrossRef]
54. Ryu, G.S.; Lee, Y.B.; Koh, K.T.; Chung, Y.S. The Mechanical Properties of Fly Ash-Based Geopolymer Concrete with Alkaline Activators. *Constr. Build. Mater.* **2013**, *47*, 409–418. [CrossRef]
55. Wang, Y.; Han, F.; Mu, J. Solidification/Stabilization Mechanism of Pb(II), Cd(II), Mn(II) and Cr(III) in Fly Ash Based Geopolymers. *Constr. Build. Mater.* **2018**, *160*, 818–827. [CrossRef]
56. Fernández-Jiménez, A.; Palomo, A. Mid-Infrared Spectroscopic Studies of Alkali-Activated Fly Ash Structure. *Microporous Mesoporous Mater.* **2005**, *86*, 207–214. [CrossRef]
57. Zhang, Z.; Provis, J.L.; Ma, X.; Reid, A.; Wang, H. Efflorescence and Subflorescence Induced Microstructural and Mechanical Evolution in Fly Ash-Based Geopolymers. *Cem. Concr. Compos.* **2018**, *92*, 165–177. [CrossRef]
58. Rees, C.A.; Provis, J.L.; Lukey, G.C.; van Deventer, J.S.J. Attenuated Total Reflectance Fourier Transform Infrared Analysis of Fly Ash Geopolymer Gel Aging. *Langmuir* **2007**, *23*, 8170–8179. [CrossRef]
59. Aiken, T.A.; Kwasny, J.; Sha, W.; Soutsos, M.N. Effect of Slag Content and Activator Dosage on the Resistance of Fly Ash Geopolymer Binders to Sulfuric Acid Attack. *Cem. Concr. Res.* **2018**, *111*, 23–40. [CrossRef]
60. Yaseri, S.; Masoomi Verki, V.; Mahdikhani, M. Utilization of High Volume Cement Kiln Dust and Rice Husk Ash in the Production of Sustainable Geopolymer. *J. Clean. Prod.* **2019**, *230*, 592–602. [CrossRef]
61. El Alouani, M.; Alehyen, S.; El Achouri, M.; Taibi, M. Preparation, Characterization, and Application of Metakaolin-Based Geopolymer for Removal of Methylene Blue from Aqueous Solution. *J. Chem.* **2019**, *2019*, e4212901. [CrossRef]

62. Wang, Y.; Alrefaei, Y.; Dai, J. Silico-Aluminophosphate and Alkali-Aluminosilicate Geopolymers: A Comparative Review. *Front. Mater.* **2019**, *6*, 106. [CrossRef]
63. Fan, Y.; Luan, H. Pore Structure in Concrete Exposed to Acid Deposit. *Constr. Build. Mater.* **2013**, *49*, 407–416. [CrossRef]

buildings

MDPI

Article

Experimental Study on the Effects of Straight and Ring-Type Steel Fibres on the Bond Behaviour of Steel Bars in Rubber-Recycled Aggregate Concrete

Honglong Ma [1,2], Huawei Li [1], Jinhu Zheng [1], Wei Wei [1,3], Shaohua He [1], Xiaopeng Tian [1,4], Xiaohui Li [1] and Feng Liu [1,*]

[1] School of Civil and Transportation Engineering, Guangdong University of Technology, Guangzhou 510006, China; crccztl7zc@163.com (H.M.); hwli3880@wuyiu.edu.cn (H.L.); 2112109162@mail2.gdut.edu.cn (J.Z.); weiwei@gdcvi.edu.cn (W.W.); 1112209006@mail2.gdut.edu.cn (S.H.); crccztl7gz@163.com (X.T.); lixh@mail2.gdut.edu.cn (X.L.)
[2] China Railway 17th Bureau Group Co., Ltd., Taiyuan 030006, China
[3] Guangdong Construction Polytechnic, Guangzhou 510440, China
[4] China Railway 17th Bureau Group (Guangzhou) Construction Co., Ltd., Guangzhou 510700, China
* Correspondence: fliu@gdut.edu.cn

Abstract: The application range of rubber-recycled aggregate concrete (RRAC), a new type of green building material, is currently limited due to performance defects, including low hardness, high water absorption, and poor adhesion. To expand its application in reinforced concrete structures, it is crucial to enhance the bonding performance between RRAC and steel bars. In this study, the effects of adding straight steel fibres (SSFs) and ring-type steel fibres (RSFs) to RRAC were investigated, in order to enhance the bonding performance. To investigate the impact of steel fibres (SFs) on the bonding properties of RRAC and steel bars, a total of 51 specimens were subjected to pull-out tests to systematically examine the impact of SSF and RSF dosages on the bonding performance. The results demonstrated that incorporating the optimal amount of SSFs and RSFs can significantly improve the bond strength and bond stiffness. Moreover, the combined use of SSFs and RSFs yielded even better enhancement effects. The RRAC exhibited remarkable performance, when the total content of SFs was 1.2% and the proportion of RSFs 75%. In this case, the bond strength and bond stiffness were enhanced by 3.7% and 53.88%, respectively. Finally, a bond–slip constitutive model for RRAC and steel bar was established. The combined use of SSFs and RSFs minimizes the limitations of poor mechanical properties in traditional RRAC and holds significant value for the widespread adoption and application of RRAC.

Keywords: bond behaviour; straight steel fibre; ring-type steel fibre; rubber; recycled aggregate concrete

Citation: Ma, H.; Li, H.; Zheng, J.; Wei, W.; He, S.; Tian, X.; Li, X.; Liu, F. Experimental Study on the Effects of Straight and Ring-Type Steel Fibres on the Bond Behaviour of Steel Bars in Rubber-Recycled Aggregate Concrete. *Buildings* **2024**, *14*, 504. https://doi.org/10.3390/buildings14020504

Academic Editor: Dan Bompa

Received: 7 January 2024
Revised: 24 January 2024
Accepted: 7 February 2024
Published: 12 February 2024

1. Introduction

Improper management of waste rubber tires and construction waste globally has led to escalating environmental problems [1–6]. The construction industry is currently experiencing a shortage of natural resources, which highlights the importance of recycling waste materials for the sustainable development of the industry [7–11]. A potential remedy involves utilizing this waste through the substitution of rubber powders (RPs) for fine aggregate and recycled coarse aggregate (RCA) for coarse aggregate in concrete. This innovative approach results in the production of rubber-recycled aggregate concrete (RRAC). Numerous studies have been conducted on the performance and application of RRAC by researchers. The addition of suitable quantities of RPs and RCA to ordinary concrete can enhance its durability, impact resistance, and noise reduction capabilities to a certain extent [12–15]. However, excessive amounts of RPs and RCA may lead to significant strength decrease in concrete [16,17]. Gholampout et al. [18] conducted an experiment where they added varying amounts of RPs to concrete. Their findings revealed that as the content of RPs increased, the compressive strength of the concrete decreased. Chen et al. [19] discovered that when RCA was used exclusively, the mechanical strength of recycled aggregate concrete (RAC) was only 77% of that of natural aggregate concrete. Furthermore, Zrar et al. [20] demonstrated that both the compressive strength and tensile strength of concrete decreased significantly with an increase in the substitution rate of RPs and RCA. The mechanical properties of RRAC are flawed, and with the growing use of waste materials in the construction industry, it is becoming challenging to promote and apply RRAC if its overall performance cannot be enhanced. This limitation restricts its potential applications. Currently, RRAC is mainly used for non-load-bearing components that have low mechanical strength requirements, such as roads and airport runways [21,22]. To broaden the application range of RRAC and increase the utilization of waste rubber tires and construction waste, it is crucial to find methods to enhance its mechanical properties.

The incorporation of SFs in concrete has proven to be an effective method for enhancing its mechanical properties. SFs act as a bridge within the concrete structure, effectively preventing the formation of micro-cracks and reducing the occurrence of macro-cracks. This results in a significant improvement in the concrete's splitting tensile strength, shear strength, and bending resistance [23–25]. Moreover, when adding a certain amount of SFs to concrete, the compressive failure mode shifted from brittleness to ductility, and its compressive performance was also improved [26]. Consequently, scholars have conducted extensive research on the use of SFs to enhance the properties of RRAC. Shahjalal et al. [17] demonstrated that the incorporation of SFs enhances the compressive strength, tensile strength, and ductility of RRAC. In a separate study, Xie et al. [27] investigated the impact of SFs on the fracture properties of RRAC. The findings indicate that the inclusion of SFs enhances the resistance to fracture toughness instability in RRAC. Although SFs have been proved to enhance the mechanical properties of RRAC, there are several shortcomings in the current research in this field. These can be summarized as outlined below. (1) The number of closed regions and the area of the ring formed by SFs in concrete have a notable impact on the mechanical properties of concrete [28]. However, the existing research has primarily focused on SF-enhanced RRAC with single closed regions, and there is a dearth of studies on the formation of SFs with multiple closed regions and larger ring areas. (2) The post-peak performance of straight steel fibre reinforced concrete (SSFRC) is determined by the ultimate tensile strength of SSFs passing through the failure surface [29]. However, the anchoring performance of SSFs and concrete is insufficient, resulting in SSFs being pulled out without yielding when the SSFRC is destroyed. As a result, the performance of SSFs has not been fully utilized. (3) SSFs only have an inhibitory effect on the cracks they pass through and have little impact on the cracks in their vicinity. Therefore, the enhancement effect of SSFs on concrete is limited. When subjected to external loads, ensuring the restraint effect on the concrete matrix solely through the use of SSFs can be challenging. This is particularly true for concrete with relatively weak mechanical properties, like RRAC.

In order to address these issues, this study proposes the use of RSFs to enhance RRAC. RSFs offer several advantages over SSFs as follow. (1) RSFs form a circular occlusal surface in concrete, which significantly improves the anchoring effect of RSFs and concrete. The mechanism of RSF reinforced concrete is similar to that inducing RSFs to break after yielding, allowing RSFs to fully demonstrate their performance [30]. (2) RSF provides a wider protection range compared to SSFs. When RSFs are incorporated into concrete, a closed system is formed. This closed system helps restrain and protect the concrete matrix when it is damaged by external forces, inhibiting the development of concrete cracks within the closed system. Choi et al. [30] demonstrated that the flexural toughness of ring-type steel fibre reinforced concrete (RSFRC) was significantly improved compared to traditional SSFRC, thus confirming the potential of RSFs in enhancing concrete performance. He et al. [31] found that incorporating an appropriate proportion of SSFs mixed with RSFs can greatly enhance the compressive performance of RRAC. This finding suggests that the combination of SSFs and RSFs could be a reliable approach to promoting RRAC.

At present, reinforced concrete structure remains the most commonly employed structural form in building structures [32–35]. The fundamental mechanical properties of concrete and the bonding properties between the steel bars and concrete are crucial factors that determine the suitability of concrete material for widespread use in reinforced concrete structures [36]. However, current research only focuses on the impact of SSFs and RSFs on the fundamental mechanical properties of RRAC. To further enhance the applicability of RRAC, it is crucial to investigate and understand the effects of SSFs and RSFs on the bonding properties between RRAC and steel bars. In this study, RRAC was used as a matrix to investigate the impact of SSFs and RSFs dosages on the bonding performance of RRAC and steel bars; the optimal content of SSFs and RSFs were also determined. Additionally, a bond–slip constitutive model for RRAC and steel bars was developed. The findings of this study are highly significant as they contribute to the promotion of RRAC, which has the potential to address environmental pollution caused by waste rubber tires and construction waste. Furthermore, it can facilitate the green transformation of the construction industry.

2. Experimental Programme

2.1. Materials Properties

The raw materials used in this study include cement, water, river sand, RCA, RPs, SSFs, and RSFs. Ordinary Portland cement (P.O 42.5) was used with an apparent density of 3112 kg/m^3. The test water used in this experiment was ordinary tap water with a density of 995 kg/m^3. For the fine aggregate of concrete, river sand with a maximum particle size of 5 mm was utilized. The density, fineness modulus, water content, and water absorption of this sand were 2642 kg/m^3, 2.17, 0.12%, and 0.18%, respectively. Additionally, the particle size of RCA ranged from 5 mm to 25 mm, and its apparent density, water absorption, and moisture content were 2536.9 kg/m^3, 2.45%, 2.10%, respectively. To prevent impurities on the surface of the aggregate from affecting the bonding performance of the interface, the aggregate was subjected to screening, cleaning, and drying after which it was ready for use. The gradation of the aggregate after mechanical screening can be seen in Figure 1. RPs, which are made from waste rubber tires, undergo a process of washing, drying, and mechanical crushing. The RPs have a particle size of 0.85 mm and an apparent density of 988 kg/m^3. The SSFs and RSFs used in this study are illustrated in Figure 2. The ends of the SSFs were processed into end hooks to improve the anchoring effect between the SSFs and concrete. The material properties of the SSFs and RSFs are presented in Table 1. The pull-out specimens were constructed using HRB400 hot-rolled ribbed bars, which had a length of 420 mm and a diameter of 10 mm. The steel bar had a yield strength of 440 MPa, a tensile strength of 550 MPa, and an elastic modulus of 200 GPa.

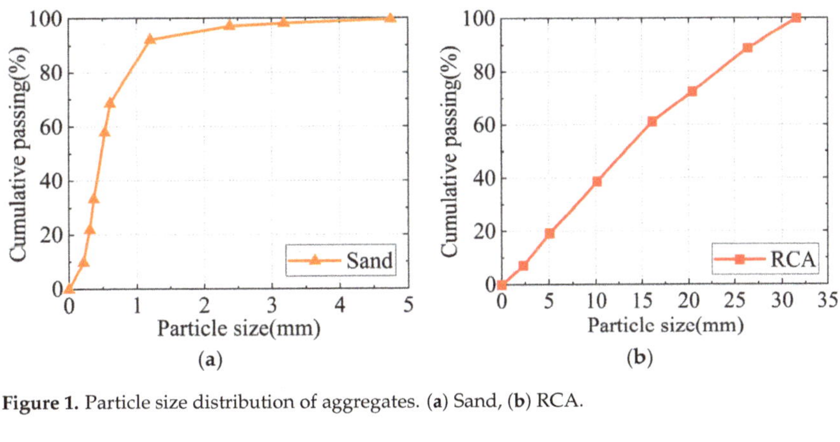

Figure 1. Particle size distribution of aggregates. (**a**) Sand, (**b**) RCA.

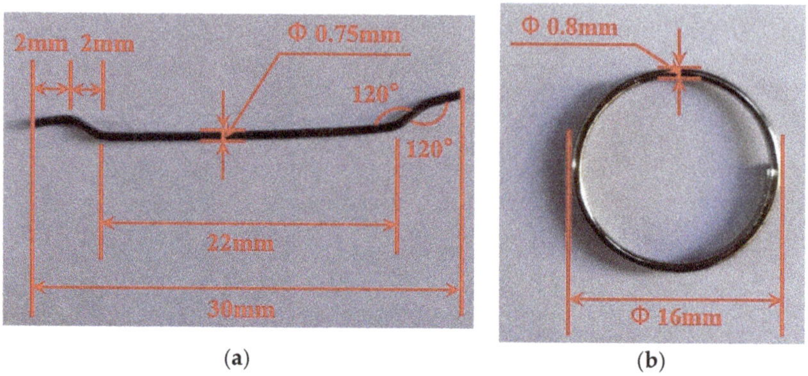

Figure 2. Geometry size of SSFs and RSFs. (**a**) SSFs, (**b**) RSFs.

Table 1. Properties of steel fibres applied in this study.

Fibre Type	Raw Material	Shape Feature	Tensile Strength f_{st} (MPa)	Apparent Density ρ (kg/m^3)	Elastic Modulus E_s (GPa)
SSFs	Carbon steel	Hooked-end steel fibre	1000	7436	200
RSFs	Stainless steel	Ring-shape steel fibre	960	7600	193

2.2. Mix Proportions

A total of 17 groups of mix proportions and their compressive strength were designed for this study, as presented in Table 2 (more relative parameters of mechanical properties can be found in our previous research [31]). Three specimens were created for each group, resulting in a total of 51 pull-out specimens. The main parameters focused on the overall amount of SFs and the ratio of RSFs in relation to the total amount of SFs. The total amount of SFs refers to the percentage of SF volume in the volume of concrete. Excessive use of steel fibres can result in fibre agglomeration and a decrease in the workability and mechanical properties of concrete. To address these problems, four different levels of SF content were selected: 0%, 0.4%, 0.8%, and 1.2%. The ratio of RSFs to the total volume of SFs is calculated by dividing the volume of RSFs by the total volume of SFs, and it is also expressed as a percentage. Five different ratios were set: 0%, 25%, 50%, 75%, and 100%. Based on our pre-experiment and taking into account the environmental advantages of utilizing waste rubber [31], we incorporated RPs by substituting a portion of the fine aggregate with an equal volume, while the incorporation ratio was consistently set at 20%. Additionally, the water–cement ratio was fixed at 0.38 (as the aggregate was dried after drying, the moisture content of the aggregate was not taken into consideration). Moreover, the objective of this study was to investigate the effect of SSFs and RSFs on the bonding properties of RRAC and steel bars. Therefore, only the reference specimen "R0S0RS0" was designed, and ordinary concrete was not considered.

Table 2. Mix proportions.

Mix Number	Cement (kg/m^3)	Water (kg/m^3)	Sand (kg/m^3)	RCA (kg/m^3)	RPs (kg/m^3)	RSFs (kg/m^3)	SSFs (kg/m^3)	Compressive Strength (MPa)
R0S0RS0	596.20	228.30	791.33	742.87	0.00	0.00	0.00	45.73
R20S0RS0	596.20	228.01	633.06	742.87	59.28	0.00	0.00	37.07
R20S0.4RS0	596.20	227.88	628.84	737.92	58.88	0.00	29.74	35.89
R20S0.4RS25	596.20	227.88	628.84	737.92	58.88	7.60	22.31	36.86
R20S0.4RS50	596.20	227.88	628.84	737.92	58.88	15.20	14.87	40.08
R20S0.4RS75	596.20	227.88	628.84	737.92	58.88	22.80	7.44	32.44
R20S0.4RS100	596.20	227.88	628.84	737.92	58.88	30.40	0.00	38.30
R20S0.8RS0	596.20	227.75	624.62	732.96	58.49	0.00	59.49	40.26
R20S0.8RS25	596.20	227.75	624.62	732.96	58.49	15.20	44.62	43.17
R20S0.8RS50	596.20	227.75	624.62	732.96	58.49	30.40	29.74	39.83
R20S0.8RS75	596.20	227.75	624.62	732.96	58.49	45.60	14.87	36.18
R20S0.8RS100	596.20	227.75	624.62	732.96	58.49	60.80	0.00	33.03
R20S1.2RS0	596.20	227.62	620.40	728.01	58.09	0.00	89.23	41.25
R20S1.2RS25	596.20	227.62	620.40	728.01	58.09	22.80	66.92	38.76
R20S1.2RS50	596.20	227.62	620.40	728.01	58.09	45.60	44.62	38.67
R20S1.2RS75	596.20	227.62	620.40	728.01	58.09	68.40	22.31	37.42
R20S1.2RS100	596.20	227.62	620.40	728.01	58.09	91.20	0.00	

Note: in specimen "R20S0.4RS25", "R20" refers to the replacement of fine aggregate by 20% RPs, "S0.4" refers to the total content of SFs being 0.4%, "RS25" refers to the RSFs ratio accounting for 25% of the total content of SFs.

2.3. Design and Preparation of Pull-Out Specimens

All the specimens for the pull-out test in this study were made of concrete cubes with a length of 150 mm and steel bars with a length of 420 mm. The steel mould and the process for making pull-out specimens are shown in Figure 3. The steel mould consists of an upper steel mould and a lower steel mould connected by bolts. The upper steel mould is made up of two L-shaped steel mould plates connected by bolts. The lower steel mould is a steel plate with a middle opening and a steel casing with a diameter of 20 mm and a length of 50 mm. To prevent stress state difference between the end of the concrete and the steel bar during the loading process, an unbonded area with a length of 50 mm was set at the bottom of the concrete, specifically at the top area of the steel casing. This was achieved by embedding a plastic pipe cap at both the upper and lower parts of the steel casing. A hole was then opened in the middle of the plastic pipe cap, allowing the steel bar to pass through. To prevent the cement slurry from seeping during pouring, the gap between the plastic pipe cap and the steel casing was sealed with lime mud. The steel bar was then inserted into the bottom of the steel mould through the lower plastic pipe cap, the steel sleeve, and finally the upper plastic pipe cap. The exposed part of the upper plastic pipe cap serves as the bonding area for the steel bar. The length of this bonding area was set to 5-times the diameter of the steel bar ($5d_0$) to maximize the bonding performance between the steel bar and concrete [37]. After the steel bar passes through, the three bolts at the lower part of the steel sleeve should be tightened to secure the steel bar in the centre of the steel sleeve. Following that, the upper and lower steel moulds were assembled and placed on the platform with the opening of the steel mould facing upwards. The RRAC was then poured into the steel mould through the upper opening, and the concrete surface was vibrated using a plug-in vibrator to remove any large bubbles. After pouring for 24 h, the mould was demoulded and cured for 28 days at a temperature of 20 ± 3 °C and a relative humidity of >95%. Finally, a pull-out test was conducted.

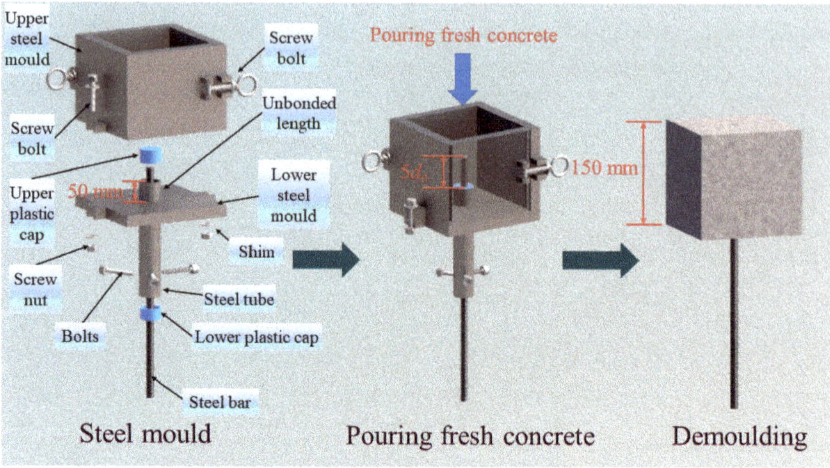

Figure 3. The steel mould and process for making pull-out specimens.

2.4. Test Setup and Machine

The tensile testing machine (MTS-370, USA) was utilized to conduct the pull-out test in this study, and the test device and machine are shown in Figure 4. The test device is connected using two steel frames and two screws in the centre. The lower clamp of the test machine clamps the steel bar of the pull-out specimen after it passes through the hole in the middle of the lower steel frame. To ensure that the test results were not affected by the uneven contact surface between the concrete and the steel plate, the bottom surface of the pull-out specimen was coated with gypsum and then placed on the steel frame. The upper steel frame has a ball hinge in the middle, and the upper pull rod of the ball hinge is clamped by the upper chuck of the testing machine. The purpose of the ball hinge is to maintain a vertical alignment between the upper and lower steel frames, preventing any eccentric tension on the steel bar. The displacement loading process was conducted at a rate of 1.2 mm/min, with the loading end positioned at the lower chuck. The tensile testing machine and the dynamic and static data acquisition instrument (JMTS-3841, China) were used to collect the pull-out force and displacement of the loading end simultaneously, at a sampling frequency of 2 Hz. The loading was halted either when the specimen was damaged or when the displacement of the loading end exceeded 25 mm (This threshold value is 2-times the rib spacing of the steel bars). It is to be noted that the slippage between the steel bar and concrete was calculated by subtracting the elongation of the steel bar from the displacement of the loading end of the steel bar. The displacement of the loading end was directly measured using a testing machine. The elongation of the steel bar was calculated based on Hooke's law and then derived, as shown in Formula (1):

$$\varepsilon = \frac{\Delta L}{L} = \frac{\sigma}{E_f} = \frac{F}{A_f E_f} \tag{1}$$

where the ε represents the strain of the steel bar, L denotes the length from the loading end of the steel bar to the midpoint of the bonding area, σ represents the stress of the steel bar, F represents the drawing force of the steel bar at the loading end, and A_f represents the cross-sectional area of the steel bar.

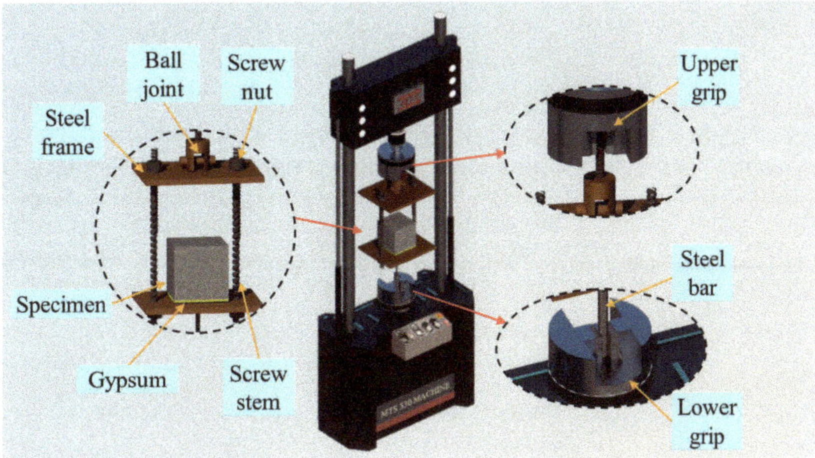

Figure 4. Test setup and machine.

3. Results and Discussion

3.1. Bond Interface Damage and Bond Mechanism

The failure mode observed in all specimens was steel bar pull-out failure. As the load increased, the slip of the steel bar gradually increased, eventually leading to the test being stopped due to excessive slip causing the pull-off of the steel bar. Throughout the entire test process, there was no occurrence of concrete splitting or steel bar breaking, and the surface integrity of the specimen remained intact. The only damage observed was at the bonding interface between the steel bar and concrete. This failure mode can be attributed to the larger thickness of the concrete protective layer on the pull-out specimen, which enhances the resistance to the splitting of the surrounding concrete.

Following the completion of the pull-out test, the concrete cubes of all specimens were divided into two pieces along the longitudinal plane of the central axis of the steel bar. This allowed for observation of the bonding damage at the interface between the steel bar and the concrete. Figure 5 illustrates the damage of the bonding interface between RRAC and the steel bars with various types of SFs. Notably, the damage to the bonding interface is consistent among all specimens. The surface of the concrete bonding interface appears rough, with noticeable peeling and scratched areas, as well as residual concrete powder. In the steel bar bonding area, the ribs of the steel bar remain intact, without any instances of rib breakage. The steel ribs are filled with concrete fragments, and some concrete fragments are adhered to the steel bars. Combined with the failure mode and the damage of the bonding interface, the bonding failure of the specimen is primarily attributed to the shear failure of the concrete in the bonding area caused by the extrusion between the steel ribs. Initially, the stress triggers the activation of the bonding mechanism between the steel bar and the concrete. During this stage, the primary bonding force at the bonding interface is the chemical bonding force generated by the cement gel on the surface of the steel bar within the concrete. As the steel bars deform and start to slip, the chemical bonding force is rapidly lost. At this stage, the bonding force is sustained by the friction between the steel bar and the concrete, as well as the mechanical bite force. As the load continues to increase, the concrete in front of the steel rib is crushed due to local extrusion. This results in the radial component of the extrusion force from the steel rib on the surrounding concrete, which generates circumferential tensile stress in the concrete. When the annular tensile stress exceeds the ultimate tensile strength of concrete, it results in the formation of radial–longitudinal cracks in the concrete. However, due to the presence of a thick protective layer of concrete and the crack resistance provided by SFs, these cracks only occur in a limited area around the bonding region and do not propagate rapidly to the surface of the concrete, thereby preventing concrete splitting. As the slip of the steel bar continues to increase, the pressure on the concrete between the ribs also increases, eventually leading to the cutting off of the concrete between the ribs. The surface of the steel bar still contains partially cut concrete blocks and concrete debris, which are visibly peeled and scratched at the interface with the concrete. As a result, the mechanical bond between the steel bar and the concrete is lost, but friction between them remains, leading to a residual pull-out force in the specimen. During this process, some concrete fragments may experience radial–longitudinal cracks due to compression, which could explain the difficulty in observing such cracks with the naked eye at the concrete interface. In the whole process of bond failure, the concrete between the ribs is compressed and severed, which is the critical factor leading to bond failure. Hence, it can be inferred that the strength of RRAC plays a significant role in the bonding performance between RRAC and steel bars. This finding aligns with previous research conclusions on the factors influencing the bond strength of concrete and steel bars [38,39]. The combined use of SSFs and RSFs has been found to have a synergistic effect in effectively controlling internal cracking of concrete. Previous studies that only used one type of SFs were limited in their ability to prevent concrete cracking [40,41].

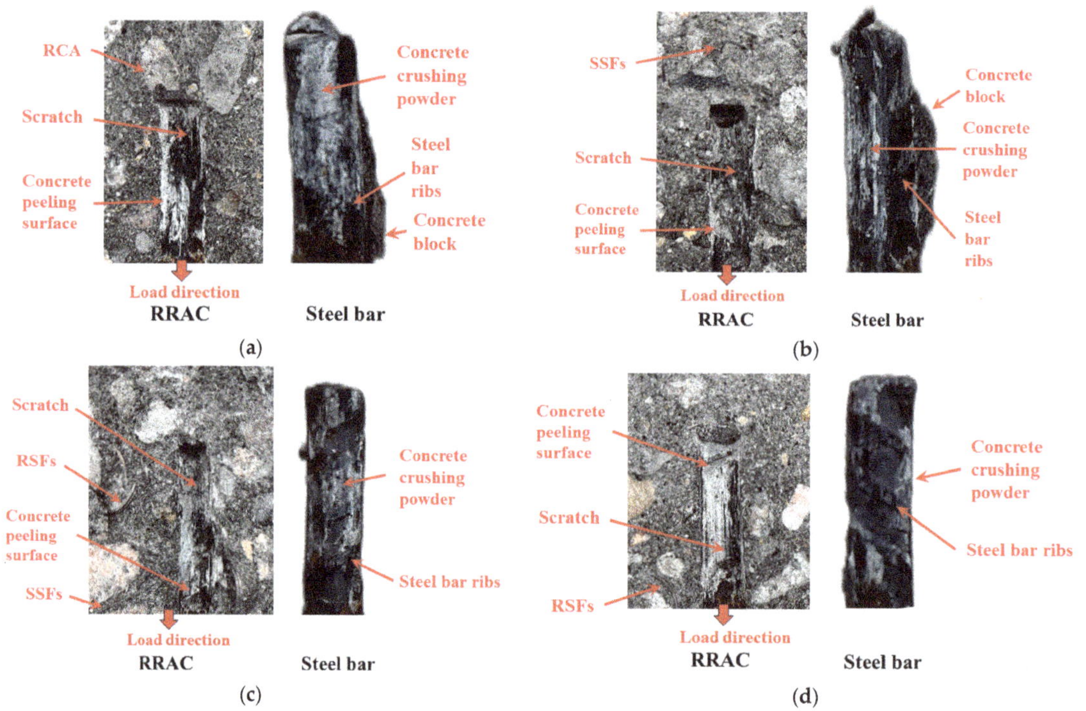

Figure 5. Bond interface damage. (**a**) 0% SFs, (**b**) 100% SSFs, (**c**) SSFs + RSFs (with any replacement ratio), (**d**)100% RSFs.

3.2. Effects of SSFs and RSFs on Bond Mechanism

In the previous section, it was mentioned that during the pull-out test of RRAC and steel bars, the concrete around the bonding interface experienced compression and developed radial–longitudinal cracks. Both SSFs and RSFs enhance the bonding effect by restricting the propagation of radial–longitudinal cracks. However, the mechanism through which SSFs and RSFs limit the development of these fractures differs and will be elaborated upon in this section.

The mechanism of SSFs and RSFs in concrete is illustrated in Figure 6. In Figure 6a, the meaning of each legend is presented. Figure 6b demonstrates that SSFs function similarly to transverse reinforcements in RRAC, enhancing the shear strength of the concrete. Furthermore, after RRAC is subjected to loads, SSFs play a bridging role between cracks. They not only alter the development path of cracks but also connect the cement matrix through which cracks pass. This increases the energy required for crack propagation, thereby delaying the progression of cracks. The presence of SSFs enhances the crack resistance and shear resistance of RRAC, improves its toughness, enhances the concrete's ability to protect steel bars embedded in the RRAC block, and enhances the bonding effect between RRAC and the steel bars. On the other hand, RSFs in concrete can be likened to sleeves. As shown in Figure 6c,d, RSFs are overlapped and embedded within the concrete, creating numerous enclosed structures of varying sizes. These RSFs serve to effectively confine the concrete within the enclosed structures, thereby reducing the rate of deformation and delaying the formation of cracks. Notably, although the concrete within the closed structure is adequately protected, the interface between the RSFs and the concrete has emerged as a vulnerable layer that influences crack propagation. Figure 6e illustrates that when SSFs and RSFs are combined with RRAC, SSFs effectively delay crack formation in the weak layer surrounding RSFs, while RSFs safeguard the concrete area that remains unaffected by SSFs. The collaborative impact of these two components proves more effective in inhibiting the development of concrete cracks.

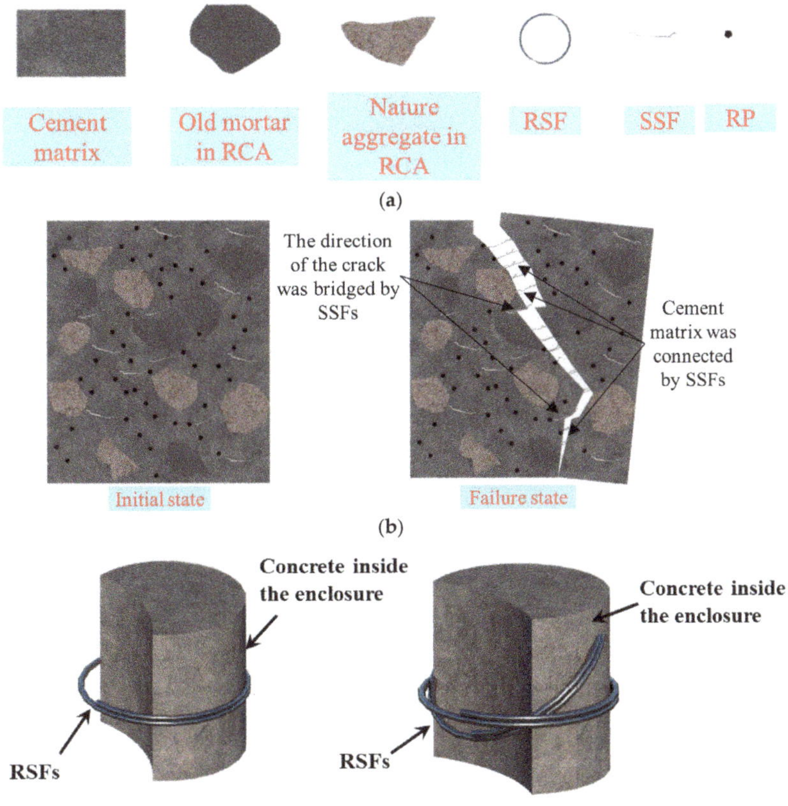

Figure 6. *Cont.*

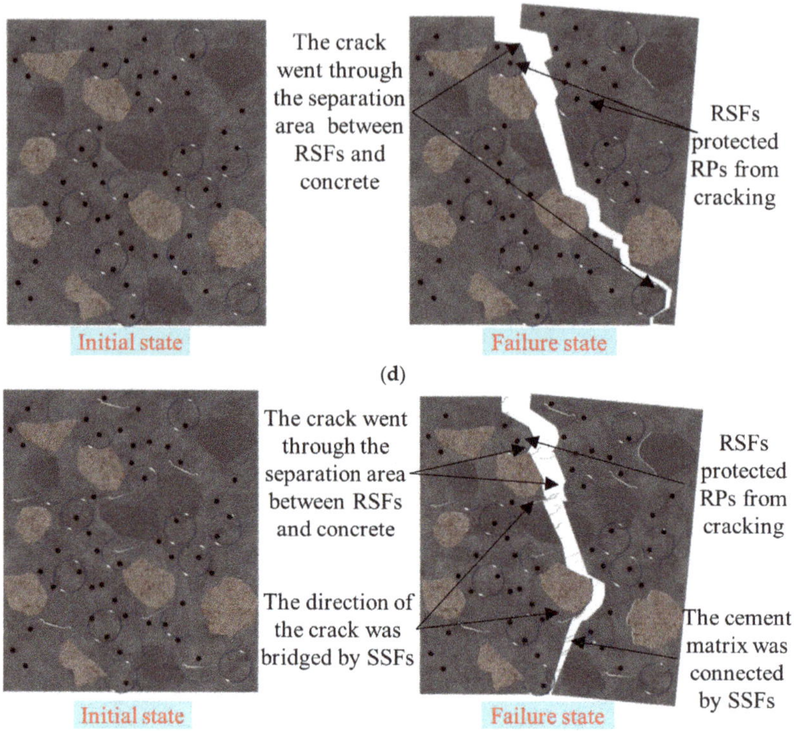

Figure 6. Effects of SSFs and RSFs on bond mechanism. (**a**) Legend, (**b**) RRAC reinforced with SSFs, (**c**) enclosure inside RSFs, (**d**) RRAC reinforced with RSFs, (**e**) RRAC reinforced with RSFs and SSFs.

3.3. Bond Stress–Slip Curves

This study assumes that the bond stress in the bond zone is uniformly distributed along the length direction. The pull-out force (F) and the displacement (s') of the loading end during the test were directly measured by the testing machine. The bond stress (τ) and slip (s) were then obtained by converting Formulas (2) and (3), respectively. Notably, the calculation of s takes into account the deformation of the steel bar:

$$\tau = \frac{F}{\pi d_o l_a} \tag{2}$$

$$s = s' - \frac{LF}{E_f A_f} \tag{3}$$

where the d_0, l_a, E_f, and A_f are the diameter, bond length, elastic modulus, and cross-sectional area of the steel bar, respectively; L is the length from the loading end of the steel bar to the midpoint of the bond area.

The bond stress–slip curves of the three specimens for each mix ratio are similar. Therefore, only the average bond stress–slip curves of the three specimens for each mix ratio are presented in Figure 7. All specimens obtained a complete bond stress–slip curve, which includes linear ascending segment, nonlinear ascending segment, descending segment, and residual segment. This indicates that the bond between steel and concrete in this study is fully utilized. From the initial slip between steel and concrete until about 70% of the peak bond stress, this stage represents the linear ascending segment of the curve. During this stage, there is a linear positive correlation between the bond stress and slip, and small radial-transverse cracks start to appear at the concrete bond interface. Then, the curve enters the nonlinear ascending segment until the bond stress reaches its peak. During this stage, cracks gradually develop in the concrete and the friction coefficient between the steel bar and the concrete reduces due to the presence of broken concrete powder in the bonding interface. As a result, the slope of the curve continuously decreases until it reaches zero. Subsequently, there is a significant increase in the slip of the steel bar and more severe concrete damage between the ribs, leading to a continuous decline in bond stress. This forms a descending segment of the curve. Eventually, the inter-ribbed concrete is completely cut off and the broken concrete powder forms a new interface at the bonding interface. At this point, there is still some friction on the interface, which tends to stabilize. Therefore, the curve in this stage becomes gentler, representing the residual segment of the curve.

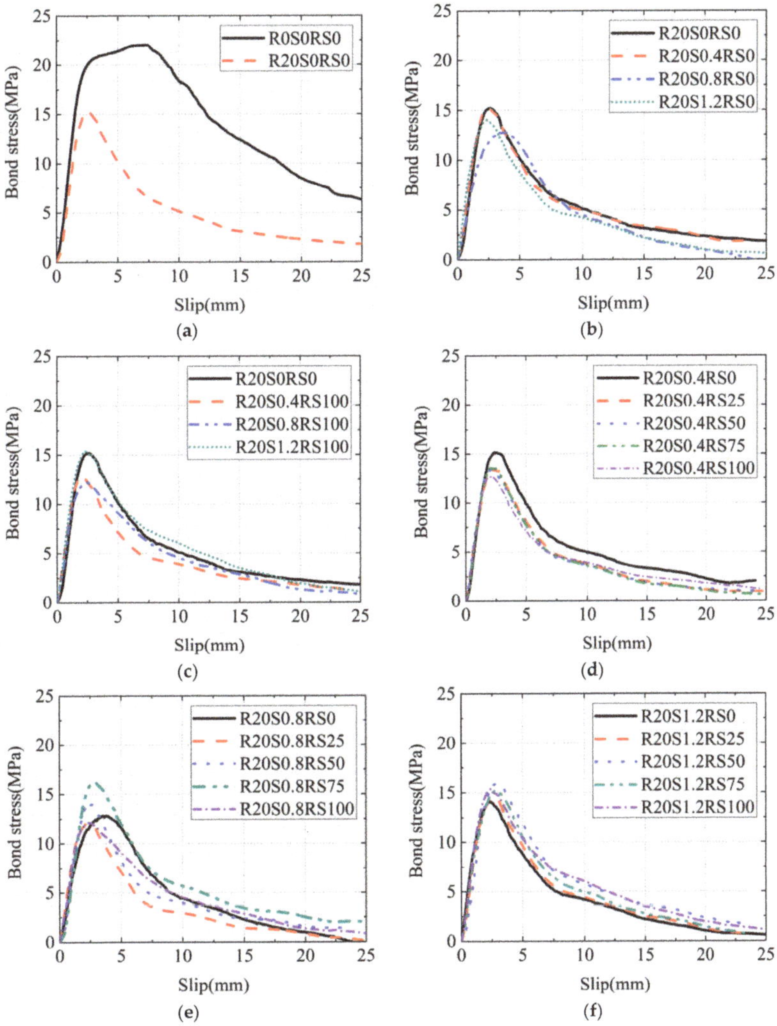

Figure 7. Bond stress–slip curves. (**a**) Effects of 20% of RPs, (**b**) effects of SSFs with different contents, (**c**) effects of RSFs with different contents, (**d**) effects of 0.4% SFs with different RSF proportions, (**e**) effects of 0.8% SFs with different RSF proportions, (**f**) effects of 1.2% SFs with different RSF proportions.

In addition, the effects of RPs, SSFs, and RSFs on the bond–slip curves are shown in Figure 7. Specifically, Figure 7a demonstrates the impact of adding 20% RPs to the bond–slip curve. It can be observed that the inclusion of 20% RPs leads to a significant decrease in the peak value of the bond stress on the curve, as well as a reduction in the slope of the rising section. This phenomenon can be attributed to the fact that the incorporation of RPs results in a decrease in the strength of concrete. RPs are hydrophobic-based materials, while concrete is hydrophilic-based. This difference in properties leads to poor bonding performance between RPs and concrete, resulting in a weak bonding interface. Additionally, RPs and concrete have different hardness and elastic modulus. Under load, stress concentration tends to occur around RPs, leading to a decrease in concrete strength. Figure 7b,c demonstrates the impact of different SF content on the bond stress–slip curve when SSFs and RSFs are added individually. When the SSF content is 0.4% and the RSF content is 1.2%, the peak value of the curve is comparable to that of the reference group (R20S0RS0). However, for other dosages, the curve tends to flatten and the peak value decreases to varying extents. This indicates that the distribution of SFs in concrete plays a crucial role in maintaining the uniformity and compactness of the internal interface of RRAC. Figure 7d,f illustrates the impact of the RSF dosage on the bond stress–slip curve at different SF content. At an SF content of 0.4%, the peak value of the curve decreases to varying degrees after the addition of RSFs, and the wrapping area of the curve is smaller compared to R20S0.4RS0. When the SF content is 0.8%, the peak value of the curve initially decreases, then increases, and finally decreases again with the increase in the proportion of RSFs. For the proportion of RSFs of 50% and 75%, the peak value of the curve is higher than that of specimen R20S0.8RS0. When the content of SF is 1.2%, the peak value and wrapping area of the curve are always greater than that of R20S1.2RS0, regardless of the proportion of RSFs. Moreover, as the proportion of RSFs increases, the overall curve initially rises and then falls. This indicates that the proportion of RSFs has a significant impact on the bonding effect between RRAC and steel bars, depending on the change in SF content. There exists an optimal SF content and RSF ratio, which provides the best enhancement effect on the bonding effect between RRAC and steel bars.

3.4. Bond Strength

The bond strength of the specimen is determined by the peak value of the bond stress–slip curve in this study. The bond strength is calculated by taking the average value of the three specimens in each mix proportions, and the bond strength and their average values are presented in Table 3. The bond strength of specimen R20S0RS0 decreased by 32.87% when compared to ordinary RAC without RPs. This decrease can be attributed to the introduction of defects in the concrete by RPs, which in turn reduces the compressive strength of the concrete [38]. Consequently, the bond strength between the concrete and the steel bars is also reduced. Figure 8 shows the impact of the SF and RSF dosages on bond strength. When solely SSFs were utilized with a 0.4% content, there was a slight enhancement in bond strength in comparison to specimen R20S0RS0. However, as the SSF content increased to 0.8% and 1.2%, the bond strength decreased by 16.41% and 7.13%, respectively, when compared to specimen R20S0RS0. The results demonstrate that the addition of an optimal amount of SSFs enhances the bond strength between RRAC and steel bars. However, excessive incorporation of SSFs can lead to varying degrees of negative effects on the bond strength. This phenomenon can be attributed to the bridging effect of SSFs in the concrete, which restricts the formation of internal cracks. Consequently, SSFs contribute to crack resistance and shear resistance, ultimately improving the bond strength. However, when the content of SSFs is too large, the distribution of SSFs in concrete is not uniform, which results in a decrease in the density of the concrete and the formation of a weak layer. As a result, the negative effect of SSFs outweighs the positive effect, leading to varying degrees of decrease in bond strength of the specimens. When RSFs are solely used at a content of 0.4% and 0.8%, the bond strength is reduced by 16.93% and 18.22% respectively, compared to specimen R20S0RS0. However, when the RSF content is increased to 1.2%, the bond strength is increased by 2.07%. This contrasting result of the sole use of SSFs suggests that the mechanisms of RSFs and SSFs differ, thus supporting the analysis presented in Section 3.2. Notably, the positive impact of RSFs on bond strength is only observed when a sufficient amount of RSFs is present to create numerous closed bodies within the concrete. Otherwise, the negative effect caused by the introduction of significant defects by RSFs continues to prevail.

Table 3. Bond strength results.

Concrete Mix No.	Bond Strength (MPa)			
	Specimen 1	Specimen 2	Specimen 3	Mean
R0S0RS0	22.15	24.67	22.10	22.97
R20S0RS0	13.85	17.31	15.10	15.42
R20S0.4RS0	15.57	15.76	15.36	15.56
R20S0.4RS25	13.59	13.36	13.97	13.64
R20S0.4RS50	12.83	13.32	14.68	13.64
R20S0.4RS75	15.28	13.75	12.47	13.83
R20S0.4RS100	13.24	12.46	12.73	12.81
R20S0.8RS0	13.02	13.98	11.68	12.89
R20S0.8RS25	13.06	11.86	12.71	12.54
R20S0.8RS50	13.53	14.57	14.90	14.33
R20S0.8RS75	16.51	18.38	14.41	16.43
R20S0.8RS100	12.12	13.17	12.55	12.61
R20S1.2RS0	12.50	14.93	15.51	14.32
R20S1.2RS25	13.70	15.11	16.21	15.01
R20S1.2RS50	15.79	16.65	16.19	16.21
R20S1.2RS75	15.42	16.81	15.75	15.99
R20S1.2RS100	16.17	16.88	14.16	15.74

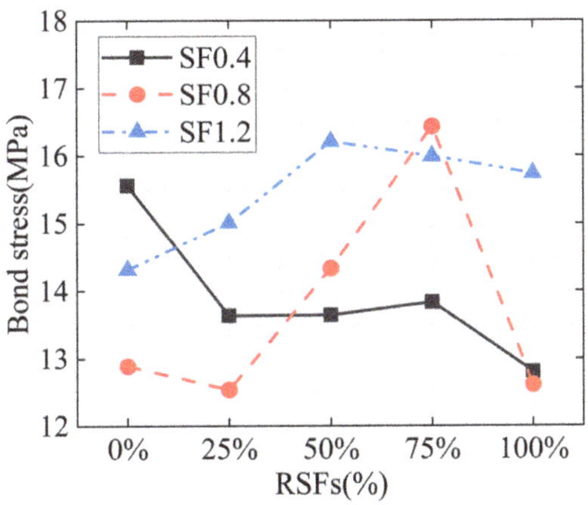

Figure 8. Bond strength results.

When the SF content is 0.4%, the bond strength decreases as the RSF ratio increases. However, when the SF content is 0.8% and 1.2%, the bond strength initially increases and then decreases with the increase in the ratio of RSFs. These findings suggest that when the amount of SFs is sufficient, the combined use of SSFs and RSFs is more effective than when using them individually to enhance the bond strength of the concrete. Additionally, incorporating too many RSFs may reduce the density of the concrete and decrease the bond strength due to uneven distribution. This suggests that there is an optimal ratio of SFs to RSFs. The maximum bond strength is achieved when the SF content is 0.8% and 1.2%, with RSF ratios of 75% and 50%, respectively. These ratios result in a bond strength that is 6.55%

and 5.12% higher than that of specimen R20S0RS0. Therefore, this study concludes that the optimal content of SSFs and RSFs to enhance bond strength synergistically is 0.8% of the total SF content, with RSFs accounting for 75%.

3.5. Bond Stiffness

The bond stiffness is a measure of the ability to resist the initial relative slip between the concrete and the steel bars. In this study, the secant stiffness at 50% peak bond stress is defined as the bond stiffness, which is the ratio of bond stress to slip at 50% peak bond stress. Table 4 shows the bond stiffness of all the specimens, with the bond stiffness of each mix proportions being the average of the bond stiffness of the three specimens in the group, placed in the last column of the table. Comparing the bonding stiffness of specimens R0S0RS0 and R20S0RS0, it is evident that the bonding stiffness decreased by 37.26% after adding 20% RPs to RAC. This decrease can be attributed to the increased likelihood of cracks forming along the interface between RPs and concrete, which accelerates the rate of concrete deformation. On the other hand, as SFs are incorporated, regardless of the SF to RSF ratio, the bond stiffness is greater than that of specimen R20S0RS0. Furthermore, the bond stiffness increases with higher SF content. This demonstrates that the inclusion of SFs can significantly enhance the bond stiffness between RRAC and steel bars. This improvement is due to the ability of SFs to resist crack development and delay the speed of concrete crushing in the concrete.

Table 4. Bond stiffness results.

Concrete Mix No.	Bond Stiffness (MPa/mm)			
	Specimen 1	Specimen 2	Specimen 3	Mean
R0S0RS0	10.81	10.72	8.02	9.85
R20S0RS0	5.05	6.64	6.85	6.18
R20S0.4RS0	7.63	7.69	7.13	7.49
R20S0.4RS25	7.60	6.50	7.99	7.36
R20S0.4RS50	7.27	6.84	7.80	7.30
R20S0.4RS75	7.34	7.03	6.67	7.01
R20S0.4RS100	6.86	7.32	7.05	7.08
R20S0.8RS0	8.12	7.72	7.61	7.82
R20S0.8RS25	6.68	9.48	7.69	7.95
R20S0.8RS50	8.44	8.10	7.43	7.99
R20S0.8RS75	7.88	7.89	8.18	7.98
R20S0.8RS100	7.39	7.18	9.00	7.85
R20S1.2RS0	7.90	8.58	8.05	8.18
R20S1.2RS25	7.90	8.75	8.44	8.36
R20S1.2RS50	8.52	9.30	8.55	8.79
R20S1.2RS75	9.77	9.18	9.57	9.51
R20S1.2RS100	7.45	8.68	8.44	8.19

Figure 9 illustrates the impact of the SF content and RSF ratio on bond stiffness. When comparing the three curves of SF content at 0.4%, 0.8%, and 1.2%, it can be observed that as the RSF ratio increases from 0% to 100%, the bonding stiffness slightly decreases when the SF content is 0.4%. Throughout this process, the bonding stiffness decreases by 5.47%. On the other hand, when the content of SFs is 0.8%, the bonding stiffness remains relatively unchanged as the proportion of RSFs increases. However, when the SF content is 1.2%, the bonding stiffness noticeably increases with the increase of RSF ratio. This is consistent with the impact of RSFs on the bond strength (mentioned in Section 3.4). It emphasizes that for RSFs to effectively improve bond stiffness, there needs to be a sufficient number of RSFs present. It is important to note that when the SF content is 1.2%, the RSF ratio increases from 75% to 100%, resulting in a 13.88% decrease in bonding stiffness. This indicates that an excessive amount of RSFs has a negative impact on bond stiffness. The reason for this is that the defects introduced by RSFs alone, as well as their excessive use, are too large.

This leads to an increase in the number of cracks in the weak layer of the interface between RSFs and concrete, accelerating their development. In addition to that, when the SF content is 1.2%, the bonding stiffness of SSFs and RSFs is considerably enhanced in comparison to that of SSFs or RSFs individually. This finding suggests the presence of a synergistic effect between SSFs and RSFs, with the combined effect of SSFs and RSFs on improving bond stiffness surpassing that of SSFs and RSFs individually. The synergistic effect of SSFs and RSFs can be attributed to the bridging effect of SSFs, which effectively delays the development of concrete cracks around RSFs. Additionally, the concrete area not crossed by SSFs is well protected due to the hooping effect of RSFs. In this study, it was found that when the SF content is 1.2% and the RSF ratio is 75%, the bond stiffness of the specimen is the highest, exhibiting a 53.88% increase compared to specimen R20S0RS0.

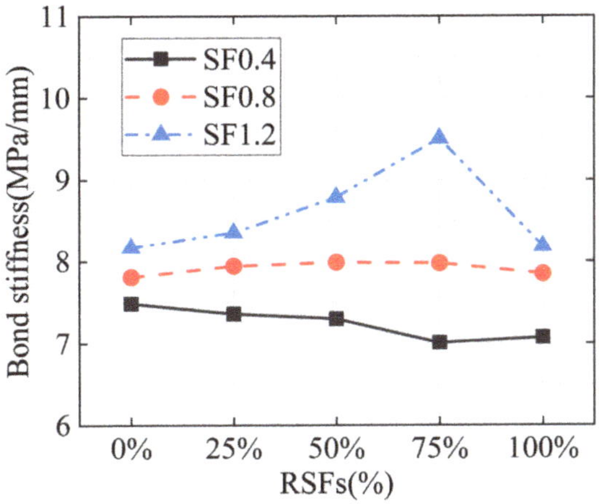

Figure 9. Bond stiffness results.

The experimental results mentioned above provide evidence of a positive synergistic effect between SSFs and RSFs, which helps in restraining the cracking of RRAC. However, it is important to note that the RSFs used in this study had a smooth surface, and it is uncertain whether this limiting effect will continue to increase with higher substitution ratios of RSFs. To address this concern, introducing textures to the surface of RSFs or exploring the use of ring fibres made from alternative materials could potentially alleviate this issue. In addition, the use of RPs in RRAC results in a weak interfacial transition zone. However, the performance of RRAC can be enhanced by the bridge effect of SSFs. On the other hand, RSFs are effective in limiting the cracking of RRAC. This strengthening effect significantly improves the bonding between RRAC and steel bars.

4. Bond–Slip Constitutive Model

As stated in Section 3.3, the bond stress–slip curves of RRAC and steel bars in this study consist of four sections (i.e., linear ascending segment, nonlinear ascending segment, descending segment, and residual segment). After comparing the bond–slip constitutive models proposed by various researchers, the bond–slip constitutive model of this study was developed based on Xiong [42], Mo [43], and Guo [44]. The curve of the bond–slip constitutive model is illustrated in Figure 10. This curve is divided into three sections: the linear ascending segment from (0, 0) to (S_r, τ_r), the nonlinear ascending segment from

(S_r, τ_r) to (S_u, τ_u), and the descending segment after (S_u, τ_u). The calculation model is as follows:

$$\tau = \begin{cases} K \cdot S & S \leq S_r \\ \tau_u \left(\frac{S}{S_u}\right)^{\alpha} & S_r \leq S \leq S_u \\ \tau_u \frac{S/S_u}{\beta(S/S_u-1)^2+S/S_u} & S_u \leq S \end{cases} \tag{4}$$

$$\alpha = \frac{\ln \tau_u - \ln \tau_r}{\ln S_u - \ln S_r} \tag{5}$$

where K is the slope of the linear ascending segment, which is defined as the secant stiffness at 50% peak bond stress (obtained from the bond stiffness values measured in Section 3.5), τ_u is the bond stress corresponding to the peak point of the bond stress–slip curve, S_u and τ_u are the bond slip during the linear ascending segment, τ_r is the bond stress corresponding to the 70% peak bond stress of the linear ascending segment of the bond stress–slip curve, S_r and τ_r are the corresponding bond slip, α is the undetermined coefficient of the nonlinear ascending segment (obtained from Formula (5)), β is the undetermined coefficient of the descending segment (obtained from the software fitting).

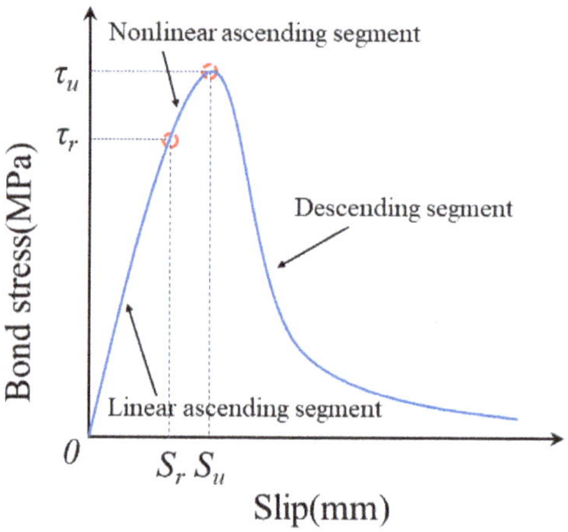

Figure 10. Bond–slip constitutive relationship model.

The bond stiffness value obtained in Section 3.5 is used as K in this study. As the linear rising section and the nonlinear rising section are continuous at (S_r, τ_r), the value of α can be determined using Formula (5). Additionally, by fitting the data of the descending curve, the value of β can be obtained. The values of relevant parameters and undetermined coefficients are presented in Table 5. The fitting correlation coefficient R2 of β in each group was found to be greater than 0.96, and the standard error *SE* was less than 0.03. The undetermined coefficients and related parameters from Table 5 are incorporated into the constitutive model (4), resulting in the bond–slip constitutive model for each group. This model is then compared with the bond–slip curve obtained from the test, as shown in Figure 11. It can be observed that the model curve aligns well with the test curve, indicating the high accuracy of the bond–slip constitutive model proposed in this study. Notably, the proposed bond–slip constitutive model is specifically designed for analysing steel bar pull-out failure, it may not be suitable for cases where the failure mode involves concrete cracking or steel breaking.

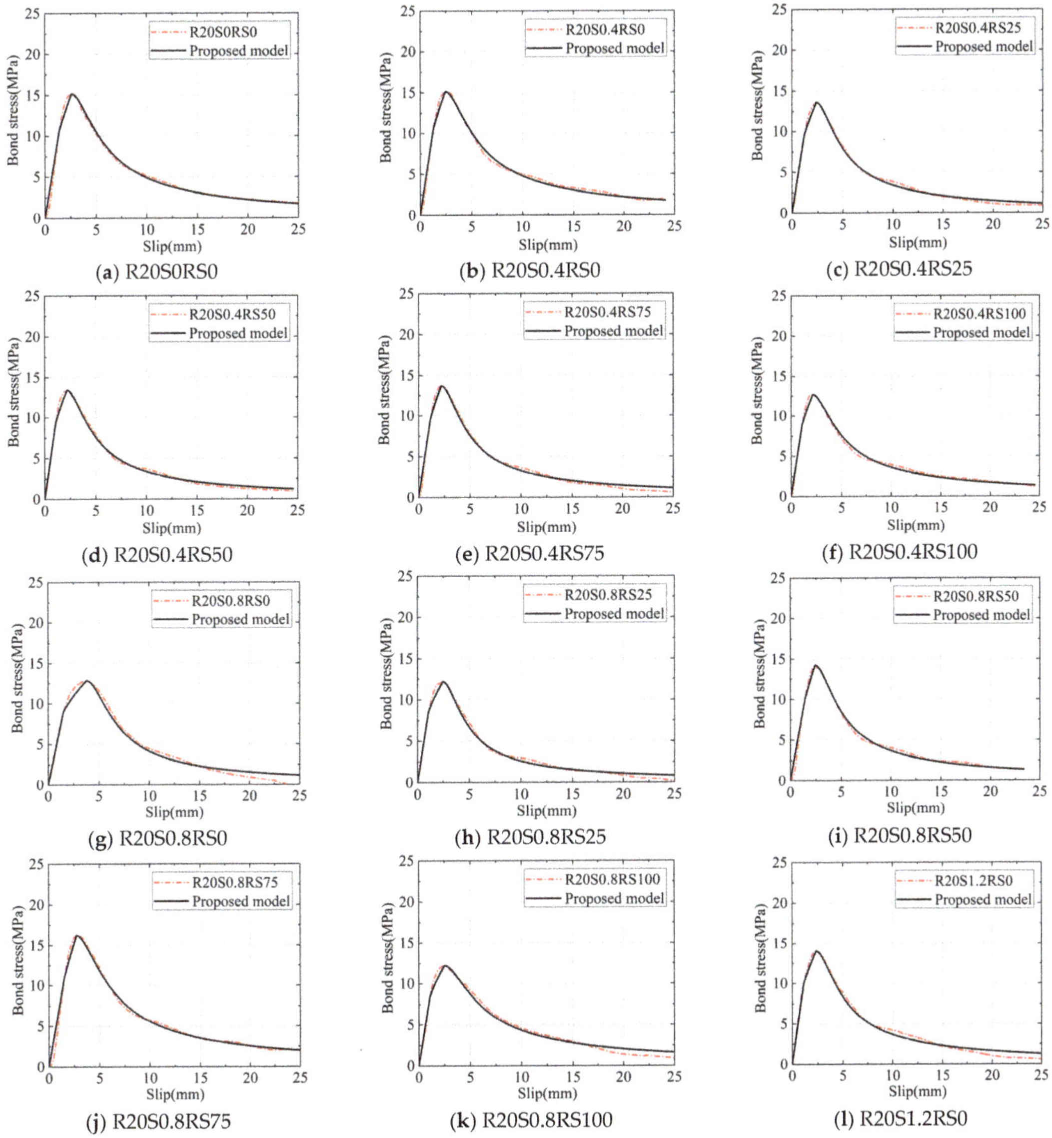

Figure 11. *Cont.*

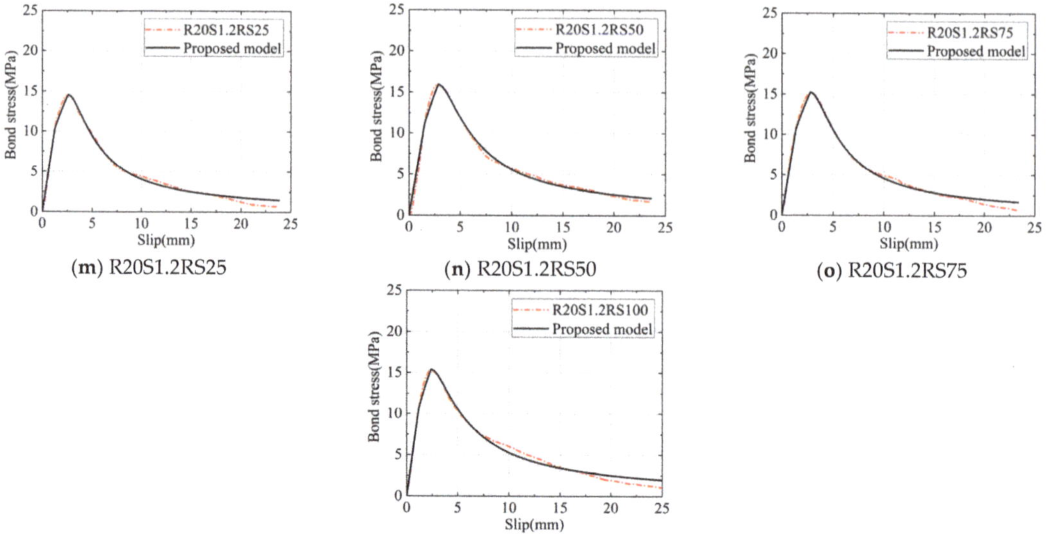

(**m**) R20S1.2RS25 (**n**) R20S1.2RS50 (**o**) R20S1.2RS75

(**p**) R20S1.2RS100

Figure 11. Comparations between the experimental curve and proposed model (the relative parameters are mentioned in Table 5).

Table 5. The values of relevant parameters and undetermined coefficients.

Concrete Mix No.	K	α	β	R^2	SE
R20S0RS0	6.18	0.5798	0.9920	0.9972	0.0024
R20S0.4RS0	7.49	0.5650	0.9231	0.9935	0.0033
R20S0.4RS25	7.36	0.5218	1.2194	0.9935	0.0051
R20S0.4RS50	7.30	0.5166	1.0699	0.9947	0.0043
R20S0.4RS75	7.01	0.5506	1.1753	0.9911	0.0058
R20S0.4RS100	7.08	0.5243	0.8547	0.9916	0.0034
R20S0.8RS0	7.82	0.3858	2.0587	0.9633	0.0235
R20S0.8RS25	7.95	0.4259	1.6187	0.9833	0.0116
R20S0.8RS50	7.99	0.6701	1.2044	0.9943	0.0042
R20S0.8RS75	7.98	0.6772	0.9801	0.9971	0.0025
R20S0.8RS100	7.85	0.4456	0.8229	0.9812	0.0053
R20S1.2RS0	8.18	0.4720	1.1265	0.9796	0.0086
R20S1.2RS25	8.36	0.4769	1.2101	0.9885	0.0064
R20S1.2RS50	8.79	0.5664	1.0161	0.9948	0.0033
R20S1.2RS75	9.51	0.5152	1.2332	0.9900	0.0061
R20S1.2RS100	8.19	0.5246	0.7804	0.9782	0.0054

R^2: coefficient of fitness; SE: standard error.

5. Conclusions

In this study, a central pull-out test was conducted on 51 pull-out specimens to investigate the impact of the content of SSFs and RSFs on the bonding properties of RRAC and steel bars. The main conclusions can be drawn as follows:

(1) The failure of reinforced inter-rib concrete after extrusion is crucial for the failure of RRAC and steel bar bonding. The combined use of SSFs and RSFs inhibits crack development of RRAC, delays damage to the inter-ribbed concrete matrix, and enhances the bonding effect between RRAC and steel bars. However, excessive SFs can reduce the compactness of concrete and create a weak layer at the interface between SFs and concrete, which negatively affects the bonding effect. In practical applications, it is rec-

ommended to limit the steel fibre content to below 1.2% and control the water–cement ratio of concrete.

(2) After the addition of 20% RPs, the bond strength experienced a decrease of 32.87%. When only SSFs or RSFs were added, the bond strength showed improvement with the addition of 0.4% SSFs or 1.2% RSFs, while other dosages resulted in varying degrees of bond-strength decrease. There is a synergistic effect between SSFs and RSFs, and the best improvement in bond strength is achieved through their combined use. When the SF content is 0.8% and the RSF ratio is 75%, the bond strength reaches its maximum value, which is 6.55% higher than that of specimen R20S0RS0.

(3) The bonding stiffness decreased by 37.26% after adding 20% RPs. However, the bonding stiffness improved when SFs were added, regardless of changes in content. The improvement effect becomes more apparent as the SF content increases. The bonding stiffness reaches its maximum at a 1.2% SF content and a 75% RSF ratio, which is 53.88% higher than that of the R20S0RS0 specimen. Considering both bonding strength and bonding stiffness, the optimal mix ratio to enhance the bonding effect between RRAC and steel bars is a 1.2% SF total content and a 75% RSF ratio.

(4) The bond–slip constitutive model of RRAC and steel bars was established based on the characteristics of the bond stress–slip curve. The model shows excellent agreement with the experimental data and can accurately predict the bond stress–slip relationship between RRAC and steel bars.

Author Contributions: Conceptualization, F.L.; methodology, S.H.; validation, J.Z.; formal analysis, H.L., W.W. and X.T.; investigation, J.Z. and X.L.; resources, F.L.; data curation, H.M.; writing—original draft preparation, H.L.; writing—review and editing, H.M.; visualization, S.H.; supervision, F.L.; project administration, W.W. All authors have read and agreed to the published version of the manuscript.

Funding: The authors gratefully acknowledge the financial support provided by the National Natural Science Foundation of China under Grant No. 12072080 and Grant No. 12032009.

Data Availability Statement: The raw data supporting the conclusions of this article will be made available by the authors on request.

Conflicts of Interest: Author Honglong Ma was employed by the company China Railway 17th Bureau Group Co., Ltd. and Xiaopeng Tian was employed by the company China Railway 17th Bureau Group (Guangzhou) Construction Co., Ltd. The remaining authors declare that the research was conducted in the absence of any commercial or financial relationships that could be construed as a potential conflict of interest.

References

1. Mai, G.H.; Li, L.J.; Lin, J.W.; Wei, W.; He, S.H.; Zhong, R.K.; Xiong, Z. Bond durability between BFRP bars and recycled aggregate seawater sea-sand concrete in freezing-thawing environment. *J. Build. Eng.* **2023**, *70*, 106422. [CrossRef]
2. Zhu, X.Y.; Jiang, Z.M. Reuse of waste rubber in pervious concrete: Experiment and DEM simulation. *J. Build. Eng.* **2023**, *71*, 106452. [CrossRef]
3. Xiong, Z.; Wei, W.; Liu, F.; Cui, C.Y.; Li, L.J.; Zou, R.; Zeng, Y. Bond behaviour of recycled aggregate concrete with basalt fibre-reinforced polymer bars. *Compos. Struct.* **2021**, *256*, 113078. [CrossRef]
4. Wang, W.D.; Lu, C.X.; Zhu, Z.M.; Zhang, Z.Y.; Liu, S.Y.; Ji, Y.C.; Xing, Z.Q. Mechanical performance of recycled aggregate concrete in green civil engineering: Review. *Case Stud. Constr. Mat.* **2023**, *19*, e02384. [CrossRef]
5. Formela, K. Waste tire rubber-based materials: Processing, performance properties and development strategies. *Adv. Ind. Eng. Polym. Res.* **2022**, *5*, 234–247. [CrossRef]
6. Tang, Y.C.; Feng, W.H.; Feng, W.X.; Chen, J.M.; Bao, D.J.; Li, L.J. Compressive properties of rubber-modified recycled aggregate concrete subjected to elevated temperatures. *Constr. Build. Mater.* **2021**, *268*, 121181. [CrossRef]
7. Zhu, H.Y.; Xiong, Z.; Song, Y.Y.; Zhou, K.T.; Su, Y. Effect of expansion agent and glass fiber on the dynamicsplitting tensile properties of seawater–sea-sand concrete. *Buildings* **2024**, *14*, 217. [CrossRef]
8. Pan, Z.Z.; Liu, F.; Li, H.W.; Li, X.H.; Wang, D.C.; Ling, Z.; Zhu, H.Y.; Zhu, Y.H. Performance evaluation of thermal insulation rubberized mortar modified by fly ash and glass fiber. *Buildings* **2024**, *14*, 221. [CrossRef]
9. Zhen, H.; Xiong, Z.; Song, Y.Y.; Li, L.J.; Qiu, Y.; Zou, X.Z.; Chen, B.; Chen, D.G.; Liu, F.; Ji, Y.D. Early mechanical performance of glass fibre-reinforced manufactured sand concrete. *J. Build. Eng.* **2024**, *83*, 108440. [CrossRef]

10. El-Zohairy, A.; Sanchez, M.; Abediniangerabi, B.; Moler, P. Performance of rubberized concrete and the effect of temperature and stainless steel fibers. *Buildings* **2023**, *12*, 280. [CrossRef]
11. Eisa, A.S.; Elshazli, M.T.; Nawar, M.T. Experimental investigation on the effect of using crumb rubber and steel fibers on the structural behavior of reinforced concrete beams. *Constr. Build. Mater.* **2020**, *252*, 119078. [CrossRef]
12. Guo, Y.C.; Zhanng, J.H.; Chen, G.; Chen, G.M.; Xie, Z.H. Fracture behaviors of a new steel fiber reinforced recycled aggregate concrete with crumb rubber. *Constr. Build. Mater.* **2014**, *53*, 32–39. [CrossRef]
13. Chen, A.J.; Han, X.Y.; Chen, M.; Wang, X.Y.; Wang, Z.H.; Guo, T.T. Mechanical and stress-strain behavior of basalt fiber reinforced rubberized recycled coarse aggregate concrete. *Constr. Build. Mater.* **2020**, *260*, 119888. [CrossRef]
14. Mohseni, E.; Koushkbaghi, M. Recycling of landfill waste tyre in construction materials: Durability of concrete made with chipped rubber. *Constr. Build. Mater.* **2023**, *409*, 134114. [CrossRef]
15. Xiong, Z.; Fang, Z.; Feng, W.H.; Liu, F.; Yang, F.; Li, L.J. Review of dynamic behaviour of rubberised concrete at material and member levels. *J. Build. Eng.* **2021**, *38*, 102237. [CrossRef]
16. Guo, Y.C.; Zhang, J.H.; Chen, G.M.; Xie, Z.H. Compressive behaviour of concrete structures incorporating recycled concrete aggregates, rubber crumb and reinforced with steel fibre, subjected to elevated temperatures. *J. Clean. Prod.* **2014**, *72*, 193–203. [CrossRef]
17. Shahjalal, M.; Islam, K.; Batool, F.; Tiznobaik, M.; Hossain, Z.F.M.; Ahmed, K.S.; Alam, S.M.; Ahsan, R. Fiber-reinforced recycled aggregate concrete with crumb rubber: A state-of-the-art review. *Constr. Build. Mater.* **2023**, *404*, 133233. [CrossRef]
18. Gholampour, A.; Ozbakkaloglu, T.; Hassanli, R. Behavior of rubberized concrete under active confinement. *Constr. Build. Mater.* **2017**, *138*, 372–382. [CrossRef]
19. Chen, F.; Wu, K.; Ren, L.J.; Xu, J.N.; Zheng, H.M. Internal curing effect and compressive strength calculation of recycled clay brick aggregate concrete. *Materials* **2019**, *12*, 1815. [CrossRef]
20. Zrar, Y.J.; Younis, K.H.; Sherwani, A.F.H. Properties of sustainable self-compacted concrete with recycled concrete and waste tire crumb rubber aggregates. *Constr. Build. Mater.* **2023**, *407*, 133524. [CrossRef]
21. Liu, F.; Meng, L.Y.; Ning, G.F.; Li, L.J. Fatigue performance of rubber-modified recycled aggregate concrete (RRAC) for pavement. *Constr. Build. Mater.* **2015**, *95*, 207–217. [CrossRef]
22. Saberian, M.; Tajaddini, A.; Li, J.; Zhang, G.M.; Wang, L.; Sun, D.A.; Maqsood, T.; Roychand, R. Mechanical properties of polypropylene fibre reinforced recycled concrete aggregate for sustainable road base and subbase applications. *Constr. Build. Mater.* **2023**, *405*, 133352. [CrossRef]
23. Peng, S.; Wu, B.; Du, X.Q.; Zhao, Y.F.; Yu, Z.P. Study on dynamic splitting tensile mechanical properties and microscopic mechanism analysis of steel fiber reinforced concrete. *Structures* **2023**, *58*, 105502. [CrossRef]
24. Han, J.H.; Zhao, M.M.; Chen, J.Y.; Lan, X.F. Effects of steel fiber length and coarse aggregate maximum size on mechanical properties of steel fiber reinforced concrete. *Constr. Build. Mater.* **2019**, *209*, 577–591. [CrossRef]
25. Li, Q.Y.; Shi, Z.M.; Zhao, F.; Yu, S.B.; Xie, K.L. Mechanical performance evaluation of steel fiber-reinforced concrete (FRC) based on multi-mechanical indicators from split hopkinson pressure bar (SHPB) test. *J. Build. Eng.* **2023**, *79*, 107898. [CrossRef]
26. Olivito, R.S.; Zuccarello, F.A. An experimental study on the tensile strength of steel fiber reinforced concrete. *Compos. Part B-Eng.* **2010**, *41*, 246–255. [CrossRef]
27. Xie, J.H.; Li, J.L.; Lu, Z.Y.; Li, Z.J.; Fang, C.; Huang, L.; Li, L.J. Combination effects of rubber and silica fume on the fracture behaviour of steel-fibre recycled aggregate concrete. *Constr. Build. Mater.* **2019**, *203*, 164–173. [CrossRef]
28. Xue, G.J.; Wang, C.L.; Zhang, J.M.; Liu, Z.P.; Zhang, T.T.; Zhang, Y.X. Influence of steel fiber shape on the performance of high-performance concrete. *Acta Mater. Compos. Sin.* **2021**, *38*, 43413–44324.
29. Lee, C.; Kim, H. Orientation factor and number of fibers at failure plane in ring-type steel fiber reinforced concrete. *Cement Concrete Res.* **2010**, *40*, 810–819. [CrossRef]
30. Choia, O.C.; Lee, C. Flexural performance of ring-type steel fiber-reinforced concrete. *Cem. Concr. Res.* **2003**, *33*, 841–849. [CrossRef]
31. He, S.H.; Li, L.J.; Xiong, Z.; Zhang, H.Q.; Zheng, J.H.; Su, Y.; Huang, J.; Liu, F. Effects of ring-type and straight steel fibres on the compressive performance of rubber-recycled aggregate concrete. *J. Build. Eng.* **2023**, *76*, 107148. [CrossRef]
32. Zheng, Y.X.; Fang, C.C.; Ma, J.J.; Wang, S.Q. Review of research on Bond-Slip of reinforced concrete structures. *Constr. Build. Mater.* **2023**, *385*, 131437. [CrossRef]
33. Li, L.G.; Chen, Z.P.; Ouyang, Y.; Zhu, J.; Chu, S.H.; Kwan, A.K.H. Synergistic effects of steel fibres and expansive agent on steel bar-concrete bond. *Cem. Concr. Comp.* **2019**, *104*, 103380. [CrossRef]
34. Vedernikov, A.; Gemi, L.; Madenci, E.; Onuralp Özkılıç, Y.; Yazman, Ş.; Gusev, S.; Sulimov, A.; Bondareva, J.; Evlashin, S.; Konev, S.; et al. Effects of high pulling speeds on mechanical properties and morphology of pultruded GFRP composite flat laminates. *Compos. Struct.* **2022**, *301*, 116216. [CrossRef]
35. Vedernikov, A.; Tucci, F.; Carlone, P.; Gusev, S.; Konev, S.; Firsov, D.; Akhatov, I.; Safonov, A. Effects of pulling speed on structural performance of L-shaped pultruded profiles. *Compos. Struct.* **2021**, *255*, 112967. [CrossRef]
36. Huang, Y.; Yu, Z.F. Experimental study of the bonding behavior between coarse aggregate ultra-high performance concrete and steel rebar. *Eng. Struct.* **2023**, *288*, 116253. [CrossRef]
37. Gao, K.; Li, H.; Liu, G.G.; Huang, Z.Y.; Wu, G.B. Bonding properties between steel-basalt hybrid fibers reinforced cementitious composites and existing concrete at high temperatures. *J. Build. Eng.* **2023**, *70*, 106371.

38. Garcia-Taengua, E.; Martí-Vargas, J.R.; Serna, P. Bond of reinforcing bars to steel fiber reinforced concrete. *Constr. Build. Mater.* **2016**, *105*, 275–284. [CrossRef]
39. Dancygier, A.N.; Katz, A.; Wexler, U. Bond between deformed reinforcement and normal and high-strength concrete with and without fibers. *Mater. Struct.* **2010**, *43*, 839–856. [CrossRef]
40. Bao, S.H.; Zhang, Y.F.; Li, C.B.; Zhang, W.J.; Zeng, K. Interfacial bonding behavior of steel fibers when using fine glass powder as partial substitution of silica fume/cement. *Constr. Build. Mater.* **2024**, *411*, 134516. [CrossRef]
41. Majain, N.; Rahman, A.B.A.; Adnan, A.; Mohamed, R.N. Bond behaviour of deformed steel bars in steel fibre high-strength self-compacting concrete. *Constr. Build. Mater.* **2022**, *318*, 125906. [CrossRef]
42. Xiong, Z.; Zeng, Y.; Li, L.G.; Kwan, A.K.H.; He, S.H. Experimental study on the effects of glass fibres and expansive agent on the bond behaviour of glass/basalt FRP bars in seawater sea-sand concrete. *Constr. Build. Mater.* **2021**, *274*, 122100. [CrossRef]
43. Mo, K.H.; Visintin, P.; Alengaram, U.J.; Jumaat, M.Z. Bond stress-slip relationship of oil palm shell lightweight concrete. *Eng. Struct.* **2016**, *127*, 319–330. [CrossRef]
44. Guo, Z.H. *The Strength and Deformation Test Basis and Constitutive Relationship of Concrete*; Tsinghua University Press: Beijing, China, 1997.

buildings

Article

Study on the Dynamic Performance and Damage Evaluation of Rubber-Modified Non-Autoclaved Concrete Pipe Piles under Axial Drop Hammer Impact

Sheng Lan [1], Feng Liu [1], Fei Yang [1,*], Wanhui Feng [2] and Dawei Chen [1]

[1] School of Civil and Transportation Engineering, Guangdong University of Technology, Guangzhou 510006, China; 1112009004@mail2.gdut.edu.cn (S.L.); fliu@gdut.edu.cn (F.L.); 2112009178@mail2.gdut.edu.cn (D.C.)
[2] College of Urban and Rural Construction, Zhongkai University of Agriculture and Engineering, Guangzhou 510225, China; whfeng@zhku.edu.cn
* Correspondence: feiyang730@gdut.edu.cn

Abstract: In order to improve the weak impact resistance of non-autoclaved concrete pipe piles, this study replaced sand in the concrete with rubber particles of different volume contents to obtain rubber-modified non-autoclaved concrete pipe piles (with volume contents of 0%, 5%, 10%, and 15%). The dynamic impact response characteristics of rubber-modified non-autoclaved concrete pipe piles were obtained through large-scale axial hammer impact experiments. The results indicate the following. (1) Non-autoclaved concrete pipe piles without rubber additives were prone to expansion deformation instability under impact. When the rubber content was 10%, the expansion deformation of the piles was the weakest, and the state was the most stable. (2) When the impact energy exceeded 48 kJ, the deformation energies of piles with 5% and 10% rubber contents significantly increased. (3) The damage levels of the piles after hammer impact were classified into four grades: no damage, mild damage, moderate damage, and severe damage. When the impact energy was greater than or equal to 48 kJ, rubber-modified non-autoclaved concrete pipe piles exhibited damage. The zone with no damage for piles with 10% rubber content was the smallest, making it less prone to damage under impact loads. The rubber-modified non-autoclaved concrete pipe piles with 10% rubber content not only had excellent impact resistance but also utilized the advantages of being environmentally friendly and energy-saving. They filled a certain knowledge gap in green building materials.

Keywords: rubber-modified non-autoclaved concrete pipe piles; drop hammer impact; impact energy; damage level; three-dimensional digital speckle technology

Citation: Lan, S.; Liu, F.; Yang, F.; Feng, W.; Chen, D. Study on the Dynamic Performance and Damage Evaluation of Rubber-Modified Non-Autoclaved Concrete Pipe Piles under Axial Drop Hammer Impact. *Buildings* **2024**, *14*, 489. https://doi.org/10.3390/buildings14020489

Academic Editor: Binsheng (Ben) Zhang

Received: 13 January 2024
Revised: 6 February 2024
Accepted: 7 February 2024
Published: 9 February 2024

1. Introduction

The current prestressed high-strength concrete pipe piles (PHC piles) mainly adopt steam curing and high-temperature and high-pressure secondary curing technology, which allows for the rapid attainment of PHC piles that meet the usage requirements within 24 h. However, the energy consumed during the production of pipe piles is mainly derived from coal [1]. Burning coal generates a large amount of harmful gases such as sulfur dioxide and nitrogen oxides, causing pollution to the environment. In 2020, the carbon emissions from China's construction industry accounted for 50.9% of the country's total carbon emissions, with building material production accounting for 28.2% [2]. In order to reduce carbon emissions and achieve the goals of the 'dual-carbon' strategy, all pipe pile factories were required to cease the use of coal and instead utilize natural gas, which has lower pollution levels. However, not only did natural gas incur higher costs, but its carbon emissions were also not low. The emission of CO and CO_2 after the combustion of natural gas per cubic meter can reach 21.2 kg and 6.7 kg, respectively [3]. In this context, the non-autoclaved curing technology has emerged. This technology eliminates the need for the second round of high-temperature and high-pressure curing in traditional methods and only involves steam curing. As the non-autoclaved curing technology eliminates the high-temperature and high-pressure curing steps, the production cost is significantly reduced. According to Chinese standards, the carbon emissions from the non-autoclaved curing process were reduced by 17.2% compared to autoclaved curing pipe piles (AP) [4]. In addition, compared to AP concrete, non-autoclaved pipe piles (NAP) concrete exhibited significant improvements in frost resistance, sulfate resistance, strength, and other properties [5–7].

The compressive strength of NAP concrete could reach 90 MPa [8]. However, similar to many high-strength concretes, NAP concrete also exhibited a high brittleness index with a tensile–compressive strength ratio of only 0.043 [9]. And its dynamic compression enhancement factor was mostly within the range of 1.0~1.2, which was smaller than that of AP concrete [9]. Liu et al. [10] conducted drop hammer impact tests on NAP and AP, and found that under high-energy impact, NAP experienced failure first. In practical applications, under impact loads such as hammering piles and earthquakes, NAP was prone to brittle failure. Water from the external environment could enter the pile body through cracks and corrode the steel bars, greatly affecting the performance of the pipe piles. This defect has seriously affected the widespread application of non-autoclaved curing technology. In addition, high temperatures during a fire caused significant damage to the strength of ordinary concrete [11]. To improve the mechanical performance of concrete, researchers often add glass fibers and expansion agents to concrete and even used steel structures [12–14]. However, these methods have shown limited effectiveness in enhancing the impact resistance of concrete. On the other hand, with the rapid development of China's automotive industry, the pressure caused by waste tires on the environment is increasing. In 2022 alone, the number of scrap tires in China reached approximately 350 million, and this number is still growing at an annual rate of 6% to 8% [15]. Some experts and scholars uphold the concept of green energy-saving and have attempted to crush the rubber materials in waste tires into particles and add them to the concrete. This approach had yielded recycled rubber concrete with lower brittleness and higher impact resistance [16–21]. Our team attempted to incorporate recycled rubber particles into NAP concrete to produce rubber-modified non-autoclaved pipe pile (NAPR) concrete. It was found that the addition of recycled rubber significantly enhanced the crack resistance and impact resistance of NAPR concrete [9].

Although there have been some achievements in the research on NAPR concrete materials, the research at the component level is still lacking, especially regarding its impact resistance performance, which remains unclear. In practical engineering, pipe piles are prone to impact damage during hammering pile construction. Based on this background, this study investigates the dynamic impact response characteristics of NAPR through axial falling hammer impact experiments and explores the influence of recycled rubber on the impact resistance performance of NAPR.

2. Preparation of Non-Autoclaved Concrete Pipe Piles

2.1. Concrete Materials for Non-Autoclaved Pipe Piles

Due to the omission of energy-intensive high-temperature and high-pressure curing steps in NAP, the fine grinding sand in the original concrete mix cannot react to form tobermorite [22]. As a result, the strength of the concrete cannot meet the strength requirements specified in relevant Chinese standards for prestressed high-strength concrete pipe piles, which is greater than 80 MPa [23]. To meet this standard, slag powder and silica powder were used to replace fine grinding sand to enhance the concrete strength. The specific mix proportions are listed in Table 1.

Table 1. Mix proportions of concrete (kg/m^3).

Cement	1–2 Stones	0–5 Stones	Sand	Slag Powder	Silica Powder	Water Reducer (Solid Content 40%)	Water	Water to Glue Ratio
282	910	310	705	141	47	5.64	100	0.213

As shown in Figure 1, the cement selected was P.II 42.5R Portland cement produced by China Resources Cement Limited in Jiangmen City, Guangdong Province, China. The sand chosen was medium sand with a fineness modulus of 2.6. The specific surface areas of slag powder and silica powder were 412 m^2/kg and 21 m^2/kg, respectively. The water reducer was QL-PC5 type polycarboxylate high-efficiency water reducer with a solid content of 40% produced by Jiangmen Strong Building Materials Technology Co., Ltd. in Jiangmen City, Guangdong Province, China. In order to enhance the impact resistance of NAP, rubber particles with a diameter of 0.85 mm were used for partial replacement of sand in a volume substitution method, with replacement rates of 5%, 10%, and 15%.

(**a**) Cement (**b**) Sand (**c**) Rubber particles (**d**) Water reducer (**e**) Slag powder (**f**) Silica powder

Figure 1. Raw materials for the concrete part of NAPR.

2.2. Forming of Non-Autoclaved Concrete Pipe Piles

2.2.1. Production Process

As shown in Figure 2, several important preparation steps of the NAPR fabrication process include concrete mixing, mixture filling, prestress application, centrifugal molding, and steam curing. NAPR needs to be cured in a 90 °C steam chamber for twenty-four hours, after which the cover can be removed and the mold dismantled.

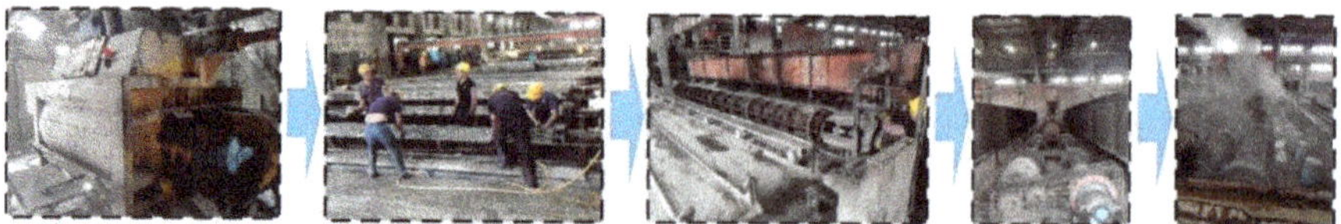

Figure 2. Preparation process of NAPR.

Figure 3 depicts the finished product of NAPR. From the figure, it can be observed that compared to AP, NAPR has an overall darker color, appearing as deep gray. The surface of the pile body is smooth and free from defects. The pipe piles are in a standard cylindrical shape without any deformation. The inner wall of the piles is smooth, with an intact protective layer and no exposed sand or gravel. The formed NAPR meets the structural requirements of high-strength prestressed piles [24].

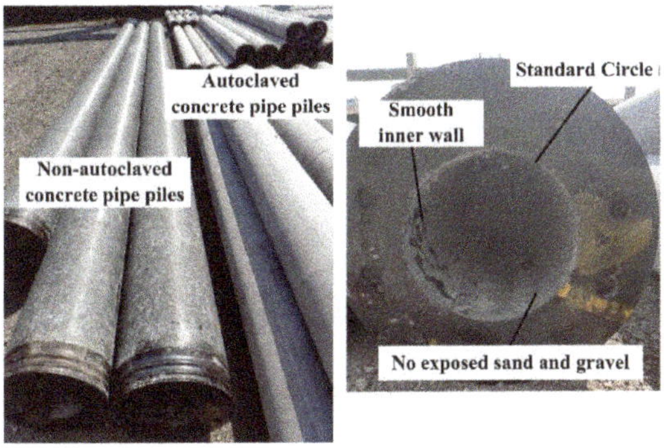

Figure 3. Appearance of NAPR.

2.2.2. Parameters of Pipe Piles

Prestressed high-strength concrete piles are divided into four major categories: A type, AB type, B type, and C type. For construction projects, A type and AB type can meet most requirements. The vertical bearing capacities of these two types of piles are the same. The non-autoclaved pipe piles prepared in this study were of the A type, and the structure and some of the reinforcing parameters of the piles are listed in Table 2. Combined with Figure 4, it can be seen that the outer diameter of the pile was 400 mm, and the wall thickness was 95 mm. Since this study focuses on the vertical impact resistance of NAPR and does not involve the bending resistance of the piles, in order to control variables, the dense area of spiral hoops was removed, and the spacing between all spiral hoops was selected as 80 mm.

Table 2. Structural parameters of NAPR (mm).

Specifications and Dimensions			Prestressed Steel Reinforcement		Stirrup	
Outside Diameter	Wall Thickness	Length	Diameter	Distribution Diameter	Diameter	Distance
400	95	12,000	9	308	4	80 ± 5

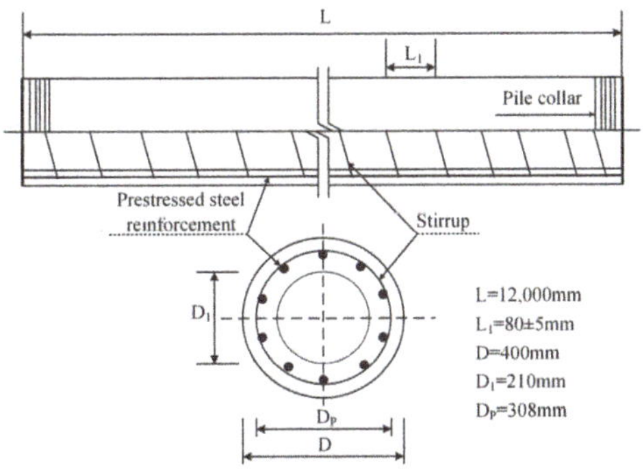

Figure 4. Schematic diagram of NAPR structure.

The study only focuses on the axial impact resistance test of NAPR, without considering the impact stability of the piles and the influence of length. As shown in Figure 5, the full-length piles were cut into one-meter-long short piles according to the experimental conditions, and the end faces of the piles were polished after cutting to ensure that the inclination unevenness of the end face was less than 0.5 mm [24]. For ease of analysis, the specimens of non-autoclaved pipe piles were named based on different rubber contents. For the name of NAPRX, "NAPR" refers to rubber-modified non-autoclaved pipe pile, and "X" indicates the rubber content in percentage. For example, NAPR5 represents the non-autoclaved pipe pile with a rubber content of 5%.

Figure 5. Specimens of NAPR.

3. Large-Scale Drop Hammer Impact Experiment

3.1. Experimental Equipment

The experimental equipment used for the axial drop hammer test on piles in this study was the Vertical Ultra-Heavy Drop Hammer Experimental Apparatus in the School of Transportation, Civil Engineering & Architecture at Foshan University of Science and Technology, China. As shown in Figure 6, the Vertical Ultra-Heavy Drop Hammer Experimental Apparatus mainly consisted of five parts: the base, support, rail, hammer head, and operating system, which could achieve numerical control and dynamic data acquisition. The maximum height of the drop hammer was 18 m, the maximum mass of the hammer head was 1200 kg, and the maximum impact energy was 170 kJ, which could meet the requirements of axial drop hammer impact tests on piles.

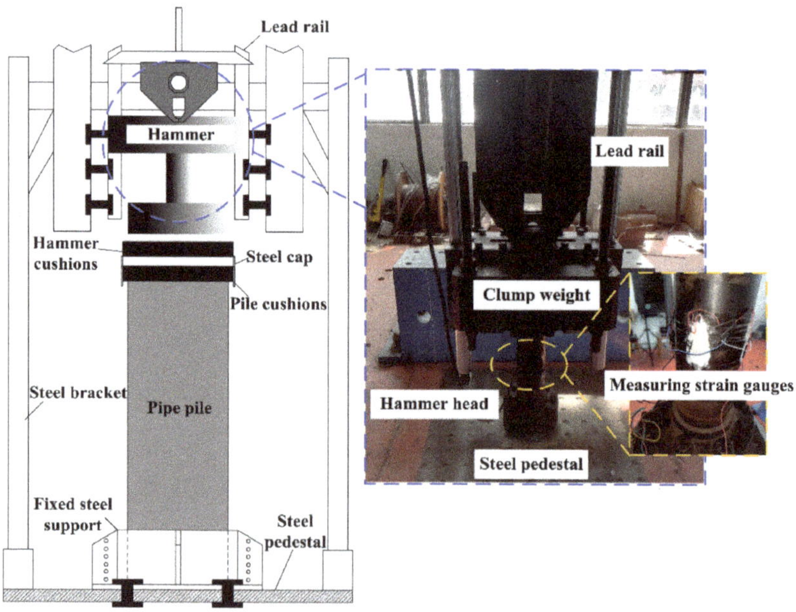

Figure 6. Vertical Ultra-Heavy Drop Hammer Experimental Apparatus.

3.2. Experimental Scheme

Before the experiment, it was necessary to fix the NAPR on the base using a custom fixture. To ensure that the impact load of the falling hammer was evenly applied to the end face of the NAPR, a steel cap with an inner diameter of 420 mm was installed at the end of the pile. In addition, to ensure the stiffness of the steel cap and reduce experimental errors caused by deformation, the thickness of the steel cap was 50 mm. In order to avoid the phenomenon of stress concentration during the drop hammer test, 30 mm thick cardboard was added as a cushion between the steel cap and the pile end. The direct impact of the hammer head on the steel cap was prone to causing damage. In order to protect the hammer head, 40 mm thick plywood was placed on the steel cap. The specific arrangement is illustrated in Figure 7. Three sets of strain gauges were pasted on the vertical surfaces of the pile body at heights of 250 mm, 500 mm, and 750 mm to obtain the axial and transverse strains of NAPR in the drop hammer test. Additionally, in order to obtain the deformation characteristics of the surface of the NAPR during the impact process of the falling hammer, a three-dimensional digital image correlation (DIC) technique was employed. Matte paint was used to spray good speckle points on the surface of the pile, and three-dimensional DIC technology could obtain the local deformation field of the pile body by calculating the displacement of the speckle points on the surface of the pile during the experimental process.

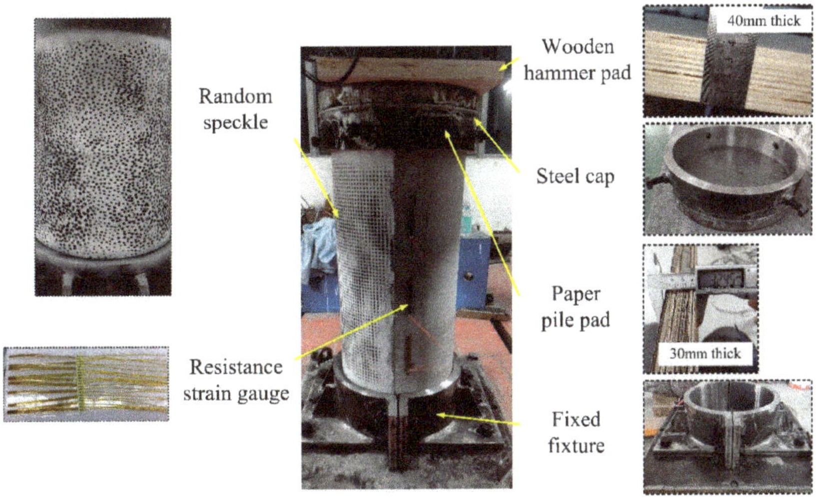

Figure 7. Preparation before the experiment.

Figure 8 is a schematic diagram of the axial drop hammer experiment for NAPR. After completing the preparation work, the high-speed cameras and drop hammer control system were calibrated and adjusted. It was found that when the two high-speed cameras were 2 m away from the piles and at an angle of 30 degrees, the calibration results were the best and the required measurement surface could be fully captured. Therefore, the two cameras were fixed at 30 degrees at a distance of 2 m from the piles [25]. Finally, the drop hammer was controlled to reach the predetermined height through the system software and complete the axial drop hammer impact experiment. During the experiment, three types of data were primarily collected: impact load, surface strains of the pile body, and high-speed speckle photos. The frequency of collecting impact load was 500 kHz, the frequency of collecting strain was 100 kHz, the resolution of the high-speed camera was 1024×1024, and the collection frequency was 12,500 frames per second.

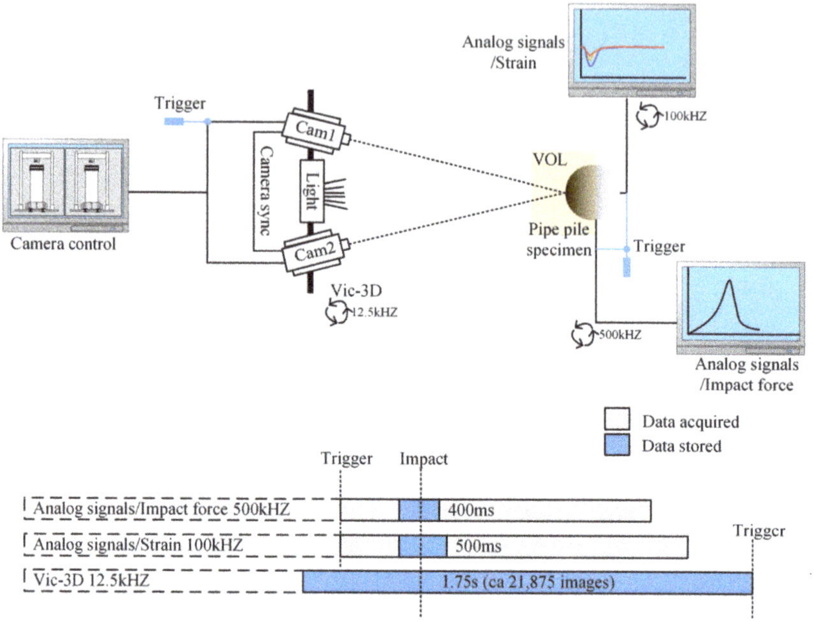

Figure 8. Schematic diagram of axial drop hammer test on NAPR.

According to Chinese standards, it is known that the maximum impact energy exerted on the piles during actual pile-driving construction will not exceed 50 kJ [26]. In this experiment, this impact energy was used as a reference to obtain the dynamic response characteristics of the piles under different impact energies by adjusting the drop hammer height. The weight of the hammer head in the drop hammer device was chosen as 1200 kg, and the experimental drop heights were 2 m, 4 m, 5 m, and 6 m, corresponding to the impact energies of 24 kJ, 48 kJ, 60 kJ, and 72 kJ, respectively.

4. Dynamic Response Characteristics of NAPR

4.1. Drop Hammer Impact Force

4.1.1. Impact Loading

The impact energies of the axial drop hammer test for NAPR with different rubber contents were controlled by the drop heights. As shown in Figure 9, the impact load on the non-steam pressure-cured piles decreased with the increase of the rubber content.

The ultimate bearing capacities of conventional precast concrete piles during pile driving ranged from 1500 kN to 3000 kN and could reach up to 4000 kN in larger cases [27–30]. However, in this experiment, the NAPR achieved a maximum dynamic bearing capacity of 5000 kN, which was significantly higher than that of conventional precast concrete piles. When the impact energy was 60 kJ, the average impact loads of NAPR with rubber contents of 0%, 5%, 10%, and 15% were 5939.7 kN, 5763.9 kN, 5702.8 kN, and 5675.7 kN, respectively. The impact loads with the addition of rubber pieces were reduced by 3%, 4%, and 4.4% compared to the impact loads without the addition of rubber pieces. When the impact energy was 72 kJ, NAPR5 exhibited the maximum impact load of 6500.4 kN.

Comparing the effect of rubber content on the impact loads of piles, it was found that when the impact energy was below 48 kJ, the impact loads were mainly influenced by the impact energy, and the effect of rubber content was relatively small. Especially when the impact energy was 24 kJ, the impact loads were not affected by the rubber content, which were the same as that of NAPR with different rubber contents. When the impact energy exceeded 48 kJ, the influence of rubber content on the impact load gradually manifested.

With the increase in rubber content, the impact loads under the same impact energy showed a trend of first increasing and then decreasing with the increase in rubber content. The impact load of NAPR was maximum when the rubber content was 5%.

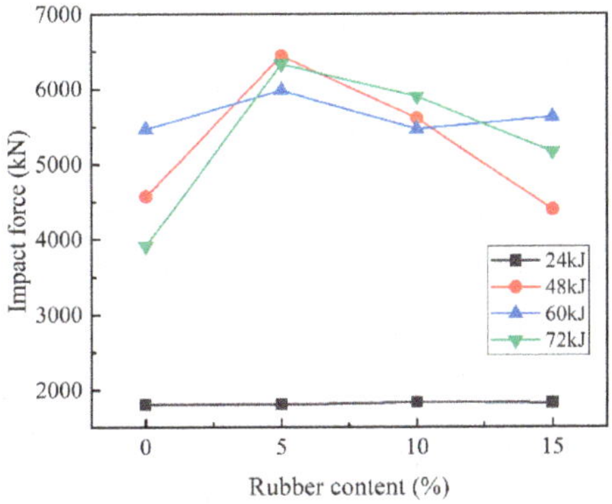

Figure 9. Relationship between impact load and rubber content for different impact energies.

4.1.2. Deformation Energy

The drop hammer impact test had a short duration, on the millisecond scale, where the drop distance of the hammer during contact with the bearing surface of the pile was used as the deformation displacement of the pile [10]. The calculation formula is shown in Equation (1).

$$z(t) = \iint \frac{F(t) - m \cdot g}{m} d^2 t \tag{1}$$

where $F(t)$ is the impact load of the hammer, m is the total mass of the hammer weight, and g is the acceleration due to gravity.

In Figure 10, the impact load–displacement curves for NAPR with different rubber contents and under different impact energies are presented. The impact load increased with the increasing impact energy. At the same time, the shorter duration of the impact resulted in an increased slope of the load–displacement curve in the linear segment. Greater impact energy and shorter impact duration meant that the pile was more likely to undergo brittle failure. However, as the impact energy increased, the postpeak segment of the load–displacement curve became smoother. From the curve, it could be observed that as the rubber content increased, the difference in the peak loads decreased for each impact energy. This indicated that the trend of increasing the impact load with the increasing impact energy was diminishing. Comparing the impact loads of 60 kJ and 72 kJ, it could be observed that with the rubber contents of 5% and 10%, the impact loads increased by 19.2% and 7.9%, respectively. Interestingly, with a higher rubber content, the increase in the impact load was relatively smaller. This indicated that rubber particles have a better cushioning effect on high impact energies.

As shown in Figure 11, the deformation energy generated during the impact process of the NAPR can be regarded as a significant indicator to measure the self-energy absorption capacity of the pile. It could be observed from the figure that when the impact energy was less than 48 kJ, the deformation energies of the steam-free piles with different rubber contents were almost the same. When the impact energy exceeded 48 kJ, the deformation energy of the NAPR increased the fastest. When the impact energy exceeded 60 kJ, the deformation energies were significant for the rubber contents of 5% and 10%, measuring 25.2 kJ and 21.9 kJ, respectively. In addition, as the impact energy increased, the proportion of the deformation energy to the impact energy increased. When the impact energies were 24 kJ, 48 kJ, and 60 kJ, the deformation energies of the piles were 3.47 kJ, 10.5 kJ, and 19.3 kJ, respectively. The proportions of the deformation energy to the impact energy were 14.46%, 21.9%, and 32.2%, respectively. The reason for this was that as the impact energy increased, especially when the impact energy exceeded 48 kJ, the pile underwent damage, resulting in significant increases in deformation. The evolution trend lines of the deformation energies of NAPR with four different rubber contents were compared. It could be observed that as the impact energy increased, the growth trends of the deformation energies of NAPR with 0% and 15% rubber contents were slowing down, while the growth trends of the deformation energies of NAPR with 5% and 10% rubber contents remained significant.

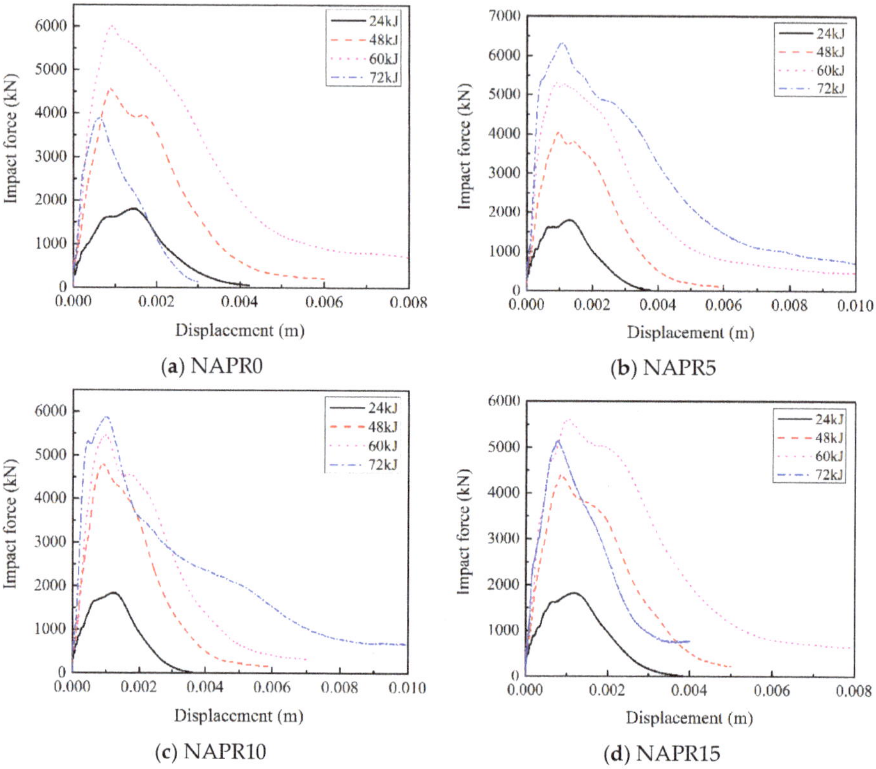

Figure 10. Impact load–displacement curves.

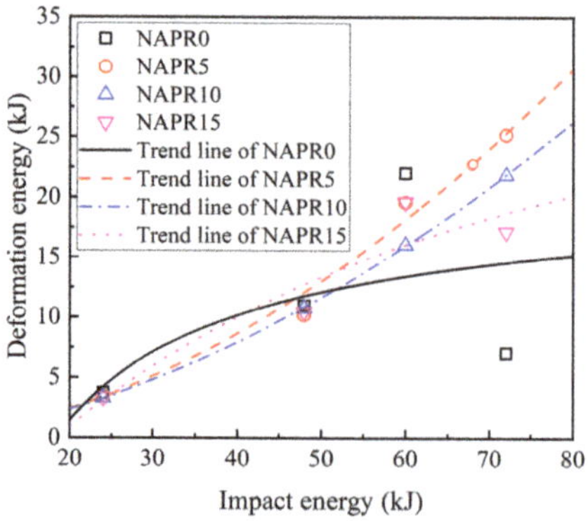

Figure 11. Relationships between the deformation energy and impact energy of NAPR with different rubber contents.

4.1.3. Strain of Pile Body

During the axial hammer impact experiment, strain gauges with a gauge length of 100 mm and a resistance of 120 Ω were attached to the pile shaft to measure the strains of the pile. As shown in Figure 12, taking the impact energies of 24 kJ and 72 kJ as examples, the axial strains, transverse strains, and lateral deformation coefficients of NAPR with different rubber contents were analyzed. Whether the impact energy was low impact energy of 24 kJ or high impact energy of 72 kJ, the axial and transverse strains of the pile both exhibited decreasing trends as the rubber content increased.

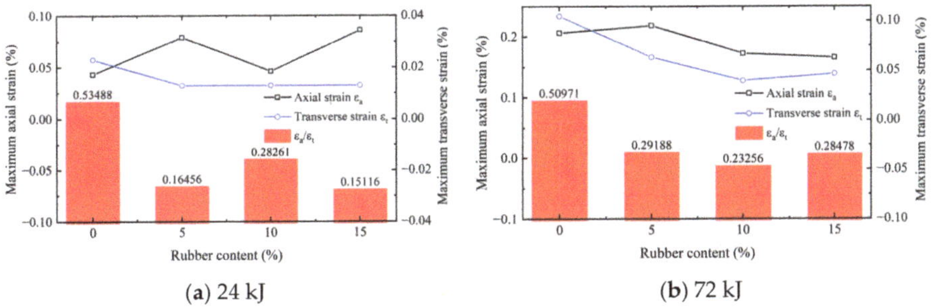

Figure 12. Strain characteristics of the pipe pile body under different impact energies.

The ratio of the transverse strain to the axial strain is called the lateral deformation coefficient, which reflects the expansion deformation of the NAPR. The lateral deformation coefficient of the pile body reflects the pile's stability under impact to some extent [31,32]. From the figure, it can be seen that whether there is low impact energy (24 kJ) or high impact energy (72 kJ), the lateral deformation coefficient of the NAPR without rubber is the largest, exceeding 0.5, indicating that the NAPR without rubber is prone to expansion deformation. When the impact energy was 24 kJ, the lateral deformation coefficients of the piles with the rubber contents of 5% and 15% were relatively small, with the values of 0.16456 and 0.15116, respectively. However, when the impact energy was 72 kJ, the lateral deformation coefficients of the piles with these two rubber contents increased to 0.29188 and 0.28478, respectively, representing increases of 77.4% and 88.4%. In comparison, the lateral deformation coefficient of the NAPR with 10% rubber content did not change significantly under low and high impact energies, indicating a more stable deformation state.

4.2. Deformation of Pile Body

The strain gauges had limited detection range and deformation area on the pile body. In order to obtain the local strain field evolution characteristics [33], three-dimensional digital speckle (DIC) technology was used to detect the deformation of a part of the pile body during the hammer impact test process. The deformation characteristics of NAPR with different rubber contents were similar during the hammer impact process. As shown in Figure 13, taking the local axial strain and displacement contours of the pile under 60 kJ energy impact without rubber as an example, it can be found that the deformation of the pile went through three stages according to the strain contour. The first stage was the stress wave generated by the hammer impact propagating downwards from the impact end. In the second stage, the stress wave reached the bottom pressure plate and partly propagated vertically in the opposite direction in the form of an emission wave. The pile still had some downward stress waves from the hammer impact. In the third stage, after the hammer impact, the pile mainly had the upward reflected wave left. In contrast to the strain, the distribution of displacement during impact was not uniform, as observed from the change in the displacement contour in Figure 13b. The displacement contour at the impacted end was concave-arc-shaped, with greater deformation on the outer side and smaller deformation on the inner side. The fixed end at the bottom was opposite to this, with a convex-arc-shaped displacement contour and greater deformation on the inner side and smaller deformation on the outer side.

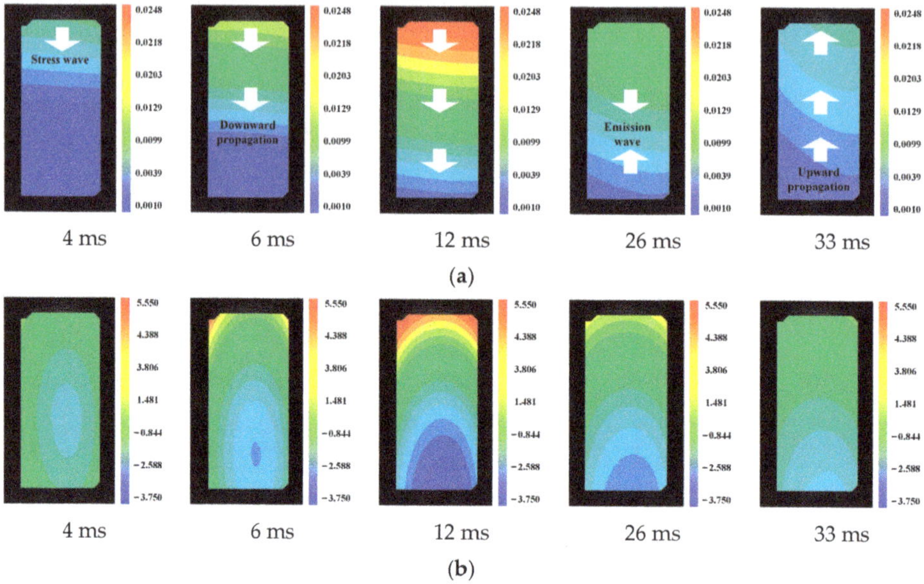

Figure 13. Deformation contours of pipe pile body under hammer impact: (**a**) Axial strains; (**b**) Axial displacements.

The deformation of NAPR bodies is mainly influenced by the impact energy. As shown in Figure 14, taking the example of NAPR without rubber, five monitoring points were arranged along the vertical direction of the pile body to detect the strains at different positions of the pile under different impact energies. The distance between adjacent monitoring points was 150 mm.

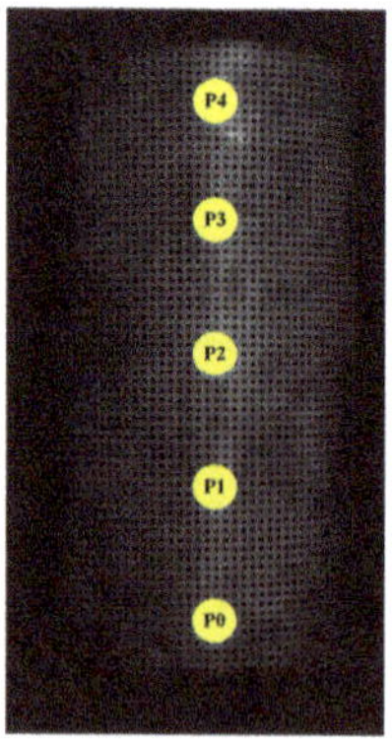

Figure 14. Layout of detection points.

As shown in Figure 15, with the increase in impact energy, the strains at various parts of the pile body increased, and the strains were greater closer to the impacted end. According to the strain values and strain contours corresponding to different time points, it could be found that before the strain reached its peak value, the deformation of the pile body was mainly in the first stage, i.e., influenced by the downward impact stress wave.

After the strain reached its peak value, it entered the second stage, where the upward reflected wave gradually affected the deformation of the pile body. Finally, when the strain decreased further, the deformation of the pile body entered the third stage, where only the upward reflected wave remained. Comparing the strain values at position P_4 under different impact energies, it was found that the strain values under four different impact energies were 0.01411, 0.02242, 0.03156, and 0.04783, respectively, and the maximum strain occurred within 30~40 ms. Comparing the strains under three different impact energies of 24 kJ, 48 kJ, and 72 kJ, it was found that their strain growth rates were 58.89% and 113.34%, respectively, indicating that the strain growth rate increased with the increasing impact energy.

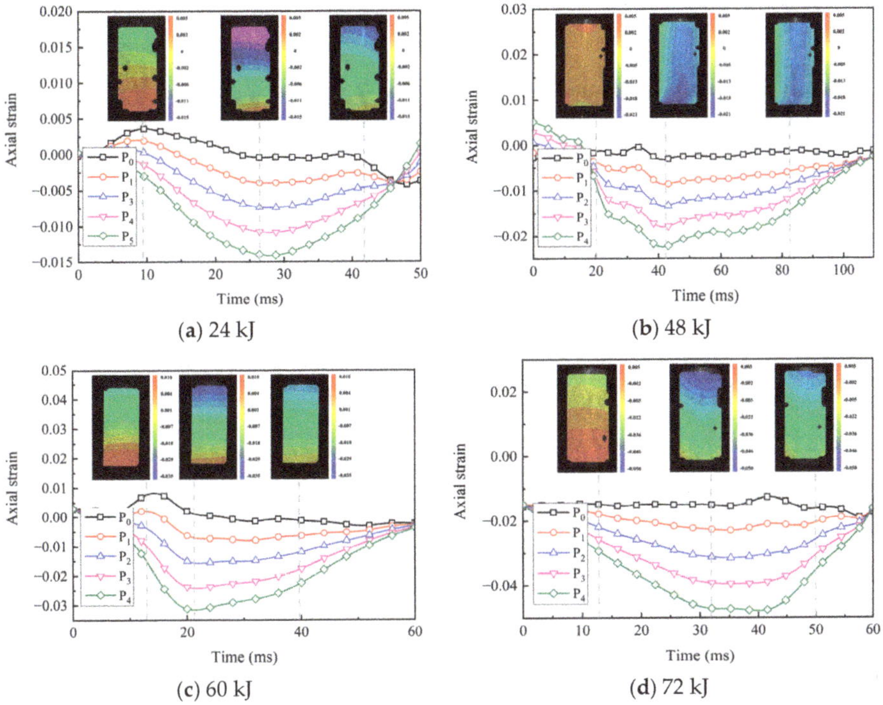

Figure 15. Strain evolutions of pipe pile body under different impact energies.

Taking the strain at point P_4 as the research object, the deformation characteristics of NAPR with different rubber contents under different impact energies were compared. As shown in Figure 16, the strain of the NAPR at the impact end generally increased with the increasing rubber content. The reason why the strain of the pile with 5% rubber content was the largest under the impact energy of 48 kJ was that all piles with different rubber contents began to show a small amount of damage under this impact load, and the pile with 5% content had the greatest brittleness. When it was damaged, brittle fracture was the main mode, and the deformation produced was larger than in other NAPR with different rubber contents. However, with the further increase in impact energy, other NAPR with different rubber contents also showed larger deformation and the measured strain once again met the rule of increasing with the increasing rubber content. In addition, when the impact energy was less than 72 kJ, the deformation time of the pile with 15% rubber content was longer, indicating that its energy absorption effect was good. However, when the impact energy was 72 kJ, its deformation time was significantly shortened, because this impact energy exceeded its absorption limit, and an obvious impact hardening phenomenon appeared in the pile.

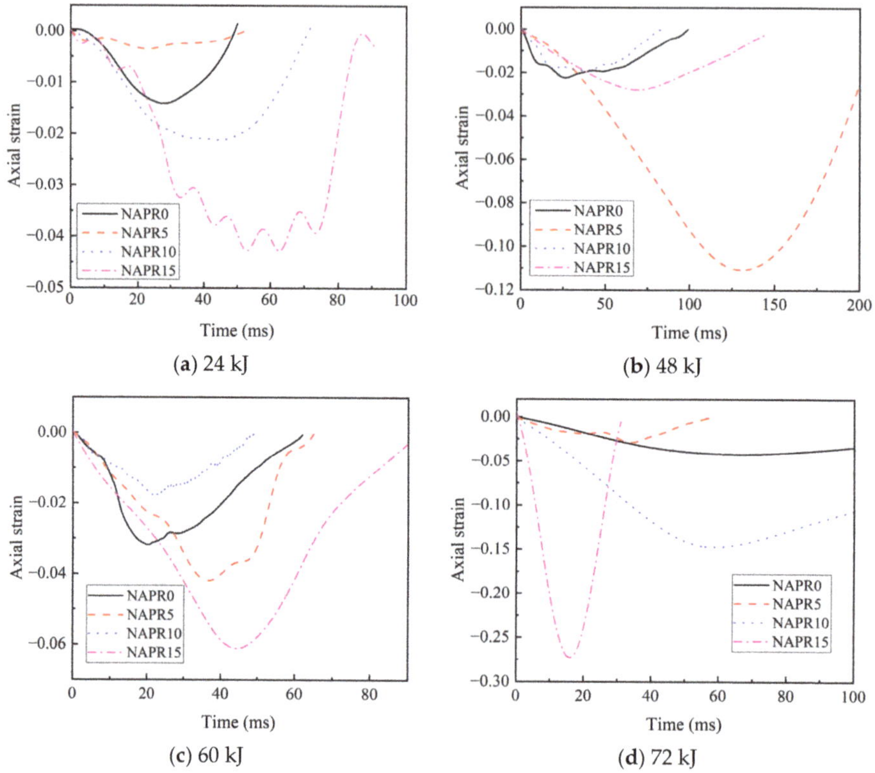

Figure 16. Strain evolutions of NAPR with different rubber contents.

4.3. Damage to Pipe Piles

4.3.1. Failure Modes

The failure mode of NAPR under axial hammer impact is also an expression of its dynamic performance. As shown in Figure 17, taking the rubber content of 15% and the impact energy of 72 kJ as an example, when subjected to a drop hammer impact, cracks began to appear at the impacted end of the pipe pile. With the further increase in the impact load, these cracks started to develop along the axial and transverse directions, eventually forming a crack network, and clear damage appeared near the end of the pipe pile.

Figure 18 illustrates the failure modes of NAPR with different rubber contents under various impact energies. When the impact energy was 24 kJ, the pile bodies of NAPR with different rubber contents remained intact, and no defects or cracks were observed on the inner and outer walls. When the impact energy was 48 kJ, the impact load caused plastic deformation of the concrete material in the NAPR, resulting in damage for piles with different rubber contents. For 0% and 5% rubber contents, large block-like fractures appeared at the bottom of the NAPR, with the most severe fracture occurring at 5% rubber content. At 10% rubber content, a long, slender crack appeared in the pile body. With 15% rubber content, only a few small microcracks appeared at the bottom of the pile. Therefore, it was evident that rubber significantly reduced pile damage under low impact energy.

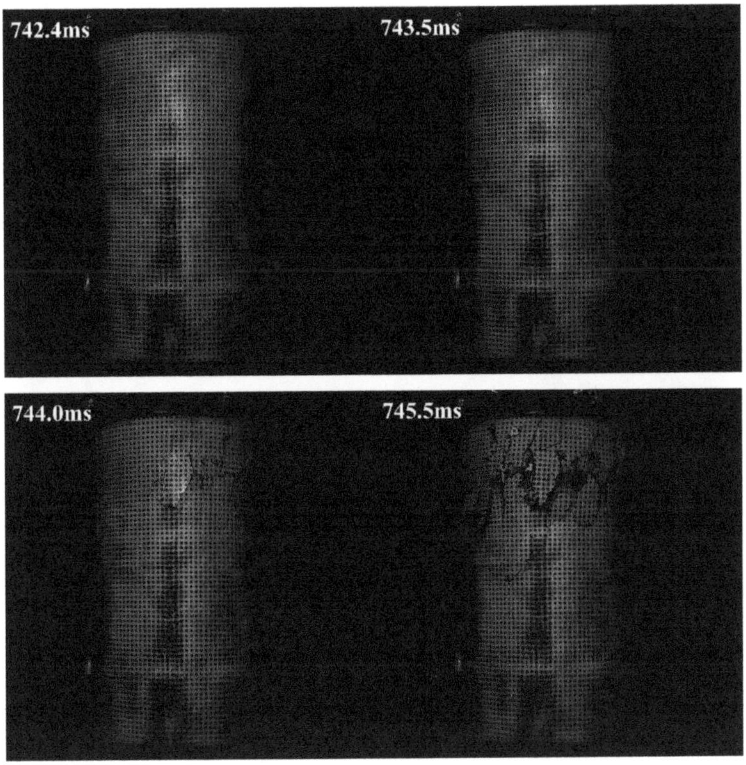

Figure 17. The process of axial hammer impact on NAPR.

When the impact energy reached 60 kJ, it fell into the high impact energy range, and NAPR with different rubber contents all experienced plastic deformation and significant damage. With 0% rubber content, three axial cracks connecting the upper and lower ends of the pile appeared on the outer wall. With 5% rubber content, two axial cracks penetrating the pile body appeared, with the least number of cracks. With 10% rubber content, the pile body of the piles showed three axial through cracks, similar to the damage observed when no rubber was added. With 15% rubber content, a large number of axial cracks appeared in the pile body, and several transverse cracks were also generated. This was because the concrete material in the NAPR with this content had weaker strength, and under the influence of reflected stress waves, the tensile stress generated exceeded the tensile strength of the concrete material.

When the impact energy was 72 kJ, the NAPR exhibited significant deformation and extensive damage. With 0% rubber content, the pile body showed a dense network of cracks, primarily consisting of numerous axial splitting cracks and some transverse cracks, and even a small area of collapse and fracture. With 5% rubber content, the pile body exhibited axial and transverse cracks, but compared to other rubber content levels, this concentration had the smallest number of cracks. With 10% rubber content, the pile body near the impact end experienced extensive cracking and damage, with the pile tilting and deforming. With 15% rubber content, the pile body near the impact end also underwent extensive cracking, and the bond between the steel reinforcement and concrete became weak, resulting in exposed reinforcement. This was because the bond strength between the concrete and steel reinforcement was insufficient to meet the normal working performance of the pipe pile.

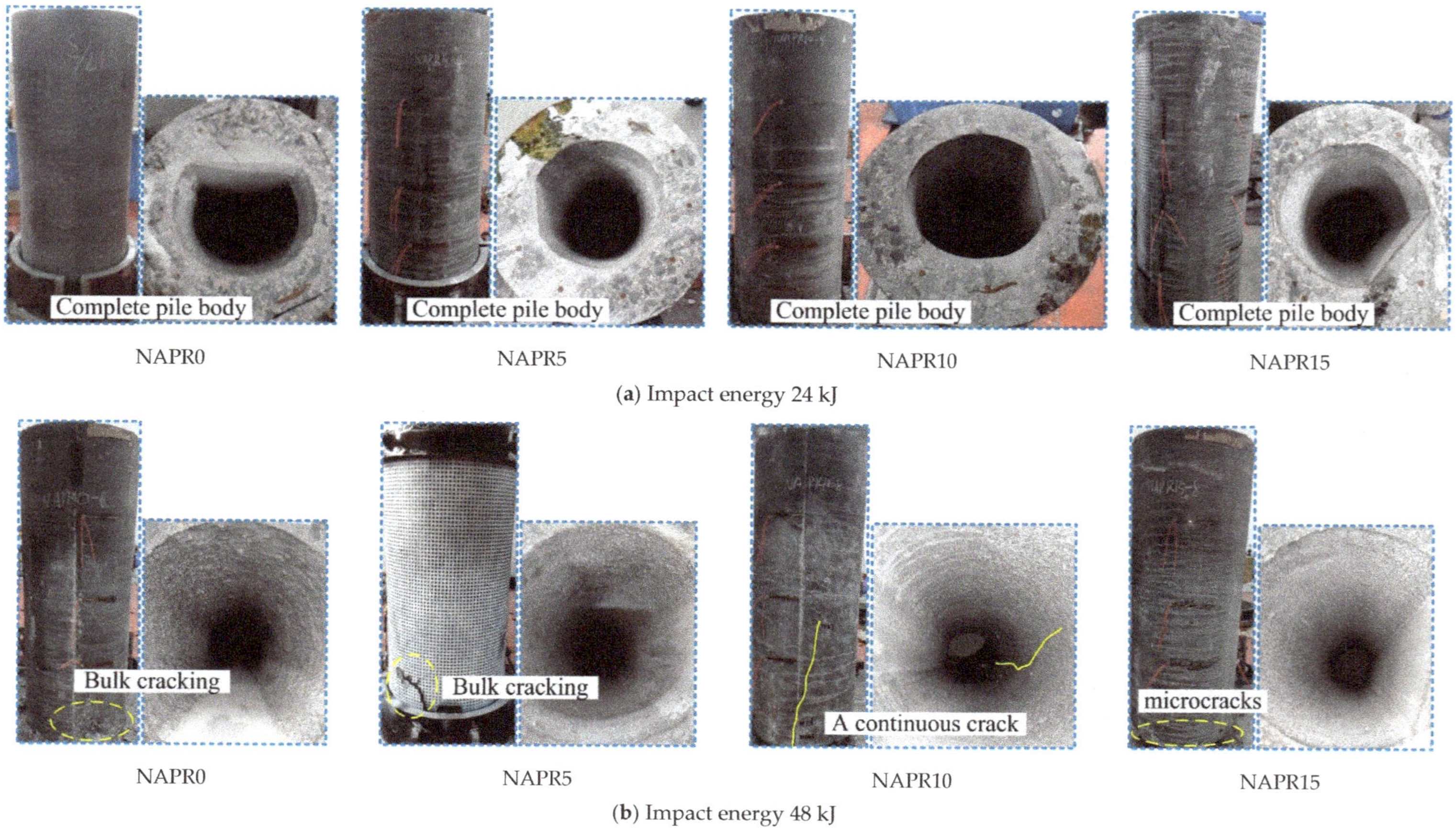

Figure 18. *Cont.*

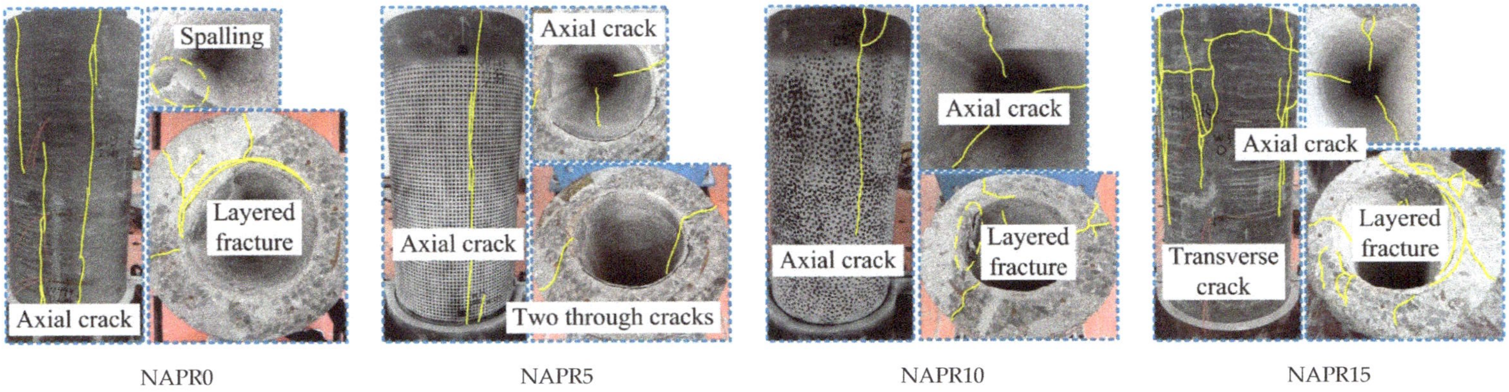

(**c**) Impact energy 60 kJ

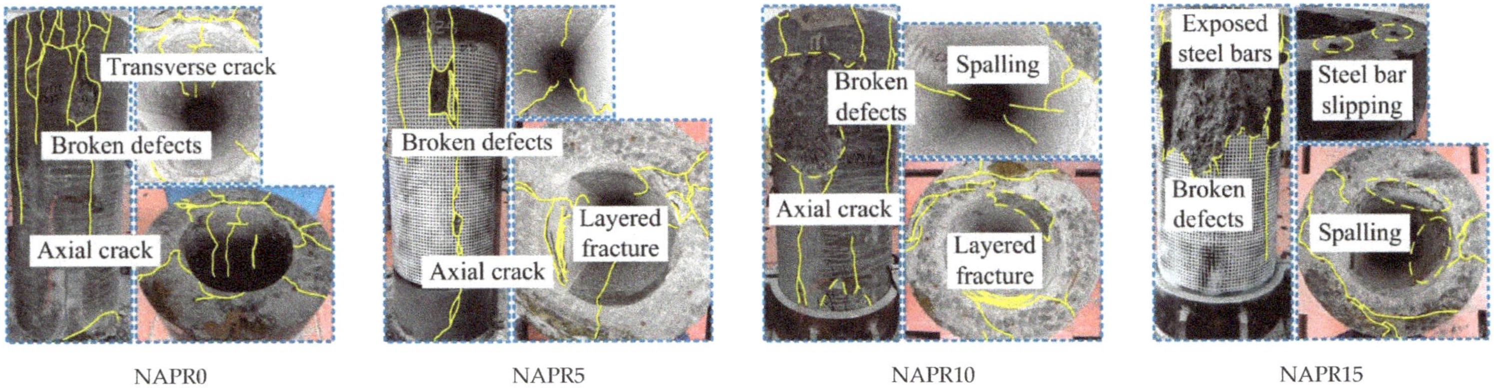

(**d**) Impact energy 72 kJ

Figure 18. Failure modes of NAPR under different impact energies.

4.3.2. Damage Levels

The damage degrees of NAPR under different impact energies were divided into four levels: no damage, mild damage, moderate damage, and severe damage. Only local damage or minor cracks constituted mild damage. The appearance of several through cracks constituted moderate damage. The appearance of a crack network, and even the occurrence of severe localized damage resulting in loss of working performance, constituted severe damage. Combining the damage degree, an analysis of the momentum and load results of NAPR yielded the corresponding damage level division area. In explosions and impact loads, the steel reinforcement ratio and concrete material were the key factors determining the damage of concrete components [34], and except for the differences in the rubber and sand contents, the four types of NAPR, including morphology, reinforcement ratio, cement matrix, etc., were all the same. Therefore, the division function of different damage levels could be approximately considered the same. Taking the division function of the severe damage area of the NAPR with 15% rubber content as the benchmark function for damage level, the divisions of the damage level areas of NAPR with different rubber contents were performed, and the benchmark function is shown in Figure 19d.

As shown in Figure 19, when the NAPR was subjected to impact from a drop hammer, the damage level corresponded to the allocated damage grade region. Without adding rubber, the NAPR would experience at least mild damage when the impact exceeded 8.9 kN·s and the load exceeded 2606.7 kN. When the impact exceeded 16.1 kN·s and the load exceeded 4390.8 kN, the pile would experience at least moderate damage. Severe damage would occur when the impact exceeded 22.0 kN·s and the load exceeded 5998.6 kN. With a rubber content of 5%, the NAPR would experience at least mild damage when the impact exceeded 10.1 kN·s and the load exceeded 2184.9 kN. Moderate damage would occur when the impact exceeded 20.5 kN·s and the load exceeded 5309.6 kN. Severe damage would occur when the impact exceeded 25.6 kN·s and the load exceeded 6256.9 kN. With a rubber content of 10%, the NAPR would experience at least mild damage when the impact exceeded 12.1 kN·s and the load exceeded 3611.7 kN. Moderate damage would occur when the impact exceeded 18.7 kN·s and the load exceeded 5561.2 kN. Severe damage would occur when the impact exceeded 24.8 kN·s and the load exceeded 7010.0 kN. With the rubber content of 15%, the NAPR would experience at least mild damage when the impact exceeded 8.3 kN·s and the load exceeded 3007.0 kN. Moderate damage would occur when the impact exceeded 14.7 kN·s and the load exceeded 4400.0 kN. Severe damage would occur when the impact exceeded 20.0 kN·s and the load exceeded 5606.7 kN. It could be observed that when the rubber content was 15%, the NAPR required the minimum impact and load to form mild, moderate, and severe damage. When the rubber content was 10%, the NAPR required the highest impulse and impact load to form mild damage, indicating its strong resistance to damage.

To further analyze the influence of rubber on the impact damage of NAPR, the areas of each damage grade region of various rubber contents at impulse of 308.3 kN.s and load of 8000.0 kN were compared. As shown in Figure 20, it could be observed that the zone of no damage was the largest for the NAPR with a rubber content of 10%. This indicated that the 10% rubber content provided better resistance to damage and optimal impact resistance performance. The zone of mild damage was the largest for the NAPR with a rubber content of 5%, indicating that this content was more prone to mild damage. The zones of mild, moderate, and severe damage were all larger for the NAPR with a rubber content of 15%, especially for severe damage, which had the highest proportion. This indicated that the NAPR with a 15% rubber content were more susceptible to severe damage. This was because the excessive rubber content significantly reduced the strength of the NAPR concrete itself and the bond strength with the steel reinforcement.

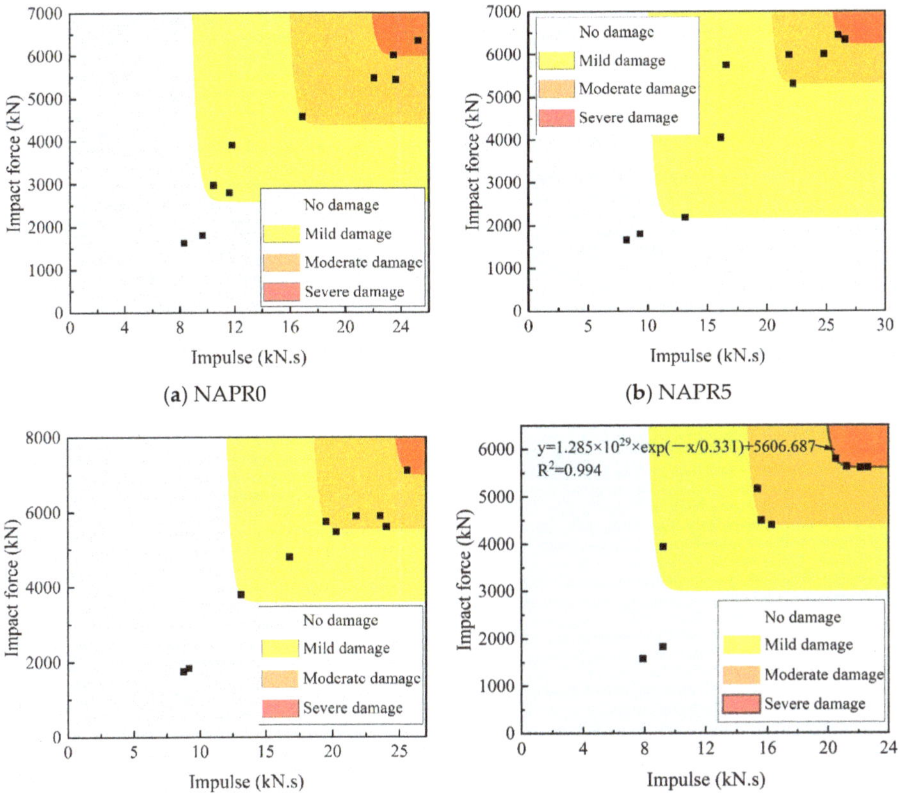

(a) NAPR0 (b) NAPR5

(c) NAPR10 (d) NAPR15

Figure 19. Classification of impact damage levels for NAPR.

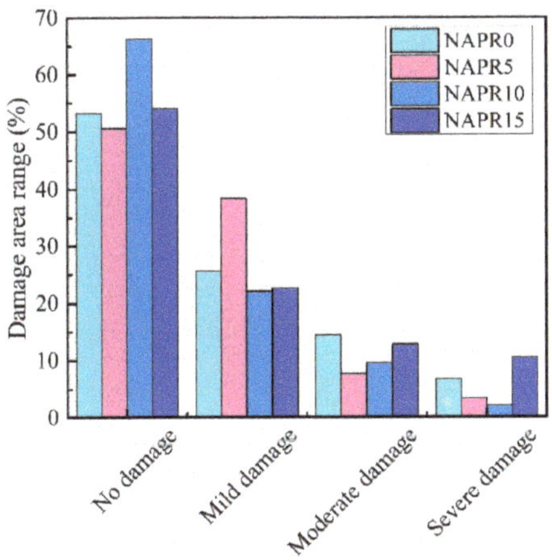

Figure 20. Damage evaluation of NAPR with different rubber contents.

5. Conclusions

Based on the results obtained from the axial drop hammer impact tests conducted on NAPR, the following conclusions can be drawn.

(1) When subjected to low impact energy (less than 48 kJ) impacts, the impact load of NAPR was minimally affected by the rubber content. When subjected to high impact energy (greater than 48 kJ) impacts, the impact load of NAPR increased initially and then decreased with the increasing rubber content, with the maximum impact load occurring at a 5% content level.

(2) When the impact energy was less than 48 kJ, the deformation energies produced by NAPR with various rubber contents during the drop hammer impact process were similar. However, as the impact energy increased, the deformation energies of NAPR with 5% and 10% rubber contents increased significantly.

(3) When no rubber particles were added, the NAPR exhibited a pronounced trend of expansion deformation under axial impact. With the increase in rubber content, the axial deformation of NAPR increased. However, at higher impact energies (72 kJ), the 5% and 15% rubber content piles also showed trends of expansion deformation, whereas only the NAPR with 10% rubber content exhibited a smaller lateral deformation coefficient, making them less susceptible to instability failure.

(4) When the impact energy exceeded 48 kJ, NAPR exhibited damage, which is categorized into four levels: no damage, mild damage, moderate damage, and severe damage. Among these, the piles with 10% rubber content had the largest undamaged area, while those with 5% rubber content were most susceptible to damage, and those with 15% rubber content were even more prone to severe damage.

(5) Based on the deformation, deformation energy, and damage degree of NAPR in the axial drop hammer tests, it can be found that the pipe pile had the best deformation and energy absorption effects when the rubber content was 10%, and it was also the most difficult to damage. It exhibited excellent impact resistance. The best rubber content, ultimate bearing capacity, and damage level evaluation obtained in the tests can provide an important basis for subsequent on-site NAPR impact tests and durability tests.

Author Contributions: Writing—original draft, S.L.; Writing—review & editing, F.L., F.Y., W.F. and D.C. All authors have read and agreed to the published version of the manuscript.

Funding: This research was funded by [the National Natural Science Foundation of China] grant number [12072080], [the National Natural Science Foundation of China] grant number [12072079], [the Science and Technology Planning Project of Guangzhou City] grant number [202002030120], [Special Foundation for Scientific and Technological Innovation Strategy of Guangdong Province] grant number [pdjh2022a0147], [Guangdong Basic and Applied Basic Research Foundation] grant number [2022A1515010008], [Guangdong Basic and Applied Basic Research Foundation] grant number [2023A1515010870].

Data Availability Statement: The data presented in this study are available in the article.

Acknowledgments: The authors gratefully acknowledge the financial support provided by the National Natural Science Foundation of China under Grant Nos. 12072080 and 12072079, the Science and Technology Planning Project of Guangzhou City under Grant No. 202002030120 (in China), Special Foundation for Scientific and Technological Innovation Strategy of Guangdong Province under Grant No. pdjh2022a0147 (in China), and Guangdong Basic and Applied Basic Research Foundation under Grant Nos. 2022A1515010008 (in China) and 2023A1515010870 (in China).

Conflicts of Interest: The authors declare no conflict of interest.

References

1. Qi, Z.; Liu, F.; Deng, M.; Hu, J. Research on cleaner production potential of pipe pile industry based on material energy flow analysis. In *IOP Conference Series: Earth and Environmental Science*; IOP Publishing: Bristol, UK, 2021; Volume 621, p. 012161.
2. Wu, Z.; Huang, H.; Chen, X.; Li, J.; He, Q.; Li, A.; Huang, J.; Lin, Y.; Liu, X.; Wang, J. Countermeasures for low-carbon transformation of construction industry in China toward the carbon peaking and carbon neutrality goals. *Strateg. Study CAE* **2023**, *25*, 202–209. [CrossRef]

3. Sonibare, J.A.; Akeredolu, F.A. A theoretical prediction of non-methane gaseous emissions from natural gas combustion. *Energy Policy* **2004**, *32*, 1653–1665. [CrossRef]
4. *T/CBMF 27-2018*; Methods and Requirements of Low-Carbon Products Evaluation for Ready-Mixed Concrete. China Building Materials Federation: Beijing, China, 2018.
5. Li, G.; Chen, X.; Zhang, Y.; Zhuang, Z.; Lv, Y. Studies of nano-SiO_2 and subsequent water curing on enhancing the frost resistance of autoclaved PHC pipe pile concrete. *J. Build. Eng.* **2023**, *69*, 106209. [CrossRef]
6. Hu, Y.; Ma, L.; He, T. Effect of mineral admixtures on the sulfate resistance of high-strength piles mortar. *Materials* **2020**, *13*, 3500. [CrossRef] [PubMed]
7. Tan, K.; Zhu, J. Influences of steam and autoclave curing on the strength and chloride permeability of high strength concrete. *Mater. Struct.* **2017**, *50*, 56. [CrossRef]
8. Zhan, Y.; Yang, Y.; Wu, X.; Xie, Y.; Luo, J.; Guo, W.; Hao, Y.; Wang, H. Experimental study on non-autoclaved PHC pile concrete applying polycarboxylate superplasticizer. *Key Eng. Mater.* **2015**, *629–630*, 345–350. [CrossRef]
9. Lan, S.; Liu, F.; Yang, F.; Li, H.; Chen, D.; Xu, K.; Zhang, H.; Kuang, J.; Fang, Z.; Feng, W. Dynamic compressive and splitting tensile characteristics of rubber-modified non-autoclaved concrete pipe piles. *J. Build. Eng.* **2023**, *69*, 106292. [CrossRef]
10. Liu, F.; Feng, W.; Xiong, Z.; Li, L.; Yang, Y.; Lin, H.; Shen, Y. Impact performance of new prestressed high-performance concrete pipe piles manufactured with an environmentally friendly technique. *J. Clean. Prod.* **2019**, *231*, 683–697. [CrossRef]
11. Thanaraj, D.P.; Anand, N.; Arulraj, P. Experimental investigation of mechanical properties and physical characteristics of concrete under standard fire exposure. *J. Eng. Des. Technol.* **2019**, *17*, 878–903. [CrossRef]
12. Zhu, H.; Xiong, Z.; Song, Y.; Zhou, K.; Su, Y. Effect of expansion agent and glass fiber on the dynamic splitting tensile properties of seawater–sea-sand concrete. *Buildings* **2024**, *14*, 217. [CrossRef]
13. Zhen, H.; Xiong, Z.; Song, Y.; Li, L.; Qiu, Y.; Zou, X.; Chen, B.; Chen, D.; Liu, F.; Ji, Y. Early mechanical performance of glass fibre-reinforced manufactured sand concrete. *J. Build. Eng.* **2024**, *83*, 108440. [CrossRef]
14. Fang, Z.; Wu, J.; Zhao, G.; Fang, S.; Ma, Y.; Jiang, H. Shear performance and design recommendations of single embedded nut bolted shear connectors in prefabricated steel–UHPC composite beams. *Steel Compos. Struct.* **2024**, *50*, 319–336.
15. Ding, Y.; Zhang, J.; Chen, X.; Wang, X.; Jia, Y.; Yan, P. Experimental study on dynamic parameter characteristics of new subgrade filler composed of granular rubber-sand mixtures. *Adv. Eng. Sci.* **2020**, *52*, 170–177.
16. Thomas, B.S.; Gupta, R.C. Properties of high strength concrete containing scrap tire rubber. *J. Clean. Prod.* **2016**, *113*, 86–92. [CrossRef]
17. Mendis, A.S.M.; Al-Deen, S.; Ashraf, M. Effect of rubber particles on the flexural behaviour of reinforced crumbed rubber concrete beams. *Constr. Build. Mater.* **2017**, *154*, 644–657. [CrossRef]
18. Bompa, D.V.; Elghazouli, A.Y. Creep properties of recycled tyre rubber concrete. *Constr. Build. Mater.* **2019**, *209*, 126–134. [CrossRef]
19. Chan, C.; Yu, T.; Zhang, S.; Xu, Q. Compressive behaviour of FRP-confined rubber concrete. *Constr. Build. Mater.* **2019**, *211*, 416–426. [CrossRef]
20. Eltayeb, E.; Ma, X.; Zhuge, Y.; Youssf, O.; Mills, J.E. Influence of rubber particles on the properties of foam concrete. *J. Build. Eng.* **2020**, *30*, 101217. [CrossRef]
21. Pan, Z.; Liu, F.; Li, H.; Li, X.; Wang, D.; Ling, Z.; Zhu, H.; Zhu, Y. Performance evaluation of thermal insulation rubberized mortar modified by fly ash and glass fiber. *Buildings* **2024**, *14*, 221. [CrossRef]
22. Klimesch, D.S.; Ray, A.; Sloane, B. Autoclaved cement-quartz pastes: The effects on chemical and physical properties when using ground quartz with different surface areas Part I: Quartz of wide particle size distribution. *Cem. Concr. Res.* **1996**, *26*, 1399–1408. [CrossRef]
23. *GB/T 13476-2009*; Pretensioned Spun Concrete Piles. General Administration of Quality Supervision, Inspection and Quarantine of the People's Republic of China and Standardization Administration of the People's Republic of China: Beijing, China, 2009.
24. *JGJ/T 406-2017*; Technical Standard for Prestressed Concrete Pipe Pile. Ministry of Housing and Urban-Rural Development of the People's Republic of China: Beijing, China, 2017.
25. Michele, G.; Mathias, F.; Williams, N.P. Low-velocity out-of-plane impact tests on double-wythe unreinforced brick masonry walls instrumented with optical measurements. *Int. J. Impact Eng.* **2023**, *178*, 104597.
26. *JGJ 94*; Technical Code for Building Pile Foundations. Ministry of Construction of the People's Republic of China: Beijing, China, 1994.
27. Fellenius, H.B.; Altaee, A.; Ismael, F.N. Analysis of load tests on piles driven through calcareous desert sands. *J. Geotech. Geoenviron. Eng.* **2001**, *127*, 200–201. [CrossRef]
28. Altaee, A.; Fellenius, B.H.; Evgin, E. Axial load transfer for piles in sand. I. Tests on an instrumented precast pile. *Can. Geotech. J.* **1992**, *29*, 11–20. [CrossRef]
29. Fellenius, B.H.; Samson, L. Testing of drivability of concrete piles and disturbance to sensitive clay. *Can. Geotech. J.* **1976**, *13*, 139–160. [CrossRef]
30. Martin, R.E.; Seli, J.J.; Powell, G.W.; Bertoulin, M. Concrete pile design in tidewater Virginia. *J. Geotech. Eng.* **1987**, *113*, 568–585. [CrossRef]
31. Zhao, C.; Lei, M.; Jia, C.; Yang, Z.; Shi, Y. An elastoplastic damage model for concrete considering the influence of mesostructure on transverse deformation. *Constr. Build. Mater.* **2023**, *406*, 133458. [CrossRef]

32. Xiao, C.; Gao, S.; Han, J.; Ding, L. Lateral facing deformation versus factor of safety for geosynthetic-reinforced soil walls in a tiered configuration. *Transp. Geotech.* **2023**, *41*, 101018. [CrossRef]
33. De Wilder, K.; Lava, P.; Debruyne, D.; Wang, Y.; de Roeck, G.; Vandewalle, L. Experimental investigation on the shear capacity of prestressed concrete beams using digital image correlation. *Eng. Struct.* **2015**, *82*, 82–92. [CrossRef]
34. Yan, Q.S. Damage assessment of subway station columns subjected to blast loadings. *Int. J. Struct. Stab. Dyn.* **2018**, *18*, 1850034. [CrossRef]

buildings

MDPI

Article

Performance Evaluation of Thermal Insulation Rubberized Mortar Modified by Fly Ash and Glass Fiber

Zezhou Pan [1], Feng Liu [1,*], Huawei Li [1], Xiaohui Li [1], Daochu Wang [2], Zao Ling [2], Huanyu Zhu [1] and Yuhao Zhu [1]

1 School of Civil and Transportation Engineering, Guangdong University of Technology, Guangzhou 510006, China; 3121003163@mail2.gdut.edu.cn (Z.P.); 1122309001@mail2.gdut.edu.cn (H.L.); lixh@mail2.gdut.edu.cn (X.L.); 3121002912@mail2.gdut.edu.cn (H.Z.); zhuyh202009@163.com (Y.Z.)

2 Guangzhou Construction Engineering Co., Ltd., Guangzhou 510180, China; wangdc202312@163.com (D.W.); lingz202106@163.com (Z.L.)

* Correspondence: fliu@gdut.edu.cn

Abstract: The utilization of waste rubber as a viable option for manufacturing building materials holds great significance for the sustainable development of the construction industry. This study explores the addition of two additives, fly ash (FA) and glass fiber (GF), to rubberized mortar in order to improve its performance. The impact of different waste rubber powder (RP) replacement rates and modified additive dosages on the performance of rubberized mortar, including fluidity, mechanical properties, drying shrinkage, impact resistance, and thermal insulation properties, was investigated. Furthermore, the analytic hierarchy process (AHP) was adopted to study the priorities of the rubberized mortar modified by FA and GF. The results indicate that the addition of RP leads to a decrease in mortar fluidity, mechanical properties, and drying shrinkage. However, it can enhance its impact resistance and thermal insulation properties. The additives, FA and GF, have a significant influence on the properties of rubberized mortar. By means of AHP method analysis, this study concludes that the optimal comprehensive properties of FA- and GF-modified rubberized mortar can be achieved by replacing 10% of sand with RP and using 10% FA and 0.4% GF. This study presents a configuration method for modified thermal insulation rubberized mortar, and it may lead to FA and GF being considered potential candidates for developing environmentally friendly building materials.

Keywords: rubberized mortar; waste rubber powder; fly ash; glass fiber; analytic hierarchy process; environmentally friendly building material

Citation: Pan, Z.; Liu, F.; Li, H.; Li, X.; Wang, D.; Ling, Z.; Zhu, H.; Zhu, Y. Performance Evaluation of Thermal Insulation Rubberized Mortar Modified by Fly Ash and Glass Fiber. *Buildings* **2024**, *14*, 221. https://doi.org/10.3390/buildings14010221

Academic Editor: Grzegorz Ludwik Golewski

Received: 6 December 2023
Revised: 9 January 2024
Accepted: 12 January 2024
Published: 14 January 2024

1. Introduction

The rapid growth of the automobile industry has resulted in a higher demand for tires [1,2]. Especially for emerging countries, growing income levels are driving the expansion of vehicle usage. As a consequence, the demand for tires is also on the rise [3]. It is estimated that approximately 1 billion tires are discarded each year, with a recycling rate of less than 50% [4]. At present, the options for disposing waste tires are limited, and one of the most commonly used methods is landfilling [4,5]. However, this approach requires a significant amount of land and poses a threat to groundwater resources in the landfill area. Moreover, the accumulation of discarded tires creates a breeding ground for mosquitoes, leading to potential health risks. Furthermore, discarded tires are highly susceptible to fires, which are notoriously difficult and costly to extinguish [6]. Thus, legislation has been introduced in many countries to restrict the disposal of waste tires in landfills and promote their reuse in other applications [3].

The construction industry is currently grappling with a shortage of natural resources, such as natural river sand, thus highlighting the need to prioritize sustainability [7,8]. One promising solution is the incorporation of waste tire rubber in the manufacturing of mortar and concrete, providing an environmentally friendly and sustainable approach to waste disposal [9,10]. In recent years, numerous researchers have explored the potential of utilizing waste tire rubber as a partial replacement for natural fine aggregate in mortar [11,12]. They have discovered that waste tire rubber can be processed into rubber powder (RP) through mechanical grinding, which can then be used in the preparation of rubberized cement-based materials. However, recent studies on rubberized mortar have highlighted a possible decrease in mechanical properties, such as compressive and flexural strengths, due to the low elastic modulus of RP and the weak bonding interface between RP and cement paste [13,14]. Despite these limitations, the elastic nature of RP offers advantages in terms of energy absorption, toughness, deformation ability, high damping, and anti-fatigue ability. As a result, it helps to minimize the brittle failure behavior typically observed in traditional mortar and concrete [15,16].

In the sustainable design of buildings, it is crucial to consider energy consumption [17]. Energy conservation in buildings encompasses various aspects such as energy consumption for building heating, air conditioning, lighting, and more. The main objective is to enhance energy efficiency while ensuring the comfort of people's lives. One way to achieve this is by improving the thermal insulation properties of building materials, which helps minimize energy dissipation. Relevant research suggests that rubber has a notably lower thermal conductivity compared to cement mortar [18,19]. This characteristic makes rubber a promising option for thermal insulation mortar. Mostofinejad et al. [20] conducted a thermal characteristics test on rubber aggregate combined with silica fume concrete. Their findings indicate that rubber particles can be utilized for thermal insulation, and fine rubber particles have a greater impact on reducing the thermal conductivity of concrete compared to coarse rubber particles. Wang et al. [21] prepared recycled aggregate crumb rubber concrete (RCC) using recycled aggregate and rubber as raw materials. The results showed that RCC had excellent thermal performance. The thermal resistance increased significantly when the replacement rate of rubber material to replace sand reached 20%. Therefore, the application of waste rubber to the field of low-energy-consumption buildings holds great promise.

Although the addition of RP can enhance certain properties of mortar or concrete and provide environmental benefits, its practical engineering application still faces limitations. This is primarily due to the lack of understanding regarding the "functional design" of such cement-based materials and the need for a comprehensive evaluation of multiple properties. For instance, the workability of rubberized cement-based materials must be taken into consideration to ensure ease of construction [22]. In addition to different environmental humidity levels, the drying shrinkage of building materials should also be taken into account. In some use environments, building materials may not necessarily require high mechanical properties to meet engineering requirements. Additionally, in specific structural engineering applications, building materials may need to withstand vibration, explosion, and impact loads. As a result, the deformation and impact resistance of building materials should also be taken into consideration [23,24]. Therefore, it is important to consider and evaluate various properties of building materials in order to select those with excellent overall performances for specific engineering environments.

To enhance specific properties of rubberized cement-based materials, incorporating additives is a direct and straightforward approach [25–27]. Fly ash (FA), a common industrial by-product, possesses smooth morphological characteristics that facilitate a beneficial "ball-bearing effect". This effect effectively enhances the interfacial behavior between the slurry and aggregate [28,29]. Additionally, FA contributes to the mechanical properties of cement-based materials. Shao et al. [30] conducted a study using FA and crumb rubber to prepare ultra-high-performance concrete (UHPC), and their results demonstrated the significant impact of FA on the flexural properties of UHPC. Similar observations were found in studies conducted by Zhao et al. and Shao et al. [31,32]. Furthermore, these researchers suggested that FA also aids in improving the ductility of cement-based materials.

The addition of fiber materials is considered an effective method to enhance the mechanical properties and deformation resistance of rubberized cement-based materials [33,34]. Yang et al. [35] conducted an experimental study on the compression–shear performance of steel fiber-reinforced rubber concrete. The results demonstrated that the combination of steel fiber and rubber exhibits a good synergistic effect, leading to the improved shear deformation toughness performance of rubber concrete. Xiong et al. [36] used a carbon fiber-reinforced polymer to prepare rubber concrete. The results show that the compressive strength of hybrid composites increases slightly, while the ductility, flexural toughness, impact resistance, and energy absorption capacity increase significantly. Su et al. [37] conducted a study on the preparation of basalt–polypropylene fiber-reinforced rubber concrete. The findings demonstrated that incorporating basalt and polypropylene fibers into rubber concrete leads to enhanced performance, particularly at higher temperatures. Glass fiber (GF) is an inorganic non-metallic material known for its insulation properties, heat resistance, tensile strength, and decay resistance [38,39]. These characteristics have made GF widely utilized in the field of constructional engineering. However, there is limited research available on the performance of glass-fiber-modified rubberized cement-based materials. Therefore, further research in this area is necessary to advance our understanding.

The objective of this work is to develop a rubberized mortar with superior comprehensive properties, including workability, mechanical properties, drying shrinkage, impact resistance, and thermal insulation properties. To achieve this, two additives, namely FA and GF, were incorporated into the rubberized mortar. This research examined the influence of RP replacement rate and FA and GF content on various specific properties of the mortar and investigated the underlying mechanisms. A representative test group was selected for evaluation of the thermal insulation properties. Additionally, the analytic hierarchy process (AHP) was utilized to determine the most suitable mix ratio for the representative test group. The findings of this study propose a configuration method for rubberized mortar with exceptional comprehensive properties, and these results are expected to contribute to the development of a new environmentally friendly building material for actual engineering projects.

2. Materials and Methods

2.1. Materials

Portland slag cement, according to the Chinese standard GB 1334-1999 [40], P.S.A 32.5, was used in this study, and its chemical composition and basic physical properties are displayed in Tables 1 and 2, respectively. The density of P.S.A 32.5 was measured according to the Chinese standard GB/T 208-2014 [41], and the setting time and flexural and compressive strength were measured according to GB/T 175-1999 [42]. RP was obtained from waste tires (with a particle size of 20 mesh) by grinding, and its density measures at 988 g/cm^3. Alkali-resistant GF was produced by Taishan Glass Fiber Co., Ltd.(Taian, China), and its physical parameters were obtained from the measurement of the company, as shown in Table 3. First-class FA used in this study was supplied by Hengyuan New Materials Co., Ltd., Huangshan, China, with a density of 2.55 kg/m^3, and its chemical composition is also listed in Table 1. Photographs of raw materials, including cement, RP, GF, and FA, are displayed in Figure 1. Well-graded river sand, with an aggregate size of 0–5 mm and density of 2.61 kg/m^3, satisfied the requirements of Chinese standard GB/T 14684-2022 [43]. A polycarboxylate superplasticizer, complying with Chinese standard JGT 223-2017 [44], was used to control the workability of the mortar.

Table 1. Chemical composition of raw materials.

Material	By Weight/%							
	CaO	SiO_2	Al_2O_3	Fe_2O_3	MgO	SO_3	TiO_2	K_2O
Cement	25.02	55.50	15.42	5.02	1.25	1.01	/	/
FA	3.50	57.03	26.12	4.23	0.72	0.72	1.81	1.34

Table 2. Physical properties of 32.5 cement.

Density/g·cm^{-3}	Setting Time		Flexural Strength		Compressive Strength	
	Initial/min	Final/min	3 d/MPa	28 d/MPa	3 d/MPa	28 d/MPa
3.6	307	373	4.6	6.3	20.6	37.6

Table 3. Physical parameters of glass fiber.

Length/mm	Diameter/μm	Tensile Strength /MPa	Elastic Modulus/GPa	Density/g·cm^{-3}
6	14	1000	72	2.68

Figure 1. Photographs of raw materials: (a) cement; (b) RP; (c) GF; (d) FA.

2.2. Mixed Proportions and Specimens Preparation

In accordance with the proposed scheme outlined in the relevant literature [45,46], we conducted a preliminary experiment to determine the mixed proportions. To evaluate the additives (FA and GF) content on the properties of rubberized mortar, 36 mixes were designed using different volume fractions (0%, 10%, 20%, and 30%) of RP as a partial replacement of fine aggregate. The different dosages of FA (volume fractions of 0%, 5%, and 10% to replace cement) and GF (volume fractions of 0%, 0.2%, and 0.4% with extra incorporation) were also considered. The mixing water/binder and binder/sand weight ratios were set at 1:2 and 2:5, respectively. The dosage of polycarboxylate superplasticizer was kept at 1 wt.%. All the mixed proportions are summarized in Table 4.

Table 4. Design parameter of mixed proportions.

0% RP		10% RP		20% RP		30% RP	
GF/%	FA/%	GF/%	FA/%	GF/%	FA/%	GF/%	FA/%
0	0	0	0	0	0	0	0
0	5	0	5	0	5	0	5
0	10	0	10	0	10	0	10
0.2	0	0.2	0	0.2	0	0.2	0
0.2	5	0.2	5	0.2	5	0.2	5
0.2	10	0.2	10	0.2	10	0.2	10
0.4	0	0.4	0	0.4	0	0.4	0
0.4	5	0.4	5	0.4	5	0.4	5
0.4	10	0.4	10	0.4	10	0.4	10

Note: in FA5GF0.2RP20 specimen, "5", "0.2", and "20" refers to volume fractions of FA, GF, and RP in the mixtures.

To ensure the uniformity of the prepared mortar specimens, the following manufacturing process was adopted to prepare the rubberized mortar modified by FA and GF. Firstly, cement and FA were dry-mixed together in a rotary mixer for 1 min, and then water was added to the mixture and mixed slowly for another 1 min to obtain the mixed slurry. Subsequently, sand, RP, and GF were added in turn to the slurry, and they were stirred at high-speed for 2 min. Finally, after allowing the mortar to stand for 90 s, it was re-stirred at high speed for 1 min, and prepared fresh mortar was obtained. After the completion of the mortar stirring, the freshly prepared mortar specimens were cast in mold and cured for corresponding hydration ages in water with 95 ± 2% relative humidity until testing.

2.3. Fluidity

The fluidity of rubberized mortar was assessed using a mini slump test following the Chinese standard GB/T 2419-2005 [47]. In order to ensure the compactness of mortar, the fresh mortar was poured into the truncated cone model with specific dimensions (height: 60 ± 0.5 mm, upper diameter: 70 ± 0.5 mm, and lower diameter: 100 ± 0.5 mm) and compacted twice by continuous ramming. The diameter of the two vertical directions was measured using a ruler, and the average value was calculated, which represents the fluidity of the mortar.

2.4. Mechanical Properties

Three series of specimens with 40 × 40 × 160 mm size were prepared for each mixed proportions, and these specimens were used to determine their mechanical properties.

The flexural strength and compressive strength were conducted using a compression testing machine (YAW-300C) in accordance with the Chinese standard GB/T 17671-2020 [48]. For the flexural strength test, three specimens with hydration ages of 7 and 28 days were measured when the loading rate was 50 ± 10 N/s. After completing the flexural strength test, a total of six broken specimens were used for the compressive strength test. The loading rate for this test was 2400 ± 200 N/s. The flexural and compressive strength of each specimen were obtained by taking the average value of the test results.

2.5. Drying Shrinkage

According to the Chinese standard JGJ/T 70-2009 [49], the specimen size of the drying shrinkage test is 40 × 40 × 160 mm. The copper shrinkage heads were embedded at both ends of the specimen, and the copper nail was exposed to the end face of the specimen by 8 ± 1 mm, as shown in Figure 2. To ensure accurate shrinkage measurement, the mortar specimens were fully vibrated. Subsequently, the drying shrinkage strain was measured at hydration ages of 14, 28, 56, and 90 days. The value of drying shrinkage (S_i) of each specimen was obtained using Equation (1). To ensure the accuracy of the experimental data, the average value of three specimens was calculated to obtain the drying shrinkage strain.

$$S_i = \frac{L_0 - L_t}{140} \tag{1}$$

where L_0 represents the initial record of the micrometer, L_t represents the record of the micrometer on corresponding days, and 140 is the effective length of the specimens.

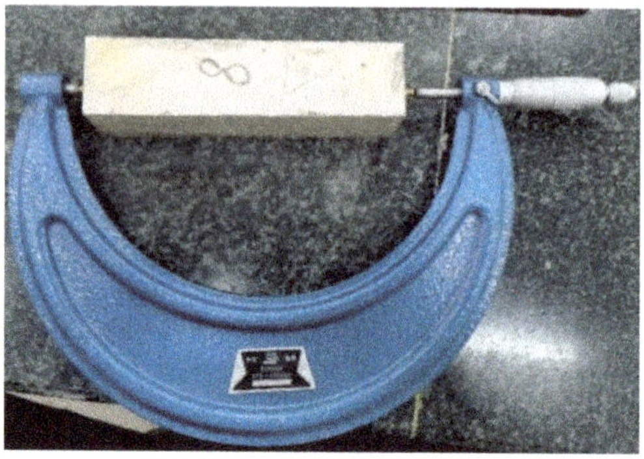

Figure 2. Test of the drying shrinkage.

2.6. Impact Resistance

The impact resistance test was conducted using a pendulum with a specified height. The non-notched sample in the cantilever beam state was struck once, and the test equipment is shown in Figure 3. In accordance with the Chinese standard GB/T 1843-2008 [50], the specimens for the impact resistance test had dimensions of 15 × 20 ×160 mm. The pendulum had an impact energy of 11 J, an angle of 50°, and a speed of 2.9 m/s. The results of the impact resistance test were determined using Equation (2):

$$\alpha_i = \frac{W}{h \times b} \quad (2)$$

where α_i is the impact strength of the specimen, W is the absorbing energy as the specimen is destroyed, and h and b represent the thickness and height of the specimen.

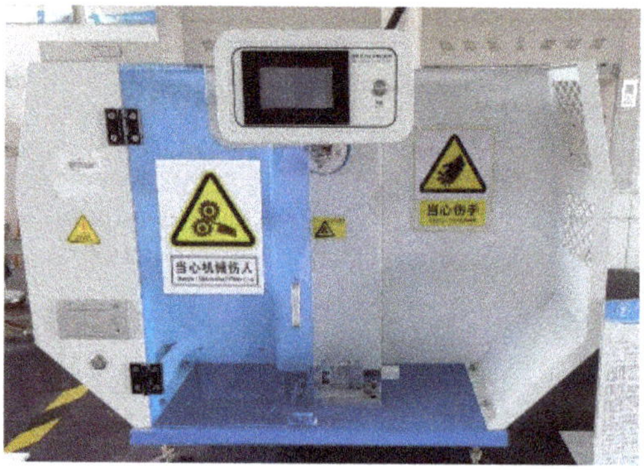

Figure 3. Cantilever beam tester.

2.7. Thermal Conductivity

The thermal conductivity of the specimens was measured by a thermal conductivity instrument (CD-DR3030), and this instrument adopts a microcomputer automatic control system, which can automatically control and complete data acquisition, as shown in Figure 4a. In accordance with the Chinese standards GB/T 6342-1996 [51] and GB/T 10294-2008 [52], the specimens, measuring 300 × 300 × 30 mm, were prepared with specific mixed proportions. Prior to the test, the specimens were dried at 105 ± 5 °C until a constant weight was reached to measure their thermal conductivity. The controlling parameters of the thermal conductivity instrument were as follows: the cold and hot plate's temperature were set at 20 and 40 °C, respectively. Finally, the thermal conductivity values were calculated using Equation (3):

$$\lambda = \frac{wd}{A(T_1 - T_2)} \quad (3)$$

where λ represents thermal conductivity, w is the power of the calorimeter plate, d and A are the average thickness and calculated area of the specimen, and T_1 and T_2 are the cold and hot plate's temperature.

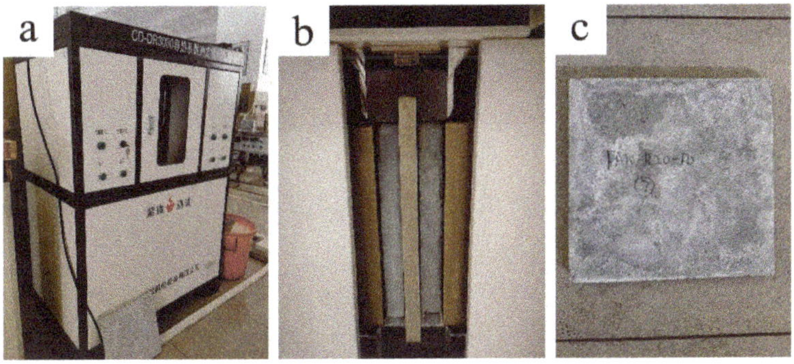

Figure 4. Characterization of thermal conductivity (**a**) measurement instrument; (**b**) thermal insulation test; (**c**) the prepared specimen.

2.8. Microstructure

Scanning electron microscopy (SEM) was conducted to observe the pore structure and interface of rubberized mortar. The specimens were soaked in anhydrous ethanol to terminate hydration and were dried at 60 °C for 2 h for testing. To obtain high-definition SEM images, the specimens were coated with gold to ensure their good electrical conductivity.

2.9. Analytic Hierarchy Process (AHP)

The AHP is a highly effective and commonly used method for analyzing problems related to multi-criterion decision making [53]. The AHP involves breaking down a complex problem into multiple factors and organizing them into a hierarchical structure based on their relationships. The overall ranking of the decision-making scheme is then determined by comparing these factors. The specific implementation process is as follows: (a) The relationship between the various factors in the system was analyzed, and the hierarchical structure of the system was established. (b) Pairwise comparisons were conducted on the nodes at each level, and the results were recorded as a_{ij} (note: *i* and *j* represent the relative importance of the node, and the importance values are shown in Table 5). Subsequently, the pairwise comparison matrix was constructed based on Equation (4). (c) Based on the pairwise comparison matrix, the relative weights of each factor were determined using the average values calculated through the arithmetic average method, geometric average method, and eigenvalue method, and the consistency test was performed using Equations (5) and (6). (d) The score was calculated based on the weight matrix, and the levels were ranked to select the best solution.

$$\begin{array}{c} \\ \mathrm{A_1\ A_2 \cdots\ A_n} \\ \\ \end{array} \begin{array}{c} \mathrm{A_1} \\ \mathrm{A_2} \\ \vdots \\ \mathrm{A_3} \end{array} \begin{bmatrix} a_{1,1} & a_{1,2} & \cdots & a_{1,n} \\ a_{2,1} & a_{2,2} & \cdots & a_{2,n} \\ \vdots & \vdots & & \vdots \\ a_{n,1} & a_{n,2} & \cdots & a_{n,n} \end{bmatrix} \begin{bmatrix} \omega_1 \\ \omega_2 \\ \vdots \\ \omega_n \end{bmatrix} = n \begin{bmatrix} \omega_1 \\ \omega_2 \\ \vdots \\ \omega_n \end{bmatrix} \quad (4)$$

$$\mathrm{CI} = \frac{\lambda_{\max} - n}{n - 1} \quad (5)$$

$$\mathrm{CR} = \frac{\mathrm{CI}}{\mathrm{RI}} < 0.1 \tag{6}$$

where A_n is the alternative, ω_n is the associated eigenvector, n is the order of matrix, CI is the consistency index, λ_{max} is the maximum eigen value of the matrix, and RI is the random consistency index.

Table 5. The importance value of step (b) in AHP.

Scales	Importance value of *i* and *j*
1	Equal importance
3	Slightly higher importance
5	Normal importance
7	Strong importance
9	Extreme importance
2, 4, 6, 8	Intermediate value of two adjacent scales

3. Results and Discussions

3.1. Fluidity

Fluidity is an important index reflecting the workability of mortar, which is mainly used to evaluate the flow ability of fresh mortar [54]. Figure 5 demonstrates the impact of varying FA and GF content on the fluidity of rubberized mortar. The figure indicates that the fluidity of fresh rubberized mortars falls within the range of 100–250 mm, satisfying the requirements of the Chinese standard GB/T 2419-2005 [47]. Additionally, the content of FA, GF, and RP has a significant impact on the fluidity of mortar. Consequently, the impact of these additives on fluidity was examined separately, as outlined in the following analysis.

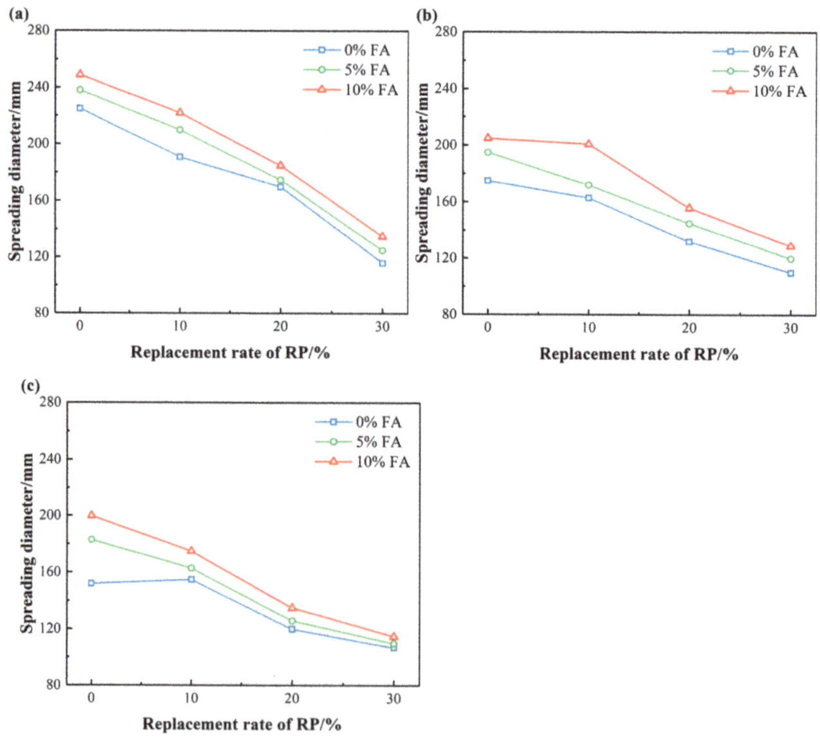

Figure 5. Fluidity of rubberized mortar modified by FA and GF: (**a**) 0% GF; (**b**) 0.2% GF; (**c**) 0.4% GF.

Figure 5a–c represent the change in fluidity of rubberized mortar with different RP contents (0%, 10%, 20%, 30%) under the influence of FA (0%, 5%, 10%) and GF (0%, 0.2%, 0.4%). As we can see from that figure, the addition of RP significantly reduces the fluidity of the mortar. In the specimens with only RP mixed, the fluidity of the mortar decreases from 225 mm to 116 mm as the RP content increases from 0% to 30%. This decrease in fluidity can be attributed to the high water absorption capacity of RP and its rough jagged-like edges. Similar findings have been reported in studies conducted by Abd-Elaty et al., Mostofinejad et al., and Zhai et al. [20,22,55]. The addition of GF also leads to a decrease in fluidity, as the surface of GF has a strong friction with paste, increasing the viscosity of the mixture [56]. Besides that, the increase in the GF dosage may increase the risk of agglomeration. However, the addition of FA improves the fluidity of mortar, particularly at higher levels of FA substitution. This is attributed to the smooth surface morphology of FA, which promotes a good "ball-bearing effect" on the contact surface between aggregate and paste [29,57]. Similar observations were observed in the research conducted by Shao et al. [30,32]. The fluidity of the rubberized mortar under the joint action of FA and GF has been altered to some extent. The addition of GF and RP to the mortar leads to a significant decrease in fluidity. For instance, the fluidity of the specimens with only 0.4% GF, increasing the percentage of RP from 0% to 20%, caused an increase in the fluidity value by 32 mm (21%). However, FA can effectively mitigate the negative impact of the decreased fluidity caused by GF and RP. For the specimens with 0.4% GF and 20% RP, increasing the percentage of FA from 0% to 10% caused an increase in the fluidity value by 15 mm (12%). This result indicates that increasing the amount of FA appropriately has a more pronounced effect on improving the fluidity of rubberized mortar modified by GF.

3.2. *Mechanical Properties*

3.2.1. Flexural Strength

Figure 6a–c present the results of the flexural strength test for rubberized mortar modified by FA and GF at 7 and 28 days. At 7 days, the flexural strength decreased as the RP replacement rate increased, although the decrease was somewhat mitigated. This decrease can be attributed to the inadequate adhesion between the RP and cement matrix, as well as the impact of RP as an aggregate on the overall stiffness. The flexural strength of rubberized mortar has almost been improved by the influence of GF. Only the flexural strength of specimen FA0GF0.2RP20 (3.7 MPa) did not show an improvement compared to specimen FA0GF0RP20 (3.7 MPa). This could be attributed to the incomplete bonding between GF and mortar at 7 days, as well as the relatively small GF content. Moreover, the flexural strength exhibited a more significant increase when the GF content reached 0.4%. The experimental data also indicated a similar change rule after the addition of FA. However, the overall flexural strength was slightly reduced due to the low early hydration activity of FA and its dilution effect [58,59].

At 28 days, the addition of RP resulted in a decrease in flexural strength. When the specimens were mixed with only RP, with replacement rates of 10%, 20%, and 30%, the flexural strength decreased by 22.7%, 41.8%, and 57.3%, respectively. This decline is higher compared to that at 7 days. Additionally, the inclusion of GF enhances the 28-day flexural strength of rubberized mortar, with a more pronounced effect observed in rubberized mortar with a higher RP replacement rate. For instance, the flexural strength of specimens with only 30% RP, increasing the dosage of GF from 0% to 0.2% and 0.4%, caused an increase in the flexural strength value by 0.7 MPa (14.9%) and 1.3 MPa (29.8%), respectively. This phenomenon can still be observed even after incorporating the FA additive, with an increase of 14.3% and 22.4% (with 5% FA) and 14.3 and 28.6% (with 10% FA), respectively.

We recorded the failure modes of the two specimens with specific mixed proportions and observed their crack development during the experimental process. The failure mode of pure cement mortar is shown in Figure 7a, where obvious brittle failure characteristics were observed during the failure process, and the specimen was disconnected from the middle of the specimen. On the other hand, Figure 7b illustrates the failure mode of rubberized mortar modified by GF. It can be observed that there were no obvious cracks, and crack development did not occur in the middle of the specimen, indicating typical ductile failure characteristics. Similar phenomena and conclusions have also been reported by Ma et al., Li et al., and Xu et al. [1,2,60].

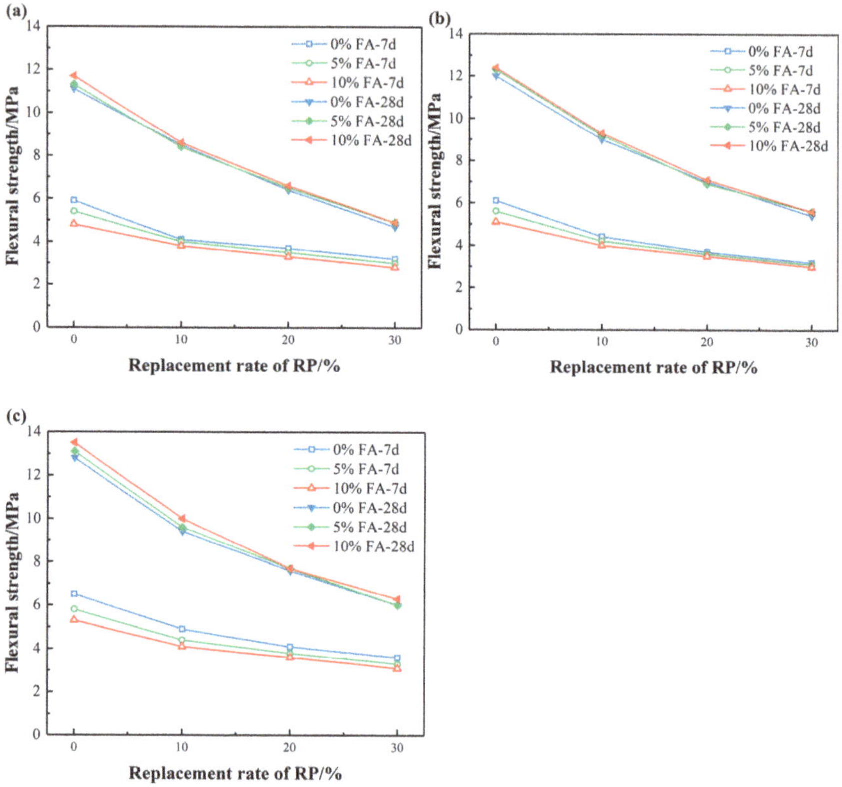

Figure 6. Flexural strength of rubberized mortar modified by FA and GF: (**a**) 0% GF; (**b**) 0.2% GF; (**c**) 0.4% GF.

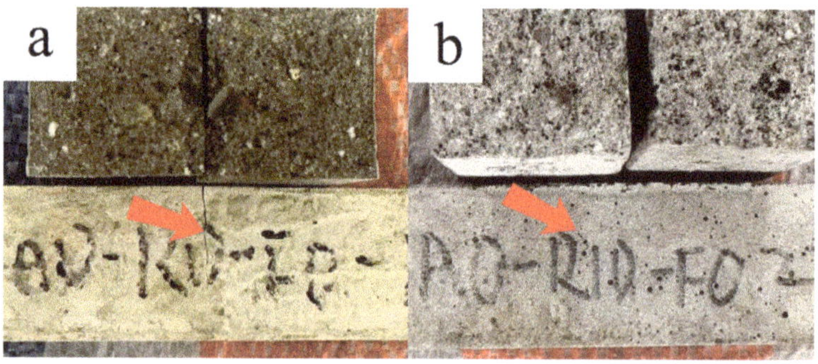

Figure 7. Failure mode of mortar: (**a**) FA0GF0RP0; (**b**) FA0GF0.2RP10.

3.2.2. Compressive Strength

The compressive strength trend of rubberized mortar modified by FA and GF is similar to the flexural strength, which is shown in Figure 8. At 7 days, when the replacement rate of RP to replace sand in the mixture increased from 0% to 30%, the compressive strength exhibited varying degrees of decline. Specifically, the rubberized mortar specimens with RP replacement rates of 10%, 20%, and 30% experienced an average reduction in compressive strength of 37.0%, 49.0%, and 65.6%, respectively, compared to the pure cement specimen. The incorporation of GF did not significantly improve the compressive strength of the sample compared to its effect on the flexural strength. In fact, in some specimens (FA5GF0.2R20 and FA5GF0.2R30), there was a decrease in compressive strength. This observation could be attributed to the relatively short curing time and the weak bonding between the fiber and the matrix. The incorporation of FA typically led to a decrease in the 7-day compressive strength of mortar. However, exceptions can be observed in certain specimens with a high RP content (FA5GF0R20 and FA5GF0R30), where a slight improvement of 4.0% and 5.9% is observed compared to the pure cement specimen. Based on these experimental findings, it can be inferred that there is a certain synergistic effect between FA and RP in enhancing compressive strength.

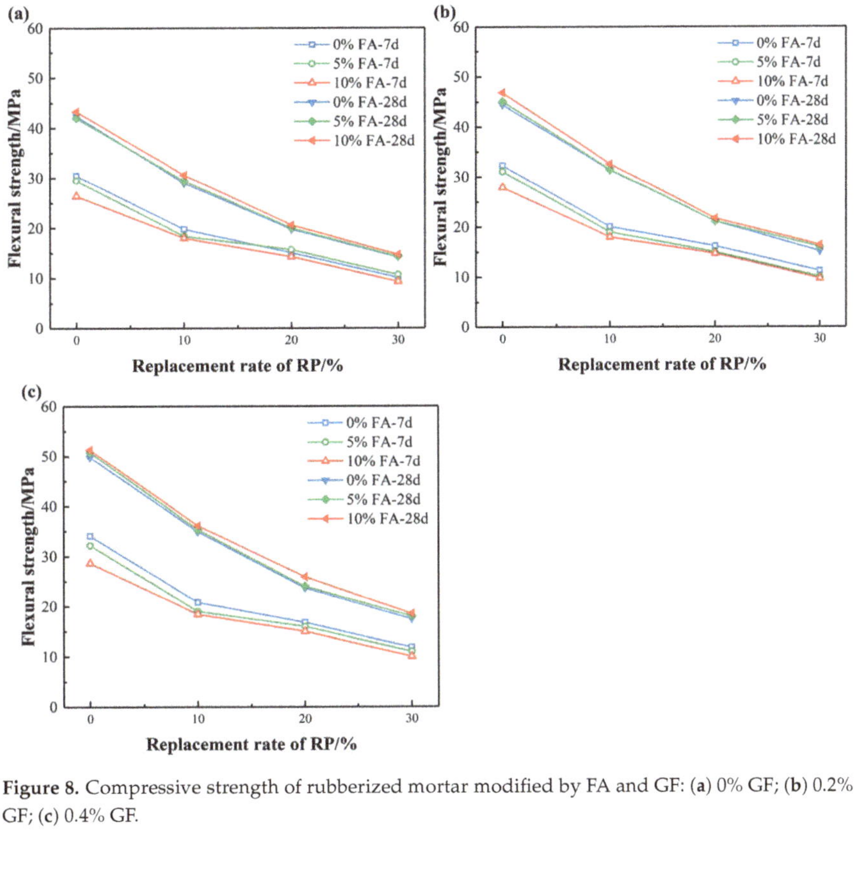

Figure 8. Compressive strength of rubberized mortar modified by FA and GF: (**a**) 0% GF; (**b**) 0.2% GF; (**c**) 0.4% GF.

At 28 days, the addition of RP also resulted in a decrease in compressive strength. When the RP replacement rate to replace sand in the mixture was 10%, 20%, and 30%, the compressive strength of the rubberized mortar was 31.6%, 53.3%, and 66.5% lower than that of the pure cement group, respectively. The decline in compressive strength gradually decreases as the RP replacement rate increases. In contrast to the 7-day results, GF has a significant positive effect on the 28-day compressive strength of mortar, and this effect becomes more pronounced with higher RP replacement rates. The addition of 0.2% GF and 10% RP to the specimens resulted in a slight increase in the 28-day compressive strength value. When the dosage of FA was increased from 0% to 5% and 10%, the 28-day compressive strength value increased by 8.3%, 7.1%, and 6.5%, respectively, compared to specimens without GF. Similarly, when the GF dosage increased to 0.4%, this ratio became 20.3%, 20.1%, and 18.0%. Additionally, the addition of 0.2% GF and 20% RP to the specimens also led to an improvement in the 28-day compressive strength. Specifically, when the dosage of FA increased from 0% to 5% and 10%, the 28-day compressive strength value increased by 6.6%, 5.5%, and 5.3%, respectively, compared to the specimens without GF. When the dosage of GF increased to 0.4%, the improvement rates changed to 19.7%, 20.0%, and 25.7%. These findings suggest that the addition of GF can effectively enhance the 28-day compressive strength of rubberized mortar, while the impact of FA is not significant.

In a previous study, Zhao et al. [31] discovered that FA decreased the strength of the rubber–cement matrix, consequently impacting the compressive strength. However, this phenomenon was not prominently observed in this study. Additionally, Li et al. and Qaidi et al. [61,62] suggested that the improved compressive strength resulting from the presence of fibers is primarily attributed to their ability to restrict crack development and provide lateral restraint. These findings align with the conclusions drawn in this study.

3.3. Drying Shrinkage

Drying shrinkage is an important phenomenon that assesses the deformation behavior in cementitious materials, which is closely related to their structural and durability properties [63]. Moreover, according to the relevant literature [64], drying shrinkage is directly influenced by the amount of free water in the system and the porosity of the cementitious materials.

Figure 9 displays the results of the drying shrinkage test for rubberized mortar modified by FA and GF at various hydration ages. As the hydration age increases, the drying shrinkage value consistently rises. However, the slope of the drying shrinkage curve for nearly all specimens noticeably decreases after 28 days. This can be attributed to the fact that the hydration process within the mortar is essentially finished within 28 days, causing the growth rate of dry shrinkage to decelerate.

The addition of RP has a varying effect on the drying shrinkage of mortar. When the RP replacement rate is low ($\leq$ 10%), as shown in Figure 9a,b, the dry shrinkage value of the mortar does not change significantly. The 28-day drying shrinkage values of specimens FA0GF0RP0 and FA0GF0RP10 were 442.9 and 450.0 $\mu\varepsilon$, respectively. When the RP replacement rate is large ($\geq$ 20%), the 28-day drying shrinkage has a significant increase, with specimen FA0GF0RP30 reaching 700.0 $\mu\varepsilon$. These phenomena may be mainly related to the porous structure of RP particles and high water absorption capacity. Under the condition of a large replacement rate, it is easy to cause the rapid growth of the dry shrinkage value of mortar. Similar conclusions have been drawn in the research conducted by Uygunoğlu et al. and Su et al. [65,66]. Additionally, Deng et al. [67] reported that the soft nature (high elasticity) characteristic of RP resulted in it being more prone to deformation behavior, which reduced the resistance to the drying shrinkage-induced strain of the matrix.

The addition of GF can effectively reduce the drying shrinkage of rubberized mortar at varying hydration ages. Especially, the drying shrinkage first decreased as the GF content increased to 0.2%, followed by a certain decrease as it further increased to 0.4%. However, this decrease is more significant when the RP replacement rate is higher ($\geq$20%), as displayed in Figure 9c,d. The main reason for this is that the binding force provided by GF effectively limits the deformation of the matrix during the evaporation of water, and the specimens with a larger RP replacement rate need more GF to provide sufficient stiffness for the matrix. In the study conducted by Chen et al. [68], they found that the addition of excessive fiber may enhance the pore connectivity inside the matrix, which has a negative impact on the drying shrinkage of the matrix. Consequently, this should be fully considered in the design of this type of material.

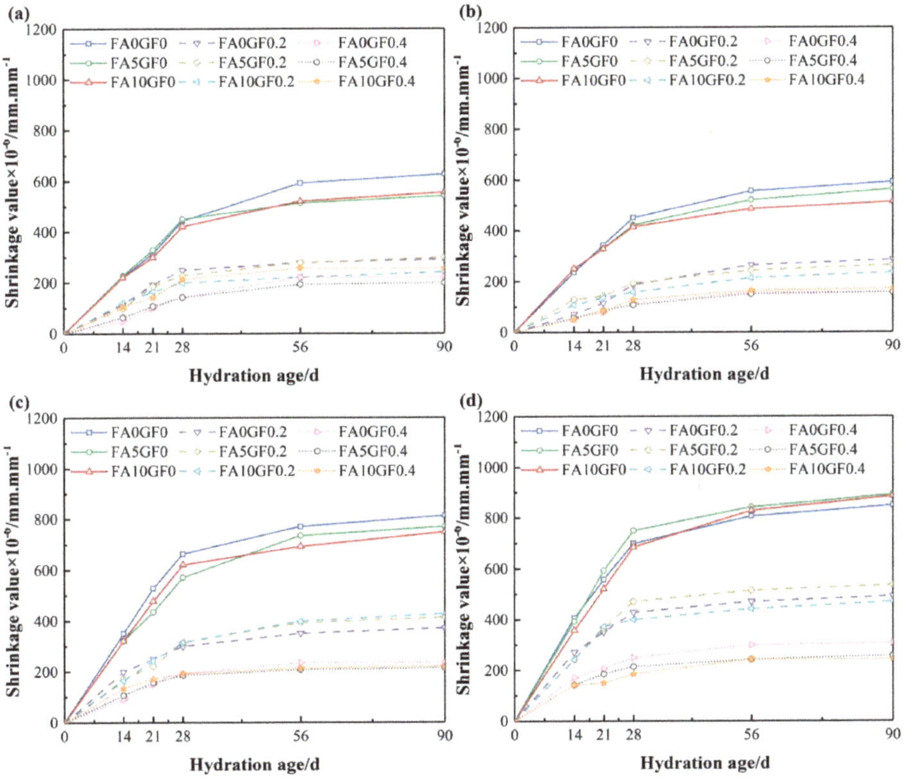

Figure 9. Drying shrinkage of rubberized mortar modified by FA and GF: (**a**) 0% RP; (**b**) 10% RP; (**c**) 20% RP; (**d**) 30% RP.

The effect of FA on the drying shrinkage of rubberized mortar is less than that of RP and GF. During the 28-day hydration age, the addition of FA generally reduces the drying shrinkage of rubberized mortar. However, after the hydration age exceeds 28 days, the specimens modified by FA generally have relatively high drying shrinkage values. In the relevant literature [69–71], the reasons for this phenomenon are summarized as follows: (1) The early hydration activity of FA is relatively low, resulting in low drying shrinkage in early hydration ages. (2) FA changes the original composition of the cementitious materials system, which in turn changes the overall pore structure. (3) The addition of FA effects the carbon resistance ability of the cement matrix; this may affect the porosity of the cementitious materials system.

3.4. Impact Resistance

Impact resistance is an important factor in evaluating toughness. Figure 10 presents the results of the impact resistance test conducted on rubberized mortar modified by FA and GF. It can be seen from the figure that when the RP replacement rate is below 20%, the impact resistance of the mortar only shows a slight improvement. When the RP replacement rate reaches 30% (as shown in Figure 10d), there is a significant enhancement in the impact resistance of the mortar. This enhancement can be attributed to the unique physical properties of RP, which enable the formation of elastic centers within the concrete and contribute to the improved toughness of the mortar. This observation aligns with previous research findings [72,73]. The addition of GF has a significant impact on improving the impact resistance of rubberized mortar. As the dosage of GF increases from 0% to 0.4%, the impact resistance of each mixture in the test group can be increased by 26–35%. This increase in impact resistance is due to the enhancement of mortar toughness, which helps to restrict the development of cracks in the mortar under impact load. It is worth noting that the FA content does not have a significant effect on the impact resistance of rubberized mortar.

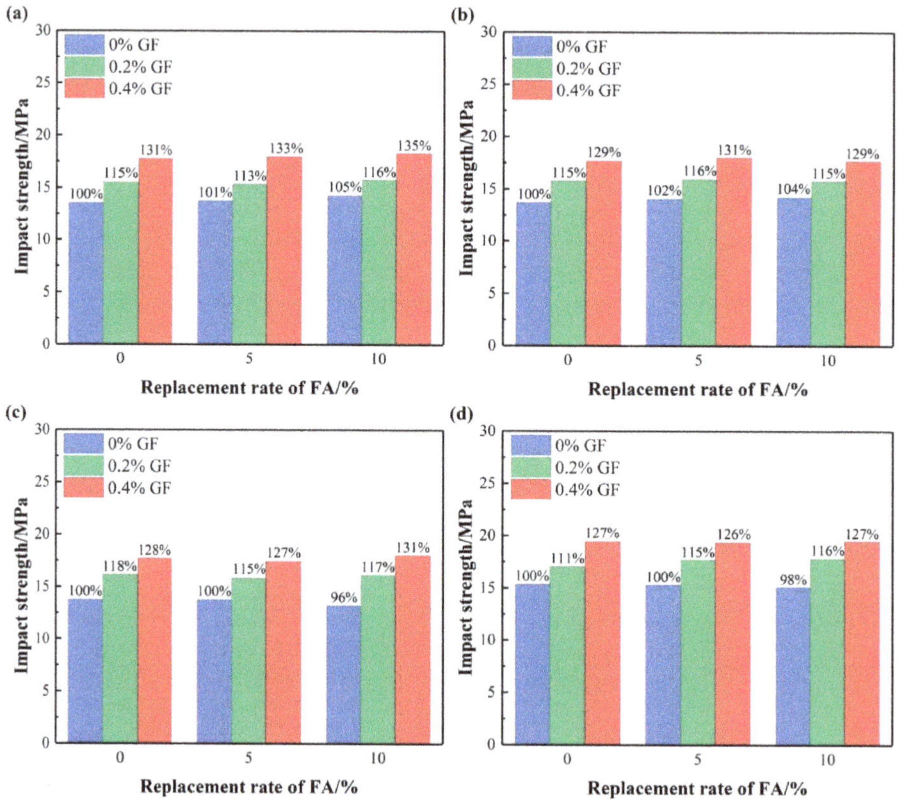

Figure 10. Impact strength of rubberized mortar modified by FA and GF: (**a**) 0% RP; (**b**) 10% RP; (**c**) 20% RP; (**d**) 30% RP.

3.5. Thermal Insulation Properties

Based on the results of the varying properties test of rubberized mortar modified by FA and GF on the above sections, it was observed that a dosage of 10% FA can enhance the workability of the mortar while preserving its mechanical and deformation resistance properties. To streamline the experimental process, we conducted the following experiments using a 10% FA dosage. Additionally, we performed a thermal conductivity test based on the mix formulation provided in Table 4.

Thermal conductivity is an important parameter for the evaluation of insulating material [74]. The lower the thermal conductivity, the better the thermal insulating properties of the material, that is, the increase in the internal insulation quality of the content. Figure 11 illustrates the thermal conductivity of rubberized mortar specimens. The figure shows that the addition of RP significantly reduces the thermal conductivity of the mortar. As the RP replacement rate increases from 0% to 30%, thermal conductivity decreases from approximately 0.36–0.39 W·(m·K)$^{-1}$ to 0.23–0.25 W·(m·K)$^{-1}$. This enhancement in the thermal insulation properties can be attributed to the relatively low density of RP and its lower thermal conductivity compared to sand. Similar findings were reported in a study conducted by Abd-Elaty et al. [55]. Furthermore, the addition of GF also enhances the thermal insulation properties of rubberized mortar, albeit to a lesser extent. This can be attributed to the thermal physical properties of GF and its impact on the internal pore structure of mortar. However, the relatively small quantity of GF present in rubberized mortar limits the extent of this improvement. These results corroborate the findings of Awoyera et al. and Guardia et al. [75,76], identifying improved thermal insulating properties with increased fiber materials.

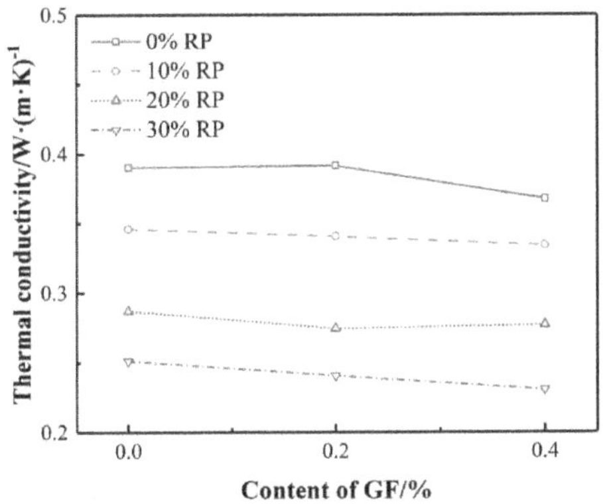

Figure 11. Thermal conductivity of rubberized mortar.

The utilization of waste materials for the production of building insulation materials can enhance thermal comfort in buildings while also offering cost-effective construction and environmental advantages. The aforementioned experimental results provide valuable insights into the potential applications of waste rubber and GF in various thermal insulation requirements.

3.6. Microstructure

Figure 12 illustrates the morphologies of rubberized mortar. In Figure 12a,b, voids and cracks can be observed around RP, suggesting that the damage in the rubberized mortar primarily occurs between the RP and matrix. These observations are consistent with previous studies [12,16,59]. However, the friction in cracks may result in energy dissipation. In Figure 12c,d, it is evident that the number of cracks in rubberized mortar is significantly reduced after the addition of GF. This improved interface has a positive impact on the strength and toughness of rubberized mortar. Although micro pores are present in rubberized mortar containing GF, the enhanced thermal insulation properties (in Section 3.5) may relate to this special observation.

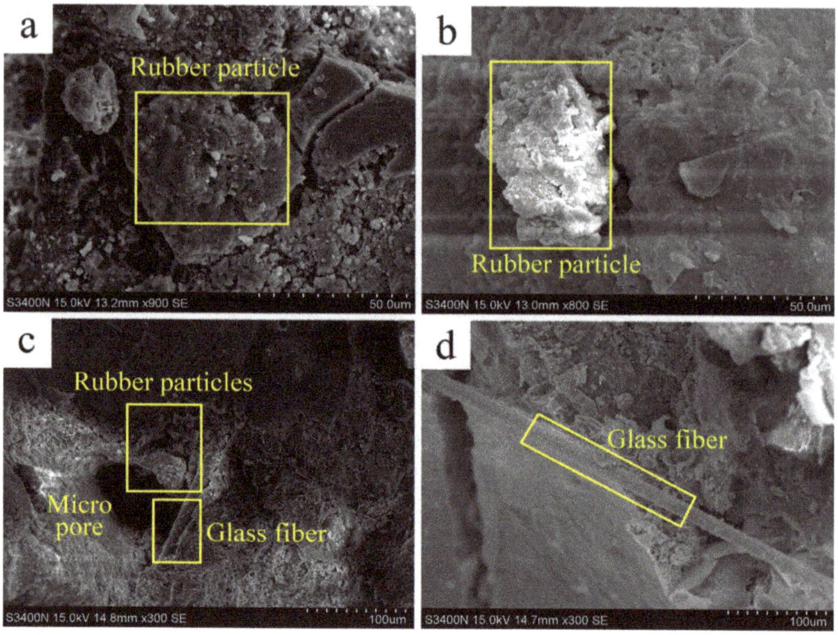

Figure 12. Microstructure: (**a**,**b**) are rubberized mortar; (**c**,**d**) are rubberized mortar containing GF.

3.7. Analytical Hierarchy Process

In this section, the AHP method was adopted to determine the optimal mixed proportion of rubberized mortar modified by FA and GF. The criteria considered were fluidity (C1), flexural strength (C2), compressive strength (C3), impact strength (C4), drying shrinkage (C5), and thermal conductivity (C6). Based on the comprehensive analysis of FA, GF, and RP on the properties of mortar in this study, six specimens (FA10GF0.2RP10, FA10GF0.4RP10, FA10GF0.2RP20, FA10GF0.4RP20, FA10GF0.2RP30, and FA10GF0.4RP30) were selected as alternatives, as shown in Figure 13.

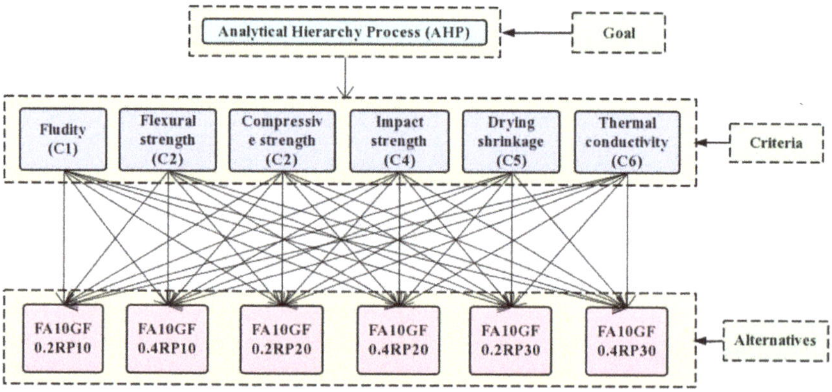

Figure 13. Flow chart of the AHP.

According to the step described in Table 5, the order of six property indicators (C1, C2, ..., C6) is sorted. Then, the pairwise comparisons matrix G was constructed as below:

$$G = \begin{bmatrix} 1 & 1/3 & 1/3 & 5 & 3 & 1/3 \\ 3 & 1 & 1 & 7 & 5 & 1 \\ 3 & 1 & 1 & 7 & 5 & 1 \\ 1/5 & 1/7 & 1/7 & 1 & 1/2 & 1/7 \\ 1/3 & 1/5 & 1/5 & 2 & 1 & 1/5 \\ 3 & 1 & 1 & 7 & 5 & 1 \end{bmatrix}$$

After calculating, the maximum eigen ($\lambda_{\max}$) of the pairwise comparisons matrix G and its corresponding eigenvector (v) were obtained as shown below:

$$\lambda_{\max} = 6.0898$$

$$v = (0.2380, 0.5560, 0.5560, 0.0667, 0.1073, 0.5560)^T$$

The consistency test was performed using Equations (5) and (6), as the following equation:

$$\mathrm{CI} = \frac{\lambda_{\max} - n}{n-1} = 0.01796$$

$$\mathrm{CR} = \frac{\mathrm{CI}}{\mathrm{RI}} = 0.0145 < 0.1$$

After satisfying the consistency test, the arithmetic average method, geometric average method, and eigenvalue method were used to calculate the weight (ω_1, ω_2, and ω_3), and their average values were calculated as the final weight vector (ω_G), as shown below:

$$\begin{aligned} \omega_1 &= (0.1155 \quad 0.2664 \quad 0.2664 \quad 0.0326 \quad 0.0525 \quad 0.2664)^T \\ \omega_2 &= (0.1111 \quad 0.2661 \quad 0.2661 \quad 0.0316 \quad 0.0512 \quad 0.2661)^T \\ \omega_3 &= (0.1144 \quad 0.2673 \quad 0.2673 \quad 0.0321 \quad 0.0516 \quad 0.2673)^T \\ \omega_G &= (0.1137 \quad 0.2666 \quad 0.2666 \quad 0.0321 \quad 0.0518 \quad 0.2666)^T \end{aligned}$$

Based on the final weight calculation result, the priority of the six properties indicators can be determined. In order to consider the quantitative differences in their importance, pairwise comparison matrices are provided for each indicator, denoted as C1, C2, ..., C6:

$$C1 = \begin{bmatrix} 1 & 2 & 3 & 4 & 5 & 6 \\ 1/2 & 1 & 2 & 3 & 4 & 5 \\ 1/3 & 1/2 & 1 & 2 & 3 & 4 \\ 1/4 & 1/3 & 1/2 & 1 & 2 & 3 \\ 1/5 & 1/4 & 1/3 & 1/2 & 1 & 2 \\ 1/6 & 1/5 & 1/4 & 1/3 & 1/2 & 1 \end{bmatrix} C2 = \begin{bmatrix} 1 & 1/2 & 5 & 4 & 8 & 7 \\ 2 & 1 & 6 & 5 & 9 & 8 \\ 1/5 & 1/6 & 1 & 1/2 & 3 & 2 \\ 1/4 & 1/5 & 2 & 1 & 4 & 3 \\ 1/8 & 1/9 & 1/3 & 1/4 & 1 & 1/2 \\ 1/7 & 1/8 & 1/2 & 1/3 & 2 & 1 \end{bmatrix}$$

$$C3 = \begin{bmatrix} 1 & 1/2 & 6 & 5 & 8 & 7 \\ 2 & 1 & 7 & 6 & 9 & 8 \\ 1/6 & 1/7 & 1 & 1/2 & 3 & 2 \\ 1/5 & 1/6 & 2 & 1 & 4 & 3 \\ 1/8 & 1/9 & 1/3 & 1/4 & 1 & 1/2 \\ 1/7 & 1/8 & 1/2 & 1/3 & 2 & 1 \end{bmatrix}$$

$$C4 = \begin{bmatrix} 1 & 1/4 & 1/2 & 1/5 & 1/4 & 1/7 \\ 4 & 1 & 2 & 1/2 & 1 & 1/4 \\ 2 & 1/2 & 1 & 1/3 & 1/2 & 1/5 \\ 5 & 2 & 3 & 1 & 2 & 1/3 \\ 4 & 1 & 2 & 1/2 & 1 & 1/4 \\ 7 & 4 & 5 & 3 & 4 & 1 \end{bmatrix}$$

$$C5 = \begin{bmatrix} 1 & 3 & 1/4 & 1 & 1/5 & 1 \\ 1/3 & 1 & 1/6 & 1/3 & 1/7 & 1/3 \\ 4 & 6 & 1 & 4 & 1/2 & 4 \\ 1 & 3 & 1/4 & 1 & 1/5 & 1 \\ 5 & 7 & 2 & 5 & 1 & 5 \\ 1 & 3 & 1/4 & 1 & 1/5 & 1 \end{bmatrix} C6 = \begin{bmatrix} 1 & 1/2 & 1/3 & 1/4 & 1/5 & 1/6 \\ 2 & 1 & 1/2 & 1/3 & 1/4 & 1/5 \\ 3 & 2 & 1 & 1/2 & 1/3 & 1/4 \\ 4 & 3 & 2 & 1 & 1/2 & 1/3 \\ 5 & 4 & 3 & 2 & 1 & 1/2 \\ 6 & 5 & 4 & 3 & 2 & 1 \end{bmatrix}$$

After calculating, their λ_{max}, CI, and CR value are displayed in Table 6. These calculation results can meet the requirements of the consistency test. Then, their weight vector (ω) was obtained as shown below:

$$\omega_{C1} = (0.3825 \quad 0.2504 \quad 0.1596 \quad 0.1006 \quad 0.0641 \quad 0.0428)^T$$

$$\omega_{C2} = (0.3071 \quad 0.4342 \quad 0.0722 \quad 0.1101 \quad 0.0312 \quad 0.0452)^T$$

$$\omega_{C3} = (0.3179 \quad 0.4430 \quad 0.0658 \quad 0.0996 \quad 0.0302 \quad 0.0435)^T$$

$$\omega_{C4} = (0.0393 \quad 0.1240 \quad 0.0702 \quad 0.2053 \quad 0.1240 \quad 0.4372)^T$$

$$\omega_{C5} = (0.0862 \quad 0.0378 \quad 0.2871 \quad 0.0862 \quad 0.4166 \quad 0.0862)^T$$

$$\omega_{C6} = (0.0416 \quad 0.0619 \quad 0.0969 \quad 0.1532 \quad 0.2401 \quad 0.4064)^T$$

Table 6. The maximum eigen, CI, and CR values of pairwise comparison matrices.

Alternatives	C1	C2	C3	C4	C5	C6
λ_{max}	6.1225	6.1605	6.2224	6.1081	6.1168	6.2306
CI	0.0245	0.0321	0.0440	0.0216	0.0234	0.0461
CR	0.0198	0.0259	0.0359	0.0174	0.0188	0.0372

Accordingly, the calculated priorities of alternatives with respect to the overall service performances of rubberized mortar modified by FA and GF are displayed in Figure 14. A higher priority indicates better overall properties. Consequently, the order of priority is as follows: C2 > C1 > C6 > C3 > C5 > C4. Overall, specimen FA10GF0.4RP10 has the highest priority of 0.381.

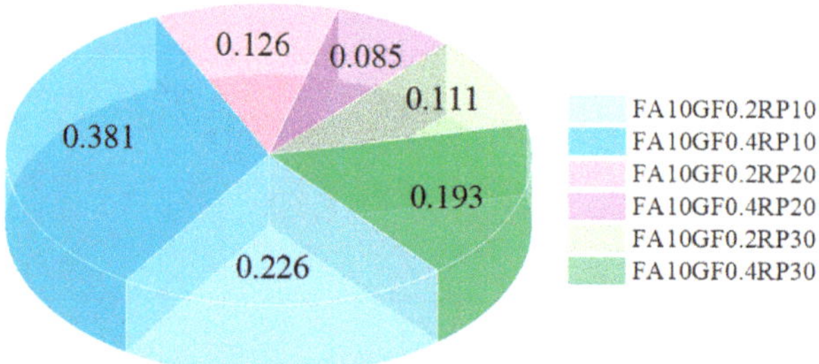

Figure 14. Priorities of rubberized mortar modified by FA and GF.

4. Conclusions

This study aimed to assess the feasibility of incorporating additives (FA and GF) into rubberized mortar. The experiment evaluated the resulting mortars based on fluidity, mechanical properties, drying shrinkage, impact resistance, and thermal insulation properties. The main findings can be drawn as follows:

1. The replacement of RP in mortar leads to a significant decrease in fluidity, and the addition of GF also contributes to reduced fluidity. Moreover, the addition of FA can appropriately improve the fluidity of the mortar. However, achieving this significant reduction in fluidity loss may necessitate the utilization of additional methods, such as incorporating chemical additives.
2. The addition of RP results in a decrease in compressive and flexural strengths. GF has a positive impact on the mechanical properties of rubberized mortar, particularly

in enhancing flexural strength. Furthermore, the influence of FA on the mechanical properties of rubberized mortar is not clearly evident.
3. The high replacement rate of RP in the mortar results in a significant increase in drying shrinkage. GF limits the deformation of the matrix during the internal water evaporation process and enhances stiffness, thereby effectively reducing the drying shrinkage of the rubberized mortar. FA initially reduces drying shrinkage within 28 days, but it subsequently leads to an increase in drying shrinkage after 28 days.
4. The addition of RP to mortar can increase its toughness, resulting in improved impact resistance of the specimens. The highest increase in impact resistance is observed when the replacement rate of RP is 30%. GF can effectively restrict the development of cracks in rubberized mortar when subjected to impact loads, while FA has a lesser effect on modifying the impact resistance of rubberized mortar.
5. The low density and low thermal conductivity of RP contribute to the reduction in thermal conductivity of rubber mortar. As the RP replacement rate increases from 0% to 30%, the thermal conductivity decreases to 0.23–0.25 $W{\cdot}(m{\cdot}K)^{-1}$. The addition of GF can further decrease the thermal conductivity of rubberized mortar.
6. In the condition of emphasizing fluidity, mechanical properties, drying shrinkage, impact resistance, and thermal insulation properties, the AHP method suggests that specimen FA10GF0.4RP10 exhibits the best comprehensive properties, with the highest priority of 0.381. The AHP method is primarily employed to identify viable options using existing indicators. To expand the applicability of this analytical tool, it may be necessary to foster creativity.

Author Contributions: Conceptualization, F.L.; methodology, Y.Z.; formal analysis, H.L. and Z.L.; data curation, Z.P., H.Z. and X.L.; writing—original draft preparation, H.L.; writing—review and editing, Z.P.; visualization, D.W.; supervision, F.L.; project administration, D.W.; funding acquisition, X.L. All authors have read and agreed to the published version of the manuscript.

Funding: This research was funded by [the National Natural Science Foundation of China] grant number [12072080] and [the Natural Science Foundation of Fujian Province] grant number [2023J011044].

Data Availability Statement: Date will be available on request.

Acknowledgments: The authors gratefully acknowledge the financial support provided by the National Natural Science Foundation of China under Grant No. 12072080 and the Natural Science Foundation of Fujian Province under Grant No. 2023J011044 (in China).

Conflicts of Interest: The authors declare that there is no conflict of interest.

References

1. Xu, J.; Houndehou, J.D.; Wang, Z.; Ma, Q. Experimental investigation on the mechanical properties and damage evolution of steel-fiber-reinforced crumb rubber concrete with porcelain tile waste. *Constr. Build. Mater.* **2023**, *370*, 130643. [CrossRef]
2. Chen, Z.; Li, L.; Xiong, Z. Investigation on the interfacial behaviour between the rubber-cement matrix of the rubberized concrete. *J. Clean. Prod.* **2019**, *209*, 1354–1364. [CrossRef]
3. Abdelmonem, A.; El-Feky, M.S.; Nasr, E.-S.A.R.; Kohail, M. Performance of high strength concrete containing recycled rubber. *Constr. Build. Mater.* **2019**, *227*, 116660. [CrossRef]
4. Roychand, R.; Gravina, R.J.; Zhuge, Y.; Ma, X.; Youssf, O.; Mills, J.E. A comprehensive review on the mechanical properties of waste tire rubber concrete. *Constr. Build. Mater.* **2020**, *237*, 117651. [CrossRef]
5. Raffoul, S.; Garcia, R.; Pilakoutas, K.; Guadagnini, M.; Medina, N.F. Optimisation of rubberised concrete with high rubber content: An experimental investigation. *Constr. Build. Mater.* **2016**, *124*, 391–404. [CrossRef]
6. Youssf, O.; Hassanli, R.; Mills, J.E. Mechanical performance of FRP-confined and unconfined crumb rubber concrete containing high rubber content. *J. Build. Eng.* **2017**, *11*, 115–126. [CrossRef]
7. Ji, Y.; Sun, Q. Experimental and numerical investigation of recycled rubber foam concrete. *Alex. Eng. J.* **2023**, *76*, 573–594. [CrossRef]
8. Assaggaf, R.A.; Ali, M.R.; Al-Dulaijan, S.U.; Maslehuddin, M. Properties of concrete with untreated and treated crumb rubber—A review. *J. Mater. Res. Technol.* **2021**, *11*, 1753–1798. [CrossRef]
9. Xiong, Z.; Fang, Z.; Feng, W.; Liu, F.; Yang, F.; Li, L. Review of dynamic behaviour of rubberised concrete at material and member levels. *J. Build. Eng.* **2021**, *38*, 102237. [CrossRef]

10. Etli, S. Evaluation of the effect of silica fume on the fresh, mechanical and durability properties of self-compacting concrete produced by using waste rubber as fine aggregate. *J. Clean. Prod.* **2023**, *384*, 135590. [CrossRef]
11. Letelier, V.; Bustamante, M.; Olave, B.; Martínez, C.; Ortega, J.M. Properties of mortars containing crumb rubber and glass powder. *Dev. Built Environ.* **2023**, *14*, 100131. [CrossRef]
12. He, S.; Li, L.; Xiong, Z.; Zhang, H.; Zheng, J.; Su, Y.; Liu, F. Effects of ring-type and straight steel fibres on the compressive performance of rubber-recycled aggregate concrete. *J. Build. Eng.* **2023**, *76*, 107148. [CrossRef]
13. Agrawal, D.; Waghe, U.; Ansari, K.; Dighade, R.; Amran, M.; Qader, D.N.; Fediuk, R. Experimental effect of pre-treatment of rubber fibers on mechanical properties of rubberized concrete. *J. Mater. Res. Technol.* **2023**, *23*, 791–807. [CrossRef]
14. Li, H.; Li, Y.; Xiong, Z.; Shu, Z.; Hong, Y.; Chen, S.; Su, Y. Experimental and theoretical study on elliptical rubber concrete filled double-skin GFRP short tubes under axial compression. *Constr. Build. Mater.* **2023**, *401*, 132796. [CrossRef]
15. Zhen, H.; Xiong, Z.; Song, Y.; Li, L.; Qiu, Y.; Zou, X.; Chen, B.; Chen, D.; Liu, F.; Ji, Y. Early mechanical performance of glass fibre-reinforced manufactured sand concrete. *J. Build. Eng.* **2024**, *83*, 108440. [CrossRef]
16. Li, X.; Zhao, Y.; Yao, W.; Wu, B.; Xia, K. Dynamic fracture behaviors of rubberized mortars with various rubber powder sizes and volume percentages. *Eng. Fract. Mech.* **2022**, *269*, 108553. [CrossRef]
17. Yu, F.; Chen, Y.; Fang, Y.; Xu, P.; Xu, B.; Liu, Q. Crack behavior of expanded polystyrene foam-ceramsite composite thermal insulation mortar. *J. Build. Eng.* **2023**, *74*, 106930. [CrossRef]
18. El-Seidy, E.; Sambucci, M.; Chougan, M.; Al-Kheetan, M.J.; Biblioteca, I.; Valente, M.; Ghaffar, S.H. Mechanical and physical characteristics of alkali-activated mortars incorporated with recycled polyvinyl chloride and rubber aggregates. *J. Build. Eng.* **2022**, *60*, 105043. [CrossRef]
19. Meshgin, P.; Xi, Y.; Li, Y. Utilization of phase change materials and rubber particles to improve thermal and mechanical properties of mortar. *Constr. Build. Mater.* **2012**, *28*, 713–721. [CrossRef]
20. Mostofinejad, D.; Aghamohammadi, O.; Bahmani, H.; Ebrahimi, S. Improving thermal characteristics and energy absorption of concrete by recycled rubber and silica fume. *Dev. Built Environ.* **2023**, *16*, 100221. [CrossRef]
21. Wang, J.; Du, B. Experimental studies of thermal and acoustic properties of recycled aggregate crumb rubber concrete. *J. Build. Eng.* **2020**, *32*, 101836. [CrossRef]
22. Zhai, S.; Zhou, X.; Zhang, Y.; Pang, B.; Liu, G.; Zhang, L.; Yang, L.; Liu, Z.; Liu, L. Effect of multifunctional modification of waste rubber powder on the workability and mechanical behavior of cement-based materials. *Constr. Build. Mater.* **2023**, *363*, 129880. [CrossRef]
23. Huang, X.; Pang, J.; Liu, G.; Chen, Y. The influence of equal amplitude high stress repeated loading on the mechanical and deformation characteristics of rubber concrete. *Constr. Build. Mater.* **2021**, *266*, 121135. [CrossRef]
24. Yu, Y.; Jin, Z.; Shen, D.; An, J.; Sun, Y.; Li, N. Microstructure evolution and impact resistance of crumb rubber concrete after elevated temperatures. *Constr. Build. Mater.* **2023**, *384*, 131340. [CrossRef]
25. Davoodi, A.; Esfahani, M.A.; Bayat, M.; Mohammadyan-Yasouj, S.E.; Rahman, A. Influence of nano-silica modified rubber mortar and EVA modified porous asphalt on the performance improvement of modified semi-flexible pavement. *Constr. Build. Mater.* **2022**, *337*, 127573. [CrossRef]
26. Xu, F.; Peng, C.; Zhu, J.; Chen, J. Design and evaluation of polyester fiber and SBR latex compound-modified perlite mortar with rubber powder. *Constr. Build. Mater.* **2016**, *127*, 751–761. [CrossRef]
27. Zhai, S.; Liu, G.; Pang, B.; Liu, C.; Zhang, Z.; Zhang, L.; Liu, L.; Yang, Y.; Liu, Z.; Zhang, Y. Investigation on the influence of modified waste rubber powder on the abrasion resistance of concrete. *Constr. Build. Mater.* **2022**, *357*, 129409. [CrossRef]
28. Luo, T.; Yuan, H.; Wang, Q. Comparison the properties of carbon fiber-based Portland cement and alkali-activated fly ash/slag conductive cementitious composites. *J. Build. Eng.* **2023**, *76*, 107134. [CrossRef]
29. Cheng, Y.; Shen, H.; Zhang, J. Understanding the effect of high-volume fly ash on micro-structure and mechanical properties of cemented coal gangue paste backfill. *Constr. Build. Mater.* **2023**, *378*, 131202. [CrossRef]
30. Shao, R.; Wu, C.; Li, J.; Liu, Z.; Wu, P.; Yang, Y. Mechanical behaviour and environmental benefit of eco-friendly steel fibre-reinforced dry UHPC incorporating high-volume fly ash and crumb rubber. *J. Build. Eng.* **2023**, *65*, 105747. [CrossRef]
31. Zhao, Y.; Chen, X.; Wang, Z.; Zhang, Y.; Leng, Y. An experimental study on the influence of fly ash and crumb rubber on the mechanical property and permeability of Engineered Cementitious Composite (ECC). *J. Build. Eng.* **2023**, *76*, 106998. [CrossRef]
32. Shao, R.; Wu, C.; Li, J.; Liu, Z. Repeated impact resistance of steel fibre-reinforced dry UHPC: Effects of fibre length, mixing method, fly ash content and crumb rubber. *Compos. Struct.* **2023**, *321*, 117274. [CrossRef]
33. Nematzadeh, M.; Hosseini, S.-A.; Ozbakkaloglu, T. The combined effect of crumb rubber aggregates and steel fibers on shear behavior of GFRP bar-reinforced high-strength concrete beams. *J. Build. Eng.* **2021**, *44*, 102981. [CrossRef]
34. Yan, Z.-W.; Bai, Y.-L.; Ozbakkaloglu, T.; Gao, W.-Y.; Zeng, J.-J. Axial impact behavior of Large-Rupture-Strain (LRS) fiber reinforced polymer (FRP)-confined concrete cylinders. *Compos. Struct.* **2021**, *276*, 114563. [CrossRef]
35. Yang, H.; Lu, X.; Gong, M.; Yang, P. Compression-shear performance of steel fiber reinforced rubber concrete. *J. Build. Eng.* **2023**, *75*, 106977. [CrossRef]
36. Xiong, C.; Li, Q.; Lan, T.; Li, H.; Long, W.; Xing, F. Sustainable use of recycled carbon fiber reinforced polymer and crumb rubber in concrete: Mechanical properties and ecological evaluation. *J. Clean. Prod.* **2021**, *279*, 123624. [CrossRef]
37. Su, Q.; Xu, J. Durability and mechanical properties of rubber concrete incorporating basalt and polypropylene fibers: Experimental evaluation at elevated temperatures. *Constr. Build. Mater.* **2023**, *368*, 130445. [CrossRef]

38. Alguhi, H.; Tomlinson, D. Crack behaviour and flexural response of steel and chopped glass fibre-reinforced concrete: Experimental and analytical study. *J. Build. Eng.* **2023**, *75*, 106914. [CrossRef]
39. Kumari, G.J.; Mousavi, S.S.; Bhojaraju, C. Influence of thermal cycles and high-temperature exposures on the residual strength of hybrid steel/glass fiber-reinforced self-consolidating concrete. *Structures* **2023**, *55*, 1532–1541. [CrossRef]
40. *GB 1334-1999*; Slag Portland Cement, Pozzolanic Portland Cement and Fly Ash Portland Cement. National Quality Supervision Bureau: Beijing, China, 1999.
41. *GB/T 208-2014*; Test method of Cement Density. National Quality Supervision Bureau: Beijing, China, 2014.
42. *GB/T 175-1999*; Portland Cement and Ordinary Portland Cement. National Quality Supervision Bureau: Beijing, China, 1999.
43. *GB/T 14684-2022*; Construction-Use Sands. National Quality Supervision Bureau: Beijing, China, 2022.
44. *JG/T 223-2017*; Polycarboxylate Superplasticizers. National Quality Supervision Bureau: Beijing, China, 2017.
45. Aziz, F.N.A.A.; Bida, S.M.; Nasir, N.A.M.; Jaafar, M.S. Mechanical properties of lightweight mortar modified with oil palm fruit fibre and tire crumb. *Constr. Build. Mater.* **2014**, *73*, 544–550. [CrossRef]
46. Benazzouk, A.; Mezreb, K.; Doyen, G.; Goullieux, A.; Quéneudec, M. Effect of rubber aggregates on the physico-mechanical behaviour of cement–rubber composites-influence of the alveolar texture of rubber aggregates. *Cem. Concr. Compos.* **2003**, *25*, 711–720. [CrossRef]
47. *GB/T 2419-2005*; Determination Method of Fluidity of Cement Mortar. National Quality Supervision Bureau: Beijing, China, 2005.
48. *GB/T 17671-2020*; Cement Mortar Strength Test Method (ISO). National Quality Supervision Bureau: Beijing, China, 2020.
49. *JGJ/T 70-2009*; Test Method for Basic Properties of Building Mortar. National Quality Supervision Bureau: Beijing, China, 2009.
50. *GB/T 1843-2008*; Determination of Impact Strength of Plastic Cantilever Beams. National Quality Supervision Bureau: Beijing, China, 2008.
51. *GB/T 6342-1996*; Determination of Linear Dimension of Foamed Plastics and Rubber. National Quality Supervision Bureau: Beijing, China, 1996.
52. *GB/T 10294-2008*; Determination of Steady-State Thermal Resistance and Related Characteristics of Thermal Insulation Materials. National Quality Supervision Bureau: Beijing, China, 2008.
53. Cao, W.; Wang, A.; Yu, D.; Liu, S.; Hou, W. Establishment and implementation of an asphalt pavement recycling decision system based on the analytic hierarchy process. *Resour. Conserv. Recycl.* **2019**, *149*, 738–749. [CrossRef]
54. Li, Y.; Mu, J.; Hao, J.; Liu, Y.; Jiang, X.; Luo, X. Research on the particle characteristics of manufactured sands affecting the flow ability of fresh mortar. *Constr. Build. Mater.* **2023**, *382*, 131287. [CrossRef]
55. Abd-Elaty, M.A.A.; Ghazy, M.F.; Khalifa, O.H. Mechanical and thermal properties of fibrous rubberized geopolymer mortar. *Constr. Build. Mater.* **2022**, *354*, 129192. [CrossRef]
56. Guler, S.; Akbulut, Z.F. Workability, physical & mechanical properties of the cement mortars strengthened with metakaolin and steel/basalt fibers exposed to freezing-thawing periods. *Constr. Build. Mater.* **2023**, *394*, 132100.
57. Feng, W.; Tang, Y.; Zhang, Y.; Qi, C.; Ma, L.; Li, L. Partially fly ash and nano-silica incorporated recycled coarse aggregate based concrete: Constitutive model and enhancement mechanism. *J. Mater. Res. Technol.* **2022**, *17*, 192–210. [CrossRef]
58. Fang, Z.; Luo, Y.; Chen, H.; Gao, Y.; Yang, W.; Wang, C. Research on mechanical properties and hydration characteristics of ultra-high performance concrete with high-volume fly ash microsphere. *J. Build. Eng.* **2023**, *78*, 107738. [CrossRef]
59. Li, H.-W.; Wang, R.; Wei, M.-W.; Lei, N.-Z.; Sun, H.-X.; Fan, J.-J. Mechanical properties and hydration mechanism of high-volume ultra-fine iron ore tailings cementitious materials. *Constr. Build. Mater.* **2022**, *353*, 129100. [CrossRef]
60. Li, D.; Zhuge, Y.; Gravina, R.; Mills, J.E. Compressive stress strain behavior of crumb rubber concrete (CRC) and application in reinforced CRC slab. *Constr. Build. Mater.* **2018**, *166*, 745–759. [CrossRef]
61. Li, H.; Long, W.-J.; Khayat, K.H. Efficient recycling of waste rubber in a sustainable fiber-reinforced mortar and its damping and energy dissipation capacity. *Cem. Concr. Compos.* **2023**, *138*, 104963. [CrossRef]
62. Qaidi, S.; Mohammed, A.S.; Ahmed, H.U.; Faraj, R.H.; Emad, W.; Tayeh, B.A.; Althoey, F.; Zaid, O.; Sor, N.H. Rubberized geopolymer composites: A comprehensive review. *Ceram. Int.* **2022**, *48*, 24234–24259. [CrossRef]
63. Chinchillas-Chinchillas, M.J.; Orozco-Carmona, V.M.; Gaxiola, A.; Alvarado-Beltrán, C.G.; Pellegrini-Cervantes, M.J.; Baldenebro-López, F.J.; Castro-Beltrán, A. Evaluation of the mechanical properties, durability and drying shrinkage of the mortar reinforced with polyacrylonitrile microfibers. *Constr. Build. Mater.* **2019**, *210*, 32–39. [CrossRef]
64. Afroughsabet, V.; Biolzi, L.; Ozbakkaloglu, T. Influence of double hooked-end steel fibers and slag on mechanical and durability properties of high performance recycled aggregate concrete. *Compos. Struct.* **2017**, *181*, 273–284. [CrossRef]
65. Uygunoğlu, T.; Topçu, I.B. The role of scrap rubber particles on the drying shrinkage and mechanical properties of self-consolidating mortars. *Constr. Build. Mater.* **2010**, *24*, 1141–1150. [CrossRef]
66. Su, P.; Dai, Q.; Li, M.; Ma, Y.; Wang, J. Investigation of the mechanical and shrinkage properties of plastic-rubber compound modified cement mortar with recycled tire steel fiber. *Constr. Build. Mater.* **2022**, *334*, 127391. [CrossRef]
67. Deng, Z.; Yang, Z.; Pan, X. Synergetic effects of recycled crumb rubber and glass cullet on the engineering properties of geopolymer mortar. *Cem. Concr. Compos.* **2023**, *137*, 104907. [CrossRef]
68. Chen, W.; Xie, Y.; Li, B.; Li, B.; Wang, J.; Thom, N. Role of aggregate and fibre in strength and drying shrinkage of alkali-activated slag mortar. *Constr. Build. Mater.* **2021**, *299*, 124002. [CrossRef]
69. Zhang, J.; Pan, G.; Yan, Y. Early hydration, mechanical strength and drying shrinkage of low-carbon alkali-activated Ti-extracted residues-fly ash cement and mortars. *Constr. Build. Mater.* **2021**, *293*, 123517. [CrossRef]

70. Wang, H.; Long, G.; Xie, Y.; Zeng, X.; Ma, K.; Dong, R.; Tang, Z.; Xiao, Q. Effects of intense ultraviolet irradiation on drying shrinkage and microstructural characteristics of cement mortar. *Constr. Build. Mater.* **2022**, *347*, 128513. [CrossRef]
71. Xiang, J.; Qiu, J.; Zhao, Y.; Zheng, P.; Peng, H.; Fei, X. Rheology, mechanical properties, and hydration of synergistically activated coal gasification slag with three typical solid wastes. *Cem. Concr. Compos.* **2023**, *147*, 105418. [CrossRef]
72. Liu, M.; Lu, J.; Ming, P.; Yin, Y. Study of fracture properties and post-peak softening process of rubber concrete based on acoustic emission. *Constr. Build. Mater.* **2021**, *313*, 125487. [CrossRef]
73. Najim, K.; Hall, M. Mechanical and dynamic properties of self-compacting crumb rubber modified concrete. *Constr. Build. Mater.* **2012**, *27*, 521–530. [CrossRef]
74. El Wardi, F.Z.; Ibaaz, K.; Bouyahyaoui, A.; Cherkaoui, M.; Ouaki, B.; Oubenmoh, S. Thermomechanical characterization and thermal simulation of a new multilayer mortar and a light-weight pozzolanic concrete for building energy efficiency. *Constr. Build. Mater.* **2022**, *346*, 128479.
75. Awoyera, P.O.; Akinrinade, A.D.; de Sousa Galdino, A.G.; Althoey, F.; Kirgiz, M.S.; Tayeh, B.A. Thermal insulation and mechanical characteristics of cement mortar reinforced with mineral wool and rice straw fibers. *J. Build. Eng.* **2022**, *53*, 104568. [CrossRef]
76. Guardia, C.; Barluenga, G.; Palomar, I.; Diarce, G. Thermal enhanced cement-lime mortars with phase change materials (PCM), lightweight aggregate and cellulose fibers. *Constr. Build. Mater.* **2019**, *221*, 586–594. [CrossRef]

buildings

MDPI

Article

Effect of Expansion Agent and Glass Fiber on the Dynamic Splitting Tensile Properties of Seawater–Sea-Sand Concrete

Huanyu Zhu, Zhe Xiong *, Yuying Song, Keting Zhou and Yue Su

School of Civil and Transportation Engineering, Guangdong University of Technology, Guangzhou 510006, China; 3121002912@mail2.gdut.edu.cn (H.Z.); 1112309005@mail2.gdut.edu.cn (Y.S.); 2112009006@mail2.gdut.edu.cn (K.Z.); 2112109012@mail2.gdut.edu.cn (Y.S.)
* Correspondence: gdgyxz263@gdut.edu.cn

Abstract: In marine structural engineering, the impact resistance of concrete holds high significance. The determination of whether the combined use of expansion agent (EA) and glass fiber (GF) has a synergistic effect on the impact resistance of seawater–sea-sand concrete (SSC) and plays a role in its performance and application. In this study, the dynamic Brazilian disc test at various strain rates was carried out with an SHPB device to investigate the effect of mixing 0% and 6% EA with 0% and 1% GF on the dynamic splitting tensile properties of SSC. The results show that strain rate effect on EA and GF-reinforced SSC during dynamic splitting tensile tests at higher strain rates, indicating strong strain rate sensitivity. The synergistic reinforcement of EA and GF consumed more energy under impact loading, thus maintaining the morphological integrity of concrete. However, the dynamic splitting tensile strength obtained in the Brazilian disc test had a significant overload effect which cannot be ignored. EA doped at 6% and GF doped at 1% showed a synergistic enhancement of SSC's dynamic splitting tensile properties.

Keywords: glass fiber; expansion agent; seawater–sea-sand concrete; dynamic splitting tensile properties; Brazilian disc splitting tensile test

Citation: Zhu, H.; Xiong, Z.; Song, Y.; Zhou, K.; Su, Y. Effect of Expansion Agent and Glass Fiber on the Dynamic Splitting Tensile Properties of Seawater–Sea-Sand Concrete. *Buildings* **2024**, *14*, 217. https://doi.org/10.3390/buildings14010217

Academic Editor: Daxu Zhang

Received: 7 December 2023
Revised: 9 January 2024
Accepted: 11 January 2024
Published: 13 January 2024

1. Introduction

Under evolving international circumstances and changing thematic priorities, countries around the world are intensifying their efforts in the development and utilization of marine resources [1,2]. In marine construction projects, the environment poses significant challenges for concrete structures, which are marked by complexity and harsh conditions [3]. Various loads, such as sea-ice impact, waves, aircraft landings, heavy falls, explosion impacts, and rock penetration, are prevalent. These loads exhibit accidental, transient, and high strain rate characteristics, posing substantial risks to concrete structures [4,5]. Analyzing statistical data from marine platform accidents revealed that incidents caused by falling objects, fire and explosion, waves, wind, ice, and currents collectively contribute to over 95% of accidents [6,7]. For marine platforms specifically, approximately one-fourth of accidents are attributable to impact loads [8]. Consequently, stringent requirements for the mechanical properties of concrete exist in the marine environment. Concrete, as a strain rate-sensitive building material, exhibits notable differences in mechanical properties between dynamic and static loading conditions. In marine environments where high-strain impact loads are prevalent, comprehensive consideration of their dynamic effects on concrete structures is imperative. The dynamic splitting tensile properties play a pivotal role in particular [9–11]. Against this engineering backdrop, the development of concrete with the dynamic splitting properties holds great significance for advancing marine engineering.

Due to the poor cracking and tensile resistance of concrete, scholars have been exploring new types of concrete to satisfy the requirements for practical engineering [12–14]. Some studies propose enhancing the overall tensile strength, impact toughness, and energy-absorbing capacity of concrete through the incorporation of fibers [15,16]. Fibers can be uniformly distributed in concrete, fostering strong bonding with the cement paste and forming an effective fiber network. This network prevents the propagation of micro-cracks, thereby improving the cracking and tensile strength of concrete, as well as enhancing its toughness [17–19]. Among various fiber-reinforced materials, glass fiber-reinforced concrete is widely employed in engineering due to its excellent impact resistance, good workability, high tensile strength, and cost-effectiveness. Mastali et al. [20] observed that the addition of glass fibers enhances the mechanical properties and impact resistance of self-compacting concrete, with notable improvements in concrete toughness. Liu et al. [21] investigated the impact of four groups of alkali-resistant glass fibers with different admixtures on the compressive strength of concrete. The test results showed that, with increasing glass fiber admixture, the concrete compressive strength followed a pattern of initial increase and subsequent decrease, reaching its optimal dosage at 2%, resulting in an 11.3% increase in compressive strength.

To enhance the mechanical properties of fiber-reinforced concrete and meet the increasing demands of infrastructure, scholars proposed the synergistic use of expansion agents and fibers as a more effective method compared to mixing different fiber types [22]. The research indicates that an appropriate quantity of expansion agent can enhance the internal structure of concrete, positively impacting its mechanical properties [23,24]. The addition of an expansion agent can lead to a larger volume of calcium hydroxide or calcium alumina, compensating for the self-shrinkage of concrete, preventing cracking, and reducing the adverse effects of micro-cracks on performance. However, Mo et al. [25] caution that excessive expansion agent admixture may result in negative expansion stress, causing microcracks and reducing concrete compactness. Therefore, the dosage of the expansion agent should be carefully controlled. Wang et al. [26] found that concrete mixed with 8% expansion agent and 1% steel fiber exhibited a 10% increase in splitting tensile strength and a 4% increase in compressive strength after 28 days of curing compared to concrete without an expansion agent. Xiong et al. [27] explored the impact of expansion agents and glass fibers on the compressive properties of sea-sand concrete. The findings revealed that the best compression properties were achieved with 18 mm long glass fibers at a 1% dosage and 6% expansion agent dosage, showcasing the synergistic reinforcing effect of combining fibers and expansion agents.

In marine engineering construction, a significant challenge appears due to the remote locations of most marine projects, resulting in higher transportation costs for essential construction materials like fresh water, sand, and stone, which are conventionally used in large quantities [28,29]. Meanwhile, the extensive use of concrete depletes crucial resources in its production process, particularly fresh water and natural river sand [30–32]. To address these challenges, a strategic approach involves leveraging marine resources, preserving the ecological environment, and substituting fresh water and river sand with seawater and sea sand in concrete raw materials. This approach not only effectively resolves the immediate raw material shortages but also reduces construction costs and shortens the construction period, providing substantial benefits to marine engineering projects [33,34]. Successful examples of utilizing sea sand as a raw material for concrete can be found in projects such as the urban expansion of Singapore and large-scale reclamation in the Middle East [35]. The research indicates that seawater–sea-sand concrete exhibits slightly higher early compressive strength than ordinary concrete, while its long-term compressive strength is comparable to that of ordinary concrete, highlighting its practical viability [36,37].

Moreover, the mechanical properties of seawater–sea-sand concrete under dynamic impact loading is vital for evaluating the long-term stability of marine construction engineering, particularly in locations with frequent saltwater washouts or airplane landings. Guan et al. [38] investigated the splitting tensile strength, flexural strength, compressive strength and stress–strain relationship of seawater–sea-sand concrete and performed microstructural analyses. However, there are some limitations in the study of tensile properties. It is well known that concrete is a brittle material with a much lower tensile strength compared to its compressive strength, and the lack of tensile strength often causes internal damage to the member, a key property that directly influences the safety of the structure [39]. Thus, understanding the dynamic tensile properties of seawater–sea-sand concrete is critical to assess the feasibility of its widespread application in marine construction projects.

There is little research on the synergistic effect of expansive agents and glass fibers. Su et al. [33] conducted three-point bending experiments using an SHPB to explore the synergy between glass fiber and expansive agent on the dynamic bending performances of seawater–sea-sand concrete at high strain rates. However, it is necessary to investigate whether the expansion agent and glass fibers can continue to maintain synergistic reinforcement under dynamic impact loading. It is difficult for brittle materials to achieve a uniaxial tensile stress state using direct tension, and the Brazilian disc test is commonly adopted to measure the tensile properties [40]. The Split Hopkinson Pressure Bar (SHPB) is currently the main device for testing the dynamic mechanics of brittle materials. Ross et al. [41] were the first to use an SHPB device to conduct Brazilian disc tests and investigate the dynamic tensile characteristics of concrete materials, which was the beginning of the application of SHPB devices in dynamic tensile strength testing of rocks. Therefore, it is essential to carry out Brazilian disc tests using an SHPB device to investigate the dynamic splitting tensile properties of expansion agent and glass fiber-reinforced seawater–sea-sand concrete.

The current investigation focused on examining the dynamic splitting tensile properties of EA and GF-reinforced SSC through the widely employed indirect Brazilian disc splitting tensile test. By conducting the test at various strain rates, the strain rate sensitivity of EA and GF-reinforced SSC was analyzed and compared with SSC. This comparative analysis aimed to elucidate the dynamic splitting tensile properties of EA and GF-reinforced SSC when subjected to impact splitting tensile forces. The results of this study provide support for the development of SSC with excellent dynamic splitting tensile properties.

2. Materials and Methods

2.1. Raw Materials

Ordinary Portland cement (P.O 42.5R, specific gravity 3.112) served as the cementing material for concrete preparation. Seawater from Zhuhai, Guangdong Province, China, (specific gravity 1.003) was used. Sea sand, characterized by a medium grain size less than 4.75 mm, specific gravity 2.60, water content 3.36%, water absorption 4.27%, and fineness modulus 2.54, was employed as the fine aggregate. Crushed granites (particle size distribution of 5-10 mm, specific gravity of 2.64, water content of 1.39%, water absorption of 1.74%) were utilized as the coarse aggregate. The particle size distribution of both the fine and coarse aggregates is illustrated in Figure 1. The expansion agent predominantly consisted of calcium sulfoaluminate. An aqueous solution of the polycarboxylic acid base mixture (specific gravity 1.03, solids content 20%) was selected as the superplasticizer. Alkali-resistant glass fibers were used for the glass fibers. The glass fibers used had a length of 18 mm, density of 2680 kg/m^3, elastic modulus of 72 GPa, and tensile strength of 1700 MPa.

2.2. Specimen Preparation

Seawater–sea-sand concrete was enhanced with an expansion agent and glass fibers, employing both single and combined methods to create reinforced concrete. According to the ISRM guidelines, the specimen thickness for the Brazilian splitting tensile test should ideally fall within the range of approximately 0.5 to 1 times the diameter [42]. Consequently, for the splitting tensile test in this study, the specimen size was specifically designed to be Ø100 × 50 mm.

To investigate the synergistic impact of the expansion agent and glass fibers under dynamic conditions, and drawing on previous research findings [27,33], specific admixture proportions were chosen. These included an expansion agent admixture (representing the ratio of expansion agent volume to the total volume of fine aggregate) set at 0% and 6%, and a glass fiber admixture (indicating the ratio of glass fiber volume to the total volume of concrete) at 0% and 1%. The water/cement (w/c) ratio was held constant at 0.40. Four distinct mix ratios of seawater–sea-sand concrete, designated as F-EA, were formulated. F0 and F1 signify glass fiber admixture proportions of 0% and 1%, while EA0 and EA6 represent expansion agent admixture proportions of 0% and 6%, as outlined in Table 1.

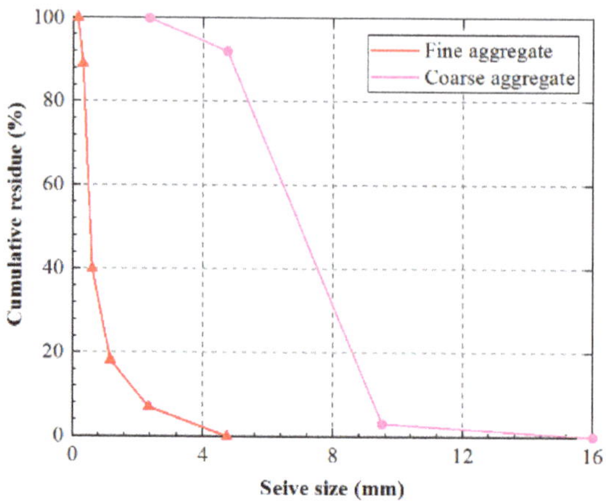

Figure 1. Particle size distribution diagram of aggregates.

Table 1. Mix proportions of sea water and sand concrete mixture.

Aggregate Type	F0EA0	F1EA0	F0EA6	F1EA6
Cement (kg/m^3)	552.27	546.74	552.08	546.56
Sea water (kg/m^3)	221.61	219.40	220.75	218.54
Sea sand (kg/m^3)	774.59	766.84	744.79	737.34
Coarse aggregate (kg/m^3)	788.60	780.71	788.60	780.71
Glass fibers (kg/m^3)	0.00	26.80	0.00	26.80
Swelling agent (kg/m^3)	0.00	0.00	33.12	32.79
Superplasticizer (kg/m^3)	11.05	10.93	11.70	11.59

2.3. *Testing Methods*

2.3.1. Static Splitting Tensile Test

The static splitting tensile test device utilized a specialized fixture comprising a circular-shaped pad and a padded strip positioned between the upper and lower platens and the specimen. This fixture was aligned with the centerline and oriented perpendicular to the top surface of the molding. The specimen was placed at the center of the lower plate of the testing machine, ensuring that the splitting surface, or the splitting bearing surface, was perpendicular to the top surface of the specimen during formation. For loading, a test rate of 0.05 MPa/s (equivalent to a strain rate of approximately $1 \times 10^{-5}\ s^{-1}$) was used. The static splitting tensile strength (f_{ts}) of the concrete was determined as follows:

$$f_{ts} = \frac{2F_{ts}}{\pi A} \quad (1)$$

where F_{ts} is the breaking load during splitting and stretching, and A is the area of the splitting surface of the specimen.

2.3.2. Dynamic Splitting Tensile Test

Brazilian disc splitting tests typically involve applying a relative linear load to the center of a cylindrical specimen. Following the principles of elastic mechanics, radial compression of the specimen induces horizontal tensile stress perpendicular to the center plane [42]. If this tensile stress surpasses the tensile strength of the material, the specimen fractures along its vertical midplane. The force analysis of the specimen in the Brazilian disc splitting test is illustrated in Figure 2.

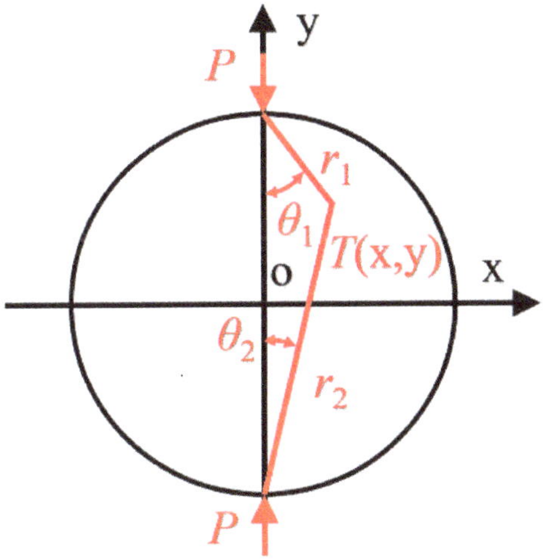

Figure 2. Force analysis of the specimen in the Brazilian disc splitting test.

The traditional Brazilian disc splitting strength formula in dynamic splitting tensile tests falls short of meeting the assumption of center crack initiation due to stress concentration at the loading end. Scholars have enhanced this formula [43,44] by introducing two platforms with central symmetry and mutual parallelism as the loading plane. This modification ensures that the crack originates from the center of the specimen during dynamic splitting tests. Following the findings of Wang et al. [45] and Huang et al. [46], the center angle of the platform was assumed to be 2α, and a correction factor (k) for the center tip of the platform was introduced. Specifically, when $2\alpha = 20°$, k was set to 0.95. Consequently, the dynamic splitting tensile strength of the Brazilian disc (ftd1) can be expressed as:

$$f_{td1} = 0.95\frac{2P}{\pi BD} \tag{2}$$

where B and D are the thickness and diameter of the specimen, respectively, and P is the destructive load of the disk.

The dynamic Brazilian disc test was conducted using a Split Hopkinson Pressure Bar (SHPB), comprising test bars (a striker, an incident bar, and a transmitted bar), specimens, and a data acquisition system. Given the low tensile strength of concrete, semiconductor strain gauges were employed for the precise measurement of incident, reflected, and transmitted pulses. These strain gauges had a resistance of 120.0 ± 0.1 Ω and a sensitivity coefficient of 110 ± 5%. Two semiconductor strain gauges were affixed in the cross-section diameter direction of the incident and transmitted rods, forming a half-bridge to eliminate bending modes. Specifically, the strain gauges on the incident rods were positioned 2880 mm away from the specimen, while those on the transmitted rods were 1890 mm away, as illustrated in Figure 3. In addition, each specimen was fitted with an additional resistive strain gauge with a sensitive grid length of 5 mm, a resistance of 120.0 ± 0.1 Ω, and a sensitivity factor of 2.22 ± 1%. This gauge was strategically placed to capture the cracking moment at the center of the specimen during each test, facilitating the computation of the true split tensile strength, as depicted in Figure 4. The force–time scale ($P(t)$) during the test is represented as:

$$P(t) = \frac{E_b A_b}{2}[\varepsilon_i(t)+\varepsilon_r(t)+\varepsilon_t(t)] \tag{3}$$

where E_b and A_b are the modulus of elasticity and cross-sectional area of the bars in the SHPB device, respectively. $\varepsilon_i(t)$ and $\varepsilon_r(t)$ are the incident and reflected wave pulses captured by the strain gauges affixed to the incident bars, respectively. $\varepsilon_t(t)$ is the transmitted wave pulse captured by the strain gauges attached to the transmission bars.

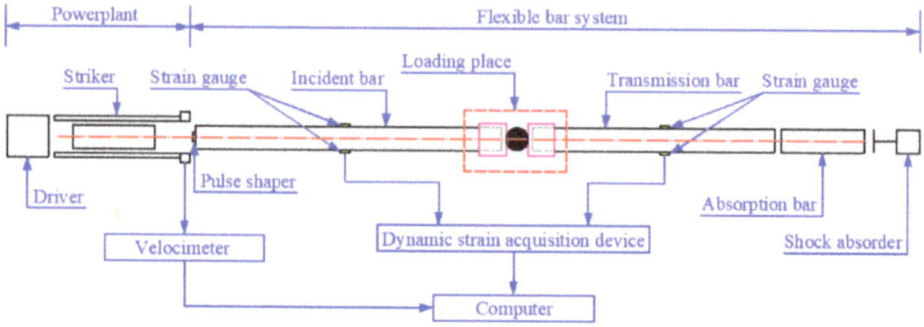

Figure 3. SHPB device for the dynamic splitting tensile tests.

Figure 4. The FBD specimen is placed between two bars.

The force sketch for the dynamic Brazilian disk splitting test is shown in Figure 5, where $\sigma_i(t)$, $\sigma_r(t)$, and $\sigma_t(t)$ are the stresses corresponding to $\varepsilon_i(t)$, $\varepsilon_r(t)$, and $\varepsilon_t(t)$, respectively. A high-precision grinder with a parallelism accuracy of 0.01/300 mm and a smoothness accuracy of Ra 0.1 μm was used for grinding and processing. From the center angle of the platform to take $2\alpha = 20°$, the grinding depth of the platform was calculated to be 0.76 mm, and the parallelism error could be controlled within 0.02 mm. The SHPB test follows the one-dimensional stress wave assumption as well as the uniformity of the force applied to the specimen during the test:

$$\sigma_i(t) + \sigma_r(t) = \sigma_t(t) \tag{4}$$

From Equations (2)–(4), the dynamic splitting tensile strength f_{td} of SHPB can be calculated as:

$$f_{td} = 0.95\frac{2\sigma_{t,max}A_0}{\pi DL} \tag{5}$$

where $\sigma_{t,max}$ is the peak value of the stress $\sigma_t(t)$ corresponding to the transmitted wave pulse.

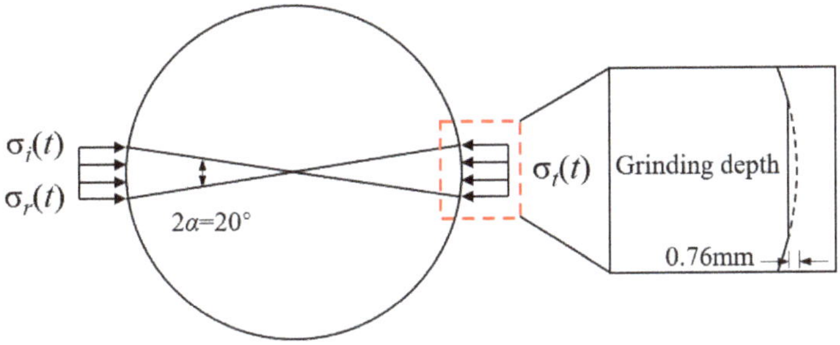

Figure 5. Stress diagram of the specimen during dynamic splitting tension.

In the dynamic splitting test, the strain rate and dynamic enhancement factor (DIF) are important parameters used to evaluate the dynamic tensile properties of concrete. Equations (6) and (7) could be used to calculate the stress rate, strain rate, and DIF values of the specimens.

$$\dot{\sigma} = \frac{f_{td}}{\tau};\ \dot{\varepsilon} = \frac{\dot{\sigma}}{E_0} \tag{6}$$

$$DIF = \frac{f_{td}}{f_{ts}} \tag{7}$$

In Equation (6), τ represents the time interval from the initial point to the peak value on the stress–time curve of the specimen, while E_0 stands for the modulus of elasticity of the specimen. Studies have indicated [47] that the modulus of elasticity in concrete exhibits low sensitivity to strain rate, with the enhancement effect notably lower than that observed in compressive and tensile strengths. Consequently, in the current study, the static modulus of elasticity was employed for estimating the strain rate of the specimens.

3. Results and Discussion

3.1. Analysis of Results of Dynamic Splitting Tensile Test

Utilizing the F1EA6 set of specimens as an illustrative case, Figure 6 shows the dynamic response to the incident stress pulse. Notably, the peak stress $\sigma_{t,max}$ corresponding to the transmitted wave pulse registered at 20.09 MPa. The stress between the starting point and the peak (τ) was approximately 315 μs, with a modulus of elasticity and thickness measuring 39.11 GPa and 50.74 mm, respectively. The SHPB device rod boasts a modulus of elasticity of 206 GPa. Referring to Equations (5)–(7) from Section 2.3, the dynamic splitting tensile strength of F1EA6-2 was calculated as 18.81 MPa, featuring a stress rate of 59.84 GPa/s and a strain rate of 1.53 s^{-1}. This methodology is applicable across all specimens for determining the dynamic splitting tensile strength, stress rate, and strain rate. The test results for the SHPB dynamic splitting tensile specimens, encompassing four groups under five distinct approximate strain rates, are tabulated in Table 2.

Table 2. Dynamic split tensile test results of F0EA0--F1EA6.

Specimen	B (mm)	$\dot{\sigma}$ (GPa/s)	$\dot{\varepsilon}$ (s^{-1})	f_{td} (MPa)	f_{td}' (MPa)	f_{ts} (MPa)	DIF	S_0
F0EA0-1	44.80	39.73	1.13	9.44	8.12	3.95	2.06	0.16
F0EA0-2	48.24	45.77	1.30	10.39	8.22	3.95	2.08	0.26
F0EA0-3	49.19	61.15	1.73	11.28	9.22	3.95	2.33	0.22
F0EA0-4	49.50	78.05	2.21	13.07	9.53	3.95	2.41	0.37
F0EA0-5	47.45	100.75	2.85	13.65	10.70	3.95	2.71	0.28
F0EA6-1	50.75	40.51	1.11	11.58	9.20	4.28	2.15	0.26
F0EA6-2	51.00	56.07	1.53	13.04	10.26	4.28	2.40	0.27

Table 2. *Cont.*

Specimen	B (mm)	σ (GPa/s)	ε (s^{-1})	f_{td} (MPa)	f_{td}' (MPa)	f_{ts} (MPa)	DIF	S_0
F0EA6-3	50.40	64.54	1.76	13.75	10.65	4.28	2.49	0.29
F0EA6-4	50.01	83.94	2.29	15.91	11.72	4.28	2.74	0.36
F0EA6-5	49.92	107.46	2.94	17.19	12.41	4.28	2.90	0.39
F1EA0-1	55.77	41.46	1.14	13.49	11.15	4.90	2.28	0.21
F1EA0-2	53.53	54.45	1.50	16.5	12.97	4.90	2.65	0.27
F1EA0-3	53.50	71.47	1.97	18.02	14.35	4.90	2.93	0.26
F1EA0-4	52.78	84.15	2.32	19.43	14.74	4.90	3.01	0.32
F1EA0-5	52.54	93.94	2.59	20.71	15.10	4.90	3.08	0.37
F1EA6-1	51.55	42.97	1.10	14.18	12.36	5.18	2.38	0.15
F1EA6-2	50.74	59.79	1.53	18.81	14.67	5.18	2.83	0.28
F1EA6-3	51.02	70.17	1.79	19.43	15.43	5.18	2.98	0.26
F1EA6-4	51.80	87.51	2.24	22.88	17.21	5.18	3.32	0.33
F1EA6-5	50.68	102.13	2.61	23.98	17.34	5.18	3.34	0.38

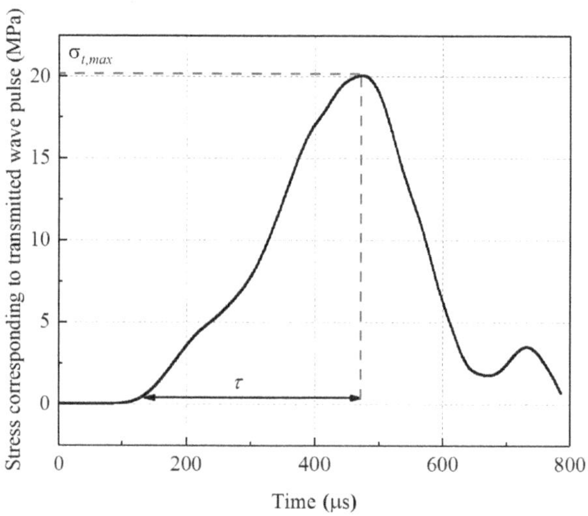

Figure 6. Dynamic response of F1EA6-2.

3.2. Failure Mode

In Figure 7, the damage modes of four groups of SSC under varying approximate strain rates are illustrated. Following the SHPB dynamic splitting test, these SSC groups displayed evident cracking damage in the horizontal radial direction along the center, resulting in the final rupture into two parts. This damage mode aligns with prior studies on dynamic splitting tensile tests [48]. At a lower strain rate (around 1.10 s^{-1}), the middle crack of the specimen was relatively small, dividing it into two halves along the developed crack. During this phase, the platform effectively served as a stress buffer. However, as the strain rate increased, the cracks of specimens gradually widened. At a strain rate of about 2.24 s^{-1}, a broad horizontal crack emerged in the middle, accompanied by the appearance of crushing triangles at the platforms on both ends. Notably, there was a more extensive crushing area at the end in contact with the projectile rod. Importantly, the damage modes across all specimens did not exhibit a "T"-shaped three-piece split, consistent with findings in dynamic Brazilian disk splitting tests conducted on rock materials [49].

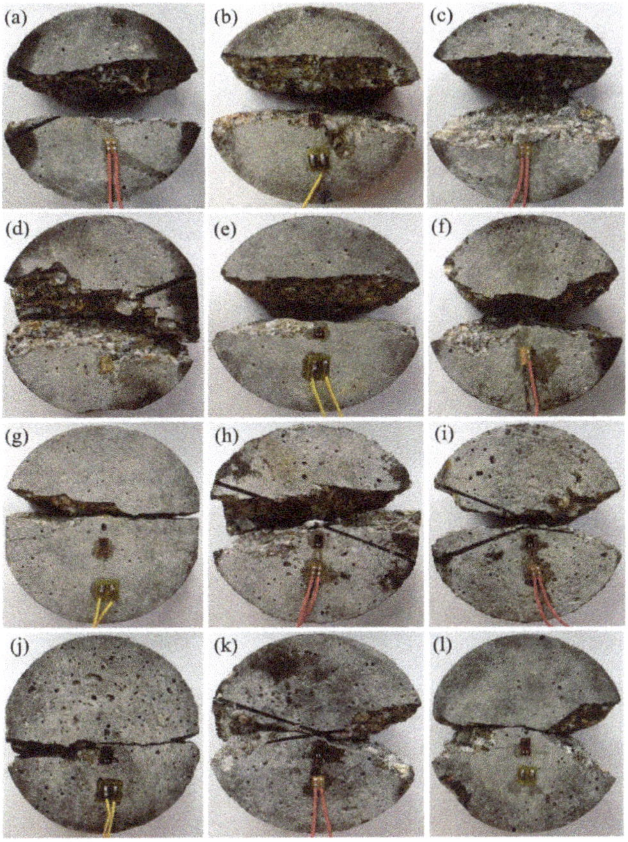

Figure 7. Failure modes at different strain rates: (**a**) F0EA0-1.13 s^{-1}, (**b**) F0EA0-1.73 s^{-1}, (**c**) F0EA0-2.21 s^{-1}, (**d**) F0EA6-1.11 s^{-1}, (**e**) F0EA6-1.76 s^{-1}, (**f**) F0EA6-2.29 s^{-1}, (**g**) F1EA0-1.14 s^{-1}, (**h**) F1EA0-1.97 s^{-1}, (**i**) F1EA0-2.32 s^{-1}, (**j**) F1EA6-1.10 s^{-1}, (**k**) F1EA6-1.79 s^{-1}, and (**l**) F1EA6-2.24 s^{-1}.

This study attributes the observed damage modes to the platforms at both ends of the specimen, which mitigate the stress concentration at the loaded contact end [50]. As the incident energy increases, the horizontal force on the platform intensifies, leading to a shorter duration of the shock wave acting on the specimen. Considering the brittle nature of concrete, rapid transmission of the stress wave and impact energy is essential near the end close to the incident pulse. Consequently, concrete produces multiple cracks to dissipate the impact energy, resulting in the eventual formation of a triangular crushing zone in the loading region after central cracking. It is noteworthy that at higher strain rates, the stress concentration at both ends of the specimen may impact the strain rate effect of the experiment.

A comparison of damage modes among the four groups of SSC at three different approximate strain rates (1.10 s^{-1}, 1.79 s^{-1}, and 2.24 s^{-1}) indicates that at lower strain rates (1.10 s^{-1}), the SSC groups with the expansion agent or glass fibers show greater damage mode integrity, in the following decreasing order of intactness: F1EA6, F1EA0, F0EA6, and F0EA0. Glass fiber incorporation was found to be more effective than the expansion agent in enhancing damage mode integrity. Conversely, at a higher strain rate (2.24 s^{-1}), the damage integrity of the two groups of SSC with either the expansion agent or glass fibers alone was similar, while the SSC with a combination of the expansion agent and glass fibers outperformed the other three. The addition of glass fibers contributed to better toughening effects on SSC. However, this toughening development seemed to weaken at high strain rates. This might be attributed to the impact high energy and short duration, causing the glass fiber to slip and not fully exert its tensile properties. Among these, the damage mode of F1EA6 exhibited greater completeness, suggesting that the expansion agent can resist shrinkage within the concrete. It can even pre-pressurize the cement paste before external loading, resulting in a more compact contact between the glass fibers and cement paste. In essence, the combination of the expansion agent and glass fibers in SSC can better leverage the excellent tensile properties of glass fibers compared to glass fiber-reinforced concrete [20,39]. This enhancement contributes to improving the cracking strength and tensile strength of the concrete.

3.3. Overload Correction

Numerous studies have utilized an SHPB device to conduct dynamic Brazilian disc tests, examining the dynamic tensile characteristics of various rock and concrete materials [51,52]. However, some scholars [53,54] have observed that the dynamic Brazilian disc tensile strength tends to surpass the dynamic direct tensile strength based on test results. In their investigation of different rock materials, Zhang et al. [55] found that the dynamic increase factor (*DIF*) obtained from the Brazilian disc test was higher than that from direct tensile tests under similar loading rates. Similarly, Xia et al. [53] measured the dynamic tensile strength of granite using various testing methods, both direct and indirect, and discovered that the dynamic splitting tensile strength exceeded the dynamic direct tensile strength at equivalent loading rates. They concluded that this phenomenon results from overload effects and internal friction effects.

Hence, in the Brazilian disc test, it was observed [54] that the load-carrying capacity of specimens continued to increase even after tensile damage occurred at the center point, a phenomenon referred to as overload. To quantitatively analyze this overload effect, a strain gauge was affixed at a distance of d = 5 mm from the center of the specimen. The time difference in the acquired signal was then employed to capture the moment of dynamic splitting onset of the Brazilian disc specimen, enabling the determination of the tensile stress at the center point of the disc at the initiation of the splitting, as depicted in Figure 8. This approach was employed in the current study to address and correct the overload effect of the specimen.

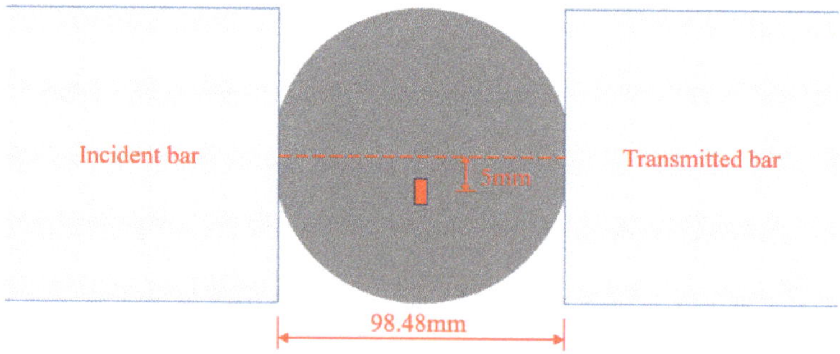

Figure 8. Arrangement of strain gauges on the specimen.

In the dynamic tensile test utilizing the SHPB device, it is imperative to achieve dynamic force balance conditions at both ends of the specimen. The forces on the incident and transmitted bars and the specimen ends are denoted as P_1 and P_2. Only when the dynamic details at both ends of the specimen are balanced, and P_1 equals P_2, can the SHPB device be deemed valid. This ensures that the data collected by the strain gauges can be simplified and processed using Equation (2). Throughout this study, all specimens met the dynamic stress equilibrium criteria. Figure 9 illustrates the dynamic stress balance verification for a representative specimen, F1EA6-5.

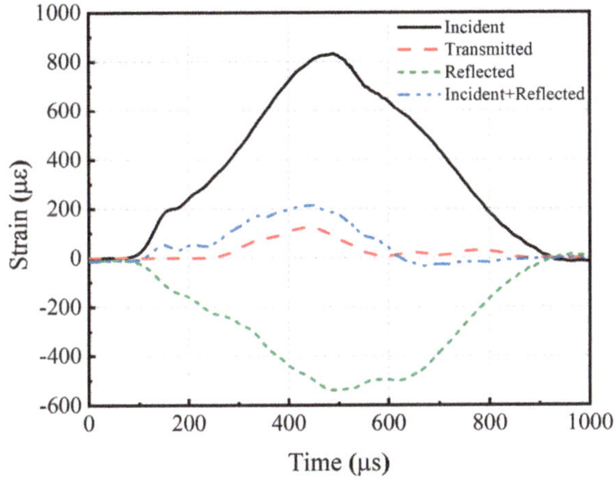

Figure 9. Dynamic stress balance verification of F1EA6-5.

According to Xia et al. [53], the initiation of cracking at the center of the disc prompts stress release in the surrounding region, leading to a notable inflection point in the strain gauge signal of the specimen. The moment corresponding to this inflection point signifies the cracking moment of the disk center. Assuming that t_c represents the starting moment of the cracking of the disk center, the elastic wave speed of the specimen is known as c_0, the elastic wave speed of the bar is c_b = 5169 m/s, the diameter of the disk is D = 98.48 mm, and the distance between the strain gauges on the transmission bar and the left end is L = 1.89 m; then, the moment of the inflection point T_1 of the captured signal by the strain gauge on the specimen is calculated as:

$$T_1 = t_c + \frac{d}{c_0} \tag{8}$$

By translating the transmitted signal according to the distance between the strain gauges on the transmission bar and the specimen center point, the corresponding moment T_2 at which the center of the specimen starts to crack can be obtained as:

$$T_2 = t_c + \frac{D}{2c_0} + \frac{L}{c_b} \tag{9}$$

Utilizing Equations (8) and (9), the strain gauge's initial signal from the specimen serves as the reference point. By shifting the transmitted wave's time-range signal to the left by $|T_2 - T_1|$, stress signals at both ends of the specimen can be computed. This process yields time-range signals of tension at the specimen ends, normalized against the cracking moment of the disc, facilitating the determination of tensile stresses at the center of disc at the point of cracking.

The calculation of a typical specimen (F1EA6-5) based on the above method yielded $|T_2 - T_1|$ = 378 μs, and its tensile strength and crack initiation time-course signals after normalizing the time axis are shown in Figure 10. The peak value of the stress time-course curve of the specimen is defined as the nominal tensile strength f_{td}, and the overload-corrected strength is the true tensile strength f_{td}'. Thus, we can calculate that the nominal tensile strength of F1EA6-5 was 23.98 MPa, and the true tensile strength was 17.34 MPa. In previous studies [50,55], it was found that the larger the loading rate, the more significant the difference between the stress at the moment of disc cracking initiation and the nominal tensile strength is, the more pronounced the overload phenomenon will be. Therefore, based on this method, the four groups of SSC were corrected for overloading, and after correction, the true tensile strengths of the specimens were determined and are shown in Table 2.

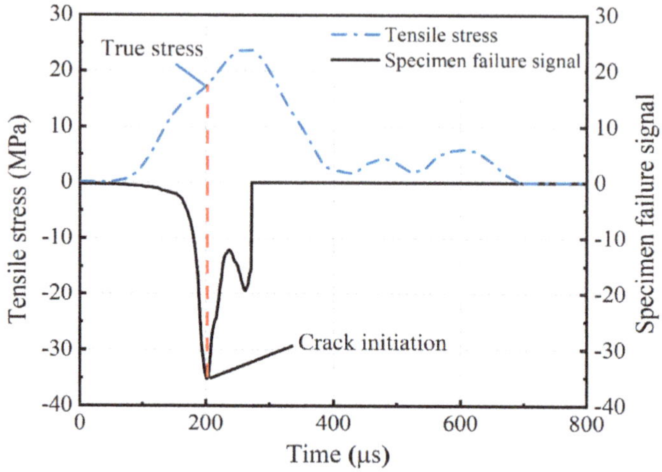

Figure 10. Tensile stress and crack initiation time history signal of F1EA6-5.

It can be found from Table 2 that there is a non-negligible difference between the nominal tensile strength f_{td} and the true tensile strength f_{td}' of the four groups of SSC. To understand this phenomenon, we plotted the nominal tensile strength versus the true tensile strength in Figure 11 and found that the higher the loading rate, the larger the deviation of f_{td} from f_{td}' is and the more pronounced the overload effect is. Therefore, the f_{td} obtained from the uncorrected overloaded Brazilian disc tests at higher strain rates is larger than f_{td}' and does not truly reflect the material properties. In order to quantitatively analyze this degree of overloading, this study introduced S_0 to denote the overload ratio of the specimen, which is calculated as:

$$S_0 = \frac{f_{td} - f'_{td}}{f'_{td}} \tag{10}$$

Using Equation (10), we calculated the overload ratios for the four groups of SSC, as detailed in Table 2. The correlation between the overload ratios and stress rates is illustrated in Figure 12. The overload ratios exhibited a logarithmic increase with the ascending stress rate, and their growth rate diminished gradually as the stress rate rose. Linear fitting of the experimental data resulted in a modified equation to represent the overload ratio of the compounded expansion agent and glass fiber-reinforced SSC, with the fitting parameters outlined in Table 3. The fitting outcomes highlight the highest linear fit correlation for F1EA6, reaching 0.928, and the trend of the overload ratio increasing with the strain rate followed a more logarithmic pattern. According to the fitting equation, it can be computed that the splitting tensile strength derived from the dynamic Brazilian disc test of compounded expansion agent and glass fiber-reinforced SSC would be overestimated by 38% when the stress rate is approximately 100 GPa/s. Therefore, its overload effect cannot be neglected.

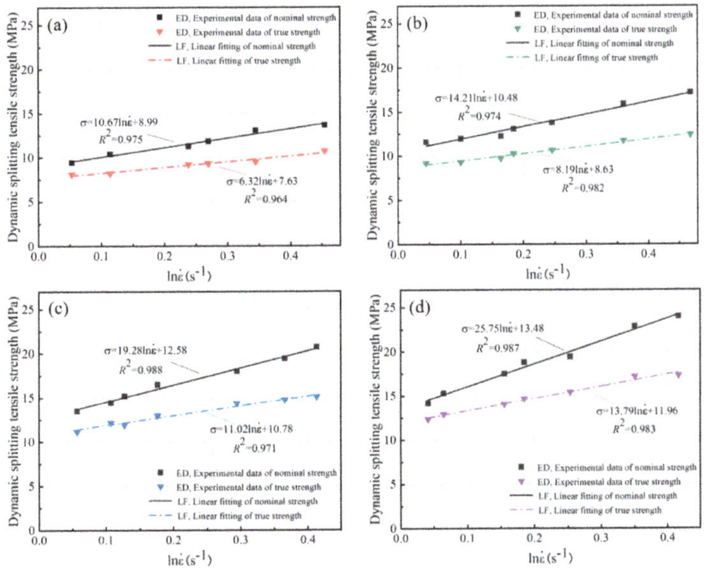

Figure 11. The relationship between nominal tensile strength and true tensile strength: (**a**) F0EA0, (**b**) F0EA6, (**c**) F1EA0, and (**d**) F1EA6.

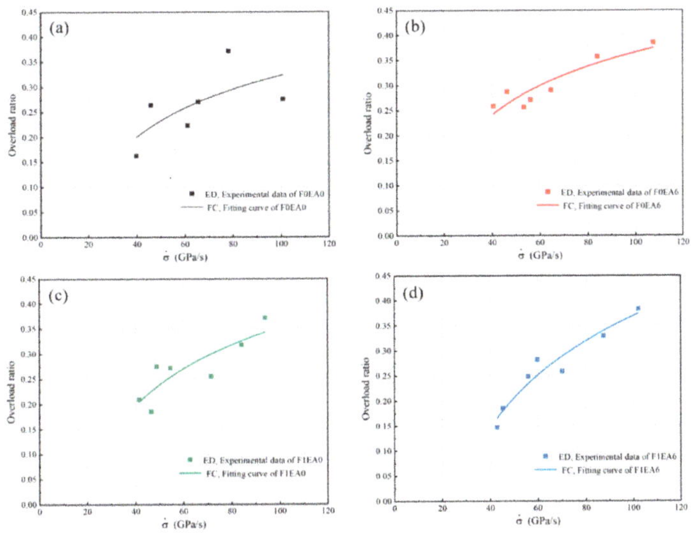

Figure 12. Overload ratio of the specimens: (**a**) F0EA0, (**b**) F0EA6, (**c**) F1EA0, and (**d**) F1EA6.

Table 3. Fitting parameters of stress rate to overload ratio.

Specimen	*a*	*b*	R^2	Fitting Equation
F0EA0	0.131	0.281	0.526	S_0 = 0.131 lnσ-0.281
F0EA6	0.136	0.259	0.849	S_0 = 0.136 lnσ-0.259
F1EA0	0.169	0.426	0.832	S_0 = 0.169 lnσ-0.426
F1EA6	0.241	0.734	0.928	S_0 = 0.241 lnσ-0.734

3.4. Strain Rate Effect

The *DIF*, serving as an indicator of the strain rate effect of concrete, exhibits an increasing trend with strain rate variations [47]. Table 2 illustrates the *DIF* values for the four SSC groups, revealing an augmentation of the *DIF* with escalating strain rates. Notably, F1EA6 boasted a higher *DIF* compared to the other groups at equivalent strain rates, underscoring the pronounced strain rate effect in the mixed expansion agent and glass fiber-reinforced SSC.

In a comprehensive synthesis of the literature and experimental findings, Zhou et al. [56] established a logarithmic relationship between the *DIF* and concrete strain rate. This relationship is divided into two segments to delineate below-critical strain rates (Equation (11)) and above-critical strain rates (Equation (12)) [55]:

$$DIF = a\log\dot{\varepsilon} + b;\ \dot{\varepsilon} \leq \dot{\varepsilon}_i \tag{11}$$

$$DIF = c\log\dot{\varepsilon} + d;\ \dot{\varepsilon} > \dot{\varepsilon}_i \tag{12}$$

where $\dot{\varepsilon}_i$ is the critical strain rate, and a, b, c, and d are the fitting parameters.

The overload-corrected Dynamic Increase Factor (DIF) plotted against the logarithmic strain rate for the four groups of SSC has been fitted using Equations (11) and (12), and is depicted in Figure 13. The fitted parameter values are detailed in Table 4. The critical strain rates for these groups fall within the range of 1.06–1.31 s^{-1}, aligning with the critical strain rate observed in ordinary concrete, approximately 1.00 s^{-1} [57]. Notably, the slopes of the fitted curves for the four SSC groups exhibited a considerable increase in the high critical strain rate stage compared to the low critical strain rate stage. This observation suggests that the strain rate effect on the dynamic splitting behavior becomes more pronounced at higher strain rates, consistent with prior research findings [58]. At below-critical strain rates, the strain rate effect on expansion agent and glass fiber-reinforced SSC was minimal, with the dynamic splitting tensile strength experiencing gradual growth with increasing strain rate. However, at above-critical strain rates, the strain rate sensitivity of the SSC significantly intensified, revealing a robust rate sensitivity. This heightened sensitivity may be attributed to the synergistic impact of the expansion agent and glass fibers. The expansion agent's compacting effect enhances the bond between glass fibers and mortar interfaces, allowing the glass fibers to optimize their properties. Simultaneously, the concrete experiences internal pre-pressure, enabling the glass fibers to withstand external forces with pre-tension. This comprehensive enhancement contributes to improved crack resistance and tensile properties for the concrete.

Table 4. The fitting parameters for DIF of SSC.

Specimen	*a*	*b*	*c*	*d*	Critical Strain Rate (s^{-1})
F0EA0	0.221	2.105	1.591	1.935	1.31
F0EA6	0.225	2.123	1.847	2.019	1.28
F1EA0	0.237	2.189	2.254	2.198	1.06
F1EA6	0.274	2.368	2.723	2.309	1.08

3.5. Energy Dissipation Analysis

In the dynamic splitting test, the splitting damage mechanism of specimens involves the absorption and release of energy. Impact loading is typically characterized by a short duration and high energy. Effective impact resistance requires the material to exhibit superior energy dissipation capabilities under such loading conditions. The axial strain observed in dynamic splitting tests is a non-uniform strain. To quantify the energy dissipation capacity of brittle specimens under dynamic splitting tensile tests using a Split Hopkinson Pressure Bar (SHPB), Song et al. [59] introduced the concept of energy dissipation, denoted as $W_s(t)$. This measure is derived by subtracting the reflected wave energy ($W_r(t)$) from the total incident wave energy ($W_i(t)$), with the transmitted wave energy ($W_t(t)$) contributing to the calculation, as shown in Equation (13). The specific calculations for $W_i(t)$, $W_r(t)$, and $W_t(t)$ are detailed in Equation (14), Equation (15), and Equation (16), respectively.

$$W_s(t) = W_i(t) - W_r(t) - W_t(t) \quad (13)$$

$$W_i(t) = E_b C_b A_b \int_0^t \varepsilon_i^2(t)dt \quad (14)$$

$$W_r(t) = E_b C_b A_b \int_0^t \varepsilon_r^2(t)dt \quad (15)$$

$$W_t(t) = E_b C_b A_b \int_0^t \varepsilon_t^2(t)dt \quad (16)$$

where E_b is the modulus of elasticity of the bar of the SHPB device, C_b is the elastic wave velocity of the bar, and A_b is the cross-sectional area of the bar.

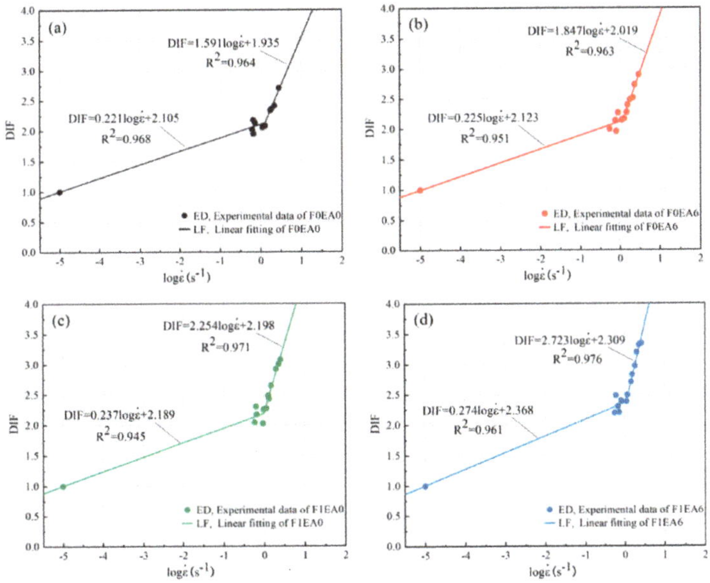

Figure 13. Relationship between DIF and logarithmic strain rate: (**a**) F0EA0, (**b**) F0EA6, (**c**) F1EA0, and (**d**) F1EA6.

Table 5 summarizes the maximum values of $W_i(t)$, $W_r(t)$, $W_t(t)$, and $W_s(t)$, denoted as W_i, W_r, W_t, and W_s, respectively, reflecting the peak energy dissipation in the dynamic splitting tensile test for the four SSC groups. The table provides details on the incident, reflected, and transmitted energies, along with the energy dissipation and absorption under various impact loads. The energy dissipation ratio (W_s/W_i), defined as the ratio of transmitted energy (W_s) to incident energy (W_i), is crucial for understanding the dynamic behavior of the specimens. At a high impact energy, the energy dissipation ratio tended to decrease with increasing incident energy (W_i). This phenomenon is attributed to the short damage time of the specimen at high strain rates, leading to increased brittleness in concrete as the impact energy rises [58]. A higher energy dissipation ratio signifies that the specimen absorbed more energy during impact.

Table 5. Energy characteristics of four groups of SSC during SHPB experiments.

Specimen	W_i (J)	W_r (J)	W_t (J)	W_s (J)	W_s/W_i
F0EA0-1	251.31	156.38	5.07	89.86	0.358
F0EA0-2	423.02	299.45	3.88	119.69	0.283
F0EA0-3	672.00	510.30	4.73	156.97	0.234
F0EA0-4	968.61	751.21	6.83	210.57	0.217
F0EA0-5	1522.71	1235.94	7.02	279.74	0.184
F0EA6-1	278.64	137.97	9.98	130.69	0.469
F0EA6-2	491.06	311.81	8.62	170.63	0.347
F0EA6-3	664.46	434.51	9.49	220.45	0.332
F0EA6-4	866.82	599.97	13.46	253.40	0.292
F0EA6-5	1512.10	1150.93	12.41	348.76	0.231
F1EA0-1	378.60	162.21	13.51	202.88	0.536
F1EA0-2	584.48	339.56	21.62	223.30	0.382
F1EA0-3	891.83	590.55	16.45	284.83	0.319
F1EA0-4	1257.96	878.58	17.71	361.66	0.288
F1EA0-5	1675.57	1204.51	21.69	449.38	0.268
F1EA6-1	363.17	144.09	7.60	211.48	0.582
F1EA6-2	888.08	517.75	17.89	352.44	0.397
F1EA6-3	1105.52	708.25	15.84	381.44	0.345
F1EA6-4	1384.49	862.06	23.29	499.14	0.361
F1EA6-5	1726.76	1206.22	17.95	502.59	0.291

The energy consumption ratio curves for the four groups of single or mixed expansion agent and glass fiber-reinforced SSC are depicted in Figure 14. At similar impact energy levels, the energy consumption ratio of the mixed expansion agent and glass fiber-reinforced SSC surpasses that of the other three groups. This outcome underscores the superior ability of the expansion agent and glass fiber-reinforced SSC to cushion and dissipate energy when exposed to impact loads in comparison to the other groups. This finding aligns with prior research indicating that fiber admixtures enhance concrete crack resistance and improve energy absorption efficiency [60]. The synergistic effect between the compounded expansion agent and glass fibers further enhances the energy dissipation capacity of the glass fiber. Moreover, the trend of the energy dissipation ratio curves for the four SSC groups revealed a decline with increasing impact energy. This suggests that at high strain rates, the damage time of the specimens is exceedingly brief, rendering the SSC more brittle when subjected to elevated impact energies.

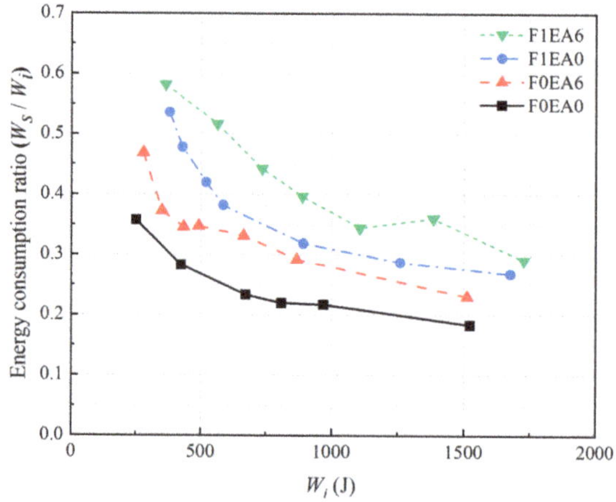

Figure 14. Energy consumption ratio curve.

3.6. Limitations and Prospects

In this work, the authors initially investigated the effects of single and mixed admixtures of expansion agent and glass fibers in seawater–sea-sand concrete on the concrete's dynamic splitting properties. Based on the results of the splitting tensile experiments under static and impact conditions, it can be concluded that a mixed admixture of expansion agent and glass fiber can enhance the internal structure of concrete and improve the overall tensile strength, impact toughness, and energy dissipation capacity. To this end, the advantages and limitations of this study are summarized in Table 6. However, there are some limitations of these research results.

Table 6. The advantages and limitations of this study.

Tensile Strength Testing	Test Method	Advantage	Limitation
Static tensile strength	Direct stretching	Simple and directly operable.	High precision requirements for experimental operation. Difficult to realize the uniaxial tensile stress state. It is easy to lead to the local stress concentration of the specimen. The appearance of a non-standard damage mode.
	Static Brazilian disc test	Avoids localized stress concentration, easy and convenient to conduct the experiments.	It is difficult to truly characterize the dynamic tensile properties by static loading.
Dynamic tensile strength	Brazilian disc test with an SHPB device	Clever measurement method to more accurately reflect the damage pattern under dynamic impact loading.	The measured intensity is anticipated to exceed the real value.

While this study delved into the dynamic splitting performance of composite seawater–sea-sand concrete, specifically focusing on damage mode, strain rate effects, and energy dissipation capacity, it did not thoroughly explore the microscopic aspects of the dynamic damage process and the changing pattern of crack formation. In essence, the analysis was macroscopic, covering the damage mode and stress changes under dynamic splitting. Future research could provide a more comprehensive understanding by employing SEM, DIC, and other technical methods to scrutinize the dynamic damage mode and crack expansion under different strain rates. This microscopic perspective would offer a more realistic and concrete portrayal of the dynamic splitting performance of the expansion agent and glass fiber-reinforced seawater–sea-sand concrete.

In this study, an SHPB device was employed to conduct dynamic Brazilian disc tests. The measured dynamic tensile strengths in the indirect Brazilian disc tests tended to surpass the dynamic direct tensile strengths, indicating an observed overloading phenomenon. This overload effect was particularly pronounced at higher strain rates. To address this, the study focused on capturing the central cracking moment by analyzing the time difference in acquired signals, allowing for the determination of tensile stress at the disc center point at the cracking moment. The overload effect was quantitatively analyzed by setting the overload ratio to obtain the true tensile strength. While the overload correction of the dynamic tensile strength is feasible experimentally and yields conclusions, certain practical deficiencies exist. These include an incomplete reflection of the real damage morphology in the damage process, the omission of friction in the simulation process, and the oversight of specimen defects. The subsequent research should delve into the dynamic damage process and crack extension laws from a microscopic perspective, supplementing the numerical analysis of microscopic models and comparing them with experimental tests. Meanwhile, the dynamic splitting performance of expansion agent and glass fiber-reinforced seawater–

sea-sand concrete remains in the laboratory research stage. Further investigations should extend the research to practical applications in engineering.

4. Conclusions

In this study, four sets of seawater–sea-sand concrete (SSC) mixtures with varying ratios were designed to explore the impact of single and combined admixtures of expansion agent and glass fibers on the dynamic splitting properties of SSC. The primary conclusions drawn from the results of the static and impact splitting tensile experiments are as follows:

(1) The damage modes of the four groups of SSC at three different approximate strain rates all exhibited clear horizontal radial cracking along the center, which was less pronounced at lower strain rates. Nevertheless, the cracks in the specimens gradually widened with increasing strain rates. At low strain rates (1.10 s^{-1}), the incorporation of glass fibers proved to be more effective than the addition of the expansion agent in enhancing damage modal integrity. However, at a high strain rate (2.24 s^{-1}), the SSC with the combined admixture of expansion agent and glass fibers demonstrated better damage modal integrity than the other three groups.

(2) The dynamic splitting tensile strength of the four SSC groups, as measured in the dynamic Brazilian disc test, exhibited significant overload, with higher strain rates amplifying the overload effect. To monitor crack initiation at the center of the disc and correct for overload, strain gauges were strategically placed. The overload ratio, denoted as S0, was introduced for the quantitative analysis. The overload ratio displayed a logarithmic increase with the stress rate. When the stress rate reached approximately 100 GPa/s, the measured dynamic splitting tensile strength of the mixed expansion agent and glass fiber-reinforced SSC was overestimated by 38%. At this point, the overload effect became substantial and cannot be overlooked.

(3) The critical strain rates for the four SSC groups fell within the range of 1.06–1.31 s^{-1}, similar to the critical strain rate of ordinary concrete, which is approximately 1.00 s^{-1}. At below-critical strain rates, the strain rate effect of the mixed expansion agent and glass fiber-reinforced SSC was negligible. However, at above-critical strain rates, the specimens exhibited a significant strain rate effect, demonstrating heightened sensitivity, particularly showcasing a robust rate sensitivity.

(4) The energy dissipation ratio of the mixed expansion agent and glass fiber-reinforced SSC surpasses that of the other three groups at comparable impact energy levels. The synergistic effect between the expansion agent and glass fiber contributed to superior cushioning and energy dissipation under impact loading. The energy consumption ratio curves for the four SSC groups showed a decreasing trend with escalating impact energy. This indicates that SSC becomes more brittle when subjected to higher impact energy levels.

(5) At present, the research on the mechanical properties of seawater–sea-sand concrete is still based on compressive strength; in comparison, the investigation of tensile strength has received less research interest. However, the research on the synergistic reinforcement effect of expansion agents and fibers is still based on static mechanical properties, with the dynamic impact property receiving less research interest. Using static loading methods may not authentically reflect the dynamic splitting tensile properties of concrete. Therefore, the SHPB device was employed in this study to conduct Brazilian disc tests. The tested nominal tensile strength was corrected for overload using a quantitative analysis to obtain the true tensile strength. Investigating the dynamic tensile properties of the expansion agent and glass fiber-reinforced seawater–sea-sand concrete under impact can support the development of seawater–sea-sand concrete with excellent dynamic splitting tensile properties.

Author Contributions: Conceptualization, H.Z., Z.X. and Y.S. (Yuying Song); Data curation, Y.S. (Yuying Song), K.Z. and Y.S. (Yue Su); Formal analysis, H.Z. and Z.X.; Funding acquisition, Z.X.; Investigation, K.Z. and Y.S. (Yue Su); Methodology, Y.S. (Yuying Song) and Z.X.; Project administration, Z.X.; Resources, Z.X.; Software, H.Z.; Supervision, Z.X.; Writing—original draft, K.Z. and Y.S. (Yuying Song); Writing—review and editing, H.Z. and Y.S. (Yuying Song). All authors have read and agreed to the published version of the manuscript.

Funding: This research was funded by the National Natural Science Foundation of China under Grants No. 12032009 and No. 12002091.

Data Availability Statement: The original data are available upon request. The data are not publicly available due to project privacy.

Acknowledgments: The authors thank all the technical personnel from the Structural Laboratory of Guangdong University of Technology for their assistance during the experiments.

Conflicts of Interest: The authors declare no conflicts of interest.

References

1. Nham, N.T.H.; Ha, L.T. The role of financial development in improving marine living resources towards sustainable blue economy. *J. Sea Res.* **2023**, *195*, 102417. [CrossRef]
2. Jiang, S.-S.; Li, J.-M. Exploring the motivation and effect of government-enterprise collusion in the utilization of marine resources: Evidence from China's coastal areas. *Ocean. Coast. Manag.* **2021**, *212*, 105822. [CrossRef]
3. Yang, Z.; Zhan, X.; Zhu, H.; Zhang, B.; Feng, P.; Li, H.; Kua, H.W. Performance-based alkali-activated seawater sea-sand concrete: Mixture optimization for mechanical, environmental, and economical objectives. *Constr. Build. Mater.* **2023**, *409*, 134156. [CrossRef]
4. Liu, X.; Asghari, V.; Lam, C.-M.; Hsu, S.-C.; Xuan, D.; Angulo, S.C.; John, V.M.; Basavaraj, A.S.; Gettu, R.; Xiao, J.; et al. Discrepancies in life cycle assessment applied to concrete waste recycling: A structured review. *J. Clean. Prod.* **2024**, *434*, 140155. [CrossRef]
5. Tayebani, B.; Said, A.; Memari, A. Less carbon producing sustainable concrete from environmental and performance perspectives: A review. *Constr. Build. Mater.* **2023**, *404*, 133234. [CrossRef]
6. Liu, F.; Zhang, J.; Sun, J.; Zhao, D.; Gao, D.; Ma, Z.; Song, H.; Wu, R.; Zhang, W. The ice resistance of self-centering jacket offshore platform with concrete filled double-skin legs: Experimental study and numerical analysis. *Ocean Eng.* **2023**, *287*, 115876. [CrossRef]
7. Jiang, S.; Chen, G.; Zhu, Y.; Li, X.; Shen, X.; He, R. Real-time risk assessment of explosion on offshore platform using Bayesian network and CFD. *J. Loss Prev. Process. Ind.* **2021**, *72*, 104518. [CrossRef]
8. Demirci, S.E.; Elçiçek, H. Scientific awareness of marine accidents in Europe: A bibliometric and correspondence analysis. *Accid. Anal. Prev.* **2023**, *190*, 107166. [CrossRef]
9. Yan, Z.-W.; Bai, Y.-L.; Ozbakkaloglu, T.; Gao, W.-Y.; Zeng, J.-J. Rate-dependent compressive behavior of concrete confined with Large-Rupture-Strain (LRS) FRP. *Compos. Struct.* **2021**, *272*, 114199. [CrossRef]
10. Bai, Y.-L.; Yan, Z.-W.; Ozbakkaloglu, T.; Gao, W.-Y.; Zeng, J.-J. Mechanical behavior of large-rupture-strain (LRS) polyethylene naphthalene fiber bundles at different strain rates and temperatures. *Constr. Build. Mater.* **2021**, *297*, 123786. [CrossRef]
11. Yan, Z.-W.; Bai, Y.-L.; Ozbakkaloglu, T.; Gao, W.-Y.; Zeng, J.-J. Axial impact behavior of Large-Rupture-Strain (LRS) fiber reinforced polymer (FRP)-confined concrete cylinders. *Compos. Struct.* **2021**, *276*, 114563. [CrossRef]
12. Xiang, J.; Qiu, J.; Zhao, Y.; Zheng, P.; Peng, H.; Fei, X. Rheology, mechanical properties, and hydration of synergistically activated coal gasification slag with three typical solid wastes. *Cem. Concr. Compos.* **2024**, *147*, 105418. [CrossRef]
13. Yang, L.; Xie, H.; Zhang, D.; Zhang, F.; Lin, C.; Fang, S. Acoustic emission characteristics and crack resistance of basalt fiber reinforced concrete under tensile load. *Constr. Build. Mater.* **2021**, *312*, 125442. [CrossRef]
14. Xiang, J.; Qiu, J.; Wu, P.; Zhang, Q.; Song, Y.; Yang, L. Autolytic capsules incorporating alkali-activated slag system for self-healing cementitious composites. *Cem. Concr. Compos.* **2024**, 105439. [CrossRef]
15. Liang, N.; Geng, S.; Mao, J.; Liu, X.; Zhou, X. Investigation on cracking resistance mechanism of basalt-polypropylene fiber reinforced concrete based on SEM test. *Constr. Build. Mater.* **2024**, *411*, 134102. [CrossRef]
16. Zhen, H.; Xiong, Z.; Song, Y.; Li, L.; Qiu, Y.; Zou, X.; Chen, B.; Chen, D.; Liu, F.; Ji, Y. Early mechanical performance of glass fibre-reinforced manufactured sand concrete. *J. Build. Eng.* **2024**, *83*, 108440. [CrossRef]
17. Ahmed, S.; Ali, M. Use of agriculture waste as short discrete fibers and glass-fiber-reinforced-polymer rebars in concrete walls for enhancing impact resistance. *J. Clean. Prod.* **2020**, *268*, 122211. [CrossRef]
18. Wang, W.; Zhang, Y.; Mo, Z.; Chouw, N.; Jayaraman, K.; Xu, Z.-D. A critical review on the properties of natural fibre reinforced concrete composites subjected to impact loading. *J. Build. Eng.* **2023**, *77*, 107497. [CrossRef]
19. Xie, C.; Cao, M.; Khan, M.; Yin, H.; Guan, J. Review on different testing methods and factors affecting fracture properties of fiber reinforced cementitious composites. *Constr. Build. Mater.* **2021**, *273*, 121766. [CrossRef]

20. Mastali, M.; Dalvand, A.; Sattarifard, A. The impact resistance and mechanical properties of reinforced self-compacting concrete with recycled glass fibre reinforced polymers. *J. Clean. Prod.* **2016**, *124*, 312–324. [CrossRef]
21. Liu, L.; Wang, X. Research on test of alkali-resistant glass fibre enhanced seawater coral aggregate concrete. In Proceedings of the 1st International Conference on Frontiers of Materials Synthesis and Processing (FMSP 2017), Changsha, China, 28–29 October 2017.
22. Sun, W.; Chen, H.; Luo, X.; Qian, H. The effect of hybrid fibers and expansive agent on the shrinkage and permeability of high-performance concrete. *Cem. Concr. Res.* **2001**, *31*, 595–601. [CrossRef]
23. Fan, Q.; Fan, L.; Quach, W.-M.; Zhang, R.; Duan, J.; Sand, W. Application of microbial mineralization technology for marine concrete crack repair: A review. *J. Build. Eng.* **2023**, *69*, 106299. [CrossRef]
24. Liu, K.; Shui, Z.; Sun, T.; Ling, G.; Li, X.; Cheng, S. Effects of combined expansive agents and supplementary cementitious materials on the mechanical properties, shrinkage and chloride penetration of self-compacting concrete. *Constr. Build. Mater.* **2019**, *211*, 120–129. [CrossRef]
25. Mo, L.; Deng, M.; Tang, M.; Al-Tabbaa, A. MgO expansive cement and concrete in China: Past, present and future. *Cem. Concr. Res.* **2014**, *57*, 1–12. [CrossRef]
26. Wang, A.G.; Deng, M.; Sun, D.S.; Mo, L.W.; Wang, J.; Tang, M.S. Effect of Combination of Steel Fiber and MgO-type Expansive Agent on Properties of Concrete. *J. Wuhan Univ. Technol. -Mater. Sci. Ed.* **2011**, *26*, 786–790. [CrossRef]
27. Xiong, Z.; He, S.; Kwan, A.; Li, L.; Zeng, Y. Compressive behaviour of seawater sea-sand concrete containing glass fibres and expansive agents. *Constr. Build. Mater.* **2021**, *292*, 123309. [CrossRef]
28. Yang, G.; Zhao, J.; Wang, Y. Durability properties of sustainable alkali-activated cementitious materials as marine engineering material: A review. *Mater. Today Sustain.* **2022**, *17*, 100099. [CrossRef]
29. Li, Y.; Chen, G.; Yang, Z.; Li, Y.; Xu, M. Experimental study on the reaction process and engineering characteristics of marine calcareous sand reinforced by eco-friendly methods. *Appl. Ocean Res.* **2023**, *138*, 103641. [CrossRef]
30. Xiao, J.; Qiang, C.; Nanni, A.; Zhang, K. Use of sea-sand and seawater in concrete construction: Current status and future opportunities. *Constr. Build. Mater.* **2017**, *155*, 1101–1111. [CrossRef]
31. Zhao, Y.; Hu, X.; Shi, C.; Zhang, Z.; Zhu, D. A review on seawater sea-sand concrete: Mixture proportion, hydration, microstructure and properties. *Constr. Build. Mater.* **2021**, *295*, 123602. [CrossRef]
32. Xiang, J.; Qiu, J.; Li, Z.; Chen, J.; Song, Y. Eco-friendly treatment for MSWI bottom ash applied to supplementary cementing: Mechanical properties and heavy metal leaching concentration evaluation. *Constr. Build. Mater.* **2022**, *327*, 127012. [CrossRef]
33. Su, Y.; Xiong, Z.; Hu, Z.; Zhu, W.; Zhou, K.; Wang, J.; Liu, F.; Li, L. Dynamic bending study of glass fiber reinforced seawater and sea-sand concrete incorporated with expansive agents. *Constr. Build. Mater.* **2022**, *358*, 129415. [CrossRef]
34. Xiong, Z.; Zeng, Y.; Li, L.; Kwan, A.; He, S. Experimental study on the effects of glass fibres and expansive agent on the bond behaviour of glass/basalt FRP bars in seawater sea-sand concrete. *Constr. Build. Mater.* **2021**, *274*, 122100. [CrossRef]
35. Chen, L.; Zhou, Q.; Yue, L.; Wu, M.; Huang, R.; Yuen, K.F.; Su, R. A theoretical model for preventing marine litter behaviour: An empirical evidence from Singapore. *J. Clean. Prod.* **2023**, *427*, 139109. [CrossRef]
36. Chen, Y.; Li, P.; Zhang, S. Experimental investigation on triaxial mechanical properties of coral coarse aggregate-sea sand seawater concrete. *Constr. Build. Mater.* **2023**, *409*, 134213. [CrossRef]
37. Zhang, B.; Zhu, H. Compressive stress–strain behavior of slag-based alkali-activated seawater coral aggregate concrete after exposure to seawater environments. *Constr. Build. Mater.* **2023**, *367*, 130294. [CrossRef]
38. Guan, H.; Hao, B.; Zhang, G. Mechanical properties of concrete prepared using seawater, sea sand and spontaneous combustion coal gangue. *Structures* **2023**, *48*, 172–181. [CrossRef]
39. Choi, D.-U.; Chun, S.-C.; Ha, S.-S. Bond strength of glass fibre-reinforced polymer bars in unconfined concrete. *Eng. Struct.* **2012**, *34*, 303–313. [CrossRef]
40. Guo, R.; Ren, H.; Zhang, L.; Long, Z.; Jiang, X.; Wu, X.; Wang, H. Direct dynamic tensile study of concrete materials based on mesoscale model. *Int. J. Impact Eng.* **2020**, *143*, 103598. [CrossRef]
41. Ross, C.A.; Thompson, P.Y.; Tedesco, J.W. Split-hopkinson pressure-bar tests on concrete and mortar in tension and compression. *Mater. J.* **1989**, *86*, 475–481.
42. Guo, H.; Tao, J.; Chen, Y.; Li, D.; Jia, B.; Zhai, Y. Effect of steel and polypropylene fibers on the quasi-static and dynamic splitting tensile properties of high-strength concrete. *Constr. Build. Mater.* **2019**, *224*, 504–514. [CrossRef]
43. Wang, Q.; Jia, X.; Kou, S.; Zhang, Z.; Lindqvist, P.-A. The flattened Brazilian disc specimen used for testing elastic modulus, tensile strength and fracture toughness of brittle rocks: Analytical and numerical results. *Int. J. Rock Mech. Min. Sci.* **2004**, *41*, 245–253. [CrossRef]
44. Wang, Q.Z.; Wu, L.Z. The flattened Brazilian disc specimen used for determining elastic modulus, tensile strength and fracture toughness of brittle rocks: Experimental results. *Int. J. Rock Mech. Min. Sci.* **2004**, *41*, 357–358. [CrossRef]
45. Wang, Q.; Li, W.; Xie, H. Dynamic split tensile test of Flattened Brazilian Disc of rock with SHPB setup. *Mech. Mater.* **2009**, *41*, 252–260. [CrossRef]
46. Huang, Y.G.; Wang, L.G.; Lu, Y.L.; Chen, J.R.; Zhang, J.H. Semi-analytical and Numerical Studies on the Flattened Brazilian Splitting Test Used for Measuring the Indirect Tensile Strength of Rocks. *Rock Mech. Rock Eng.* **2015**, *48*, 1849–1866. [CrossRef]
47. Xie, F.; Jin, Z.; Yang, T.; Han, X.; Chen, X.; Zhang, Y. Experimental study of dynamic splitting-tensile properties of precast concrete samples under different strain rates. *Constr. Build. Mater.* **2023**, *372*, 130748. [CrossRef]

48. Li, J.; Yang, L.; Xie, H.; Wei, P.; Li, D.; Xu, Y.; Zhang, F. Research on impact toughness and crack propagation of basalt fiber reinforced concrete under SHPB splitting test. *J. Build. Eng.* **2023**, *77*, 107445. [CrossRef]
49. Du, W.; Zhang, D.; Yu, B. Mechanical and fractal characteristics of sandstone with Pre-existing fissures of different lengths under varying impact loads in SHPB tests. *Theor. Appl. Fract. Mech.* **2023**, *125*, 103884. [CrossRef]
50. Wu, N.; Fu, J.; Zhu, Z.; Sun, B. Experimental study on the dynamic behavior of the Brazilian disc sample of rock material. *Int. J. Rock Mech. Min. Sci.* **2020**, *130*, 104326. [CrossRef]
51. Wu, B.; Chen, R.; Xia, K. Dynamic tensile failure of rocks under static pre-tension. *Int. J. Rock Mech. Min. Sci.* **2015**, *80*, 12–18.
52. Lv, N.; Wang, H.; Rong, K.; Chen, Z.; Zong, Q. The numerical simulation of large diameter split Hopkinson pressure bar and Hopkinson bundle bar of concrete based on mesoscopic model. *Constr. Build. Mater.* **2022**, *315*, 125728. [CrossRef]
53. Xia, K.; Yao, W.; Wu, B. Dynamic rock tensile strengths of Laurentian granite: Experimental observation and micromechanical model. *J. Rock Mech. Geotechnol. Eng.* **2017**, *9*, 116–124. [CrossRef]
54. Mellor, M.; Hawkes, I. Measurement of tensile strength by diametral compression of discs and annuli. *Eng. Geol.* **1971**, *5*, 173–225. [CrossRef]
55. Zhang, Q.B.; Zhao, J. A Review of Dynamic Experimental Techniques and Mechanical Behaviour of Rock Materials. *Rock Mech. Rock Eng.* **2014**, *47*, 1411–1478. [CrossRef]
56. Zhou, H.Y.; Chen, Y.B.; Zhang, Y.R.; Wang, H.Q. Research Progress of Strain-Rate Effect on Mechanical Properties of Concrete. *Appl. Mech. Mater.* **2014**, *638–640*, 1391–1396. [CrossRef]
57. Malvar, L.J.; Ross, C.A. Review of strain rate effects for concrete in tension. *ACI Mater. J.* **1998**, *95*, 735–739.
58. Feng, W.; Liu, F.; Yang, F.; Li, L.; Jing, L. Experimental study on dynamic split tensile properties of rubber concrete. *Constr. Build. Mater.* **2018**, *165*, 675–687. [CrossRef]
59. Song, B.; Chen, W. Energy for Specimen Deformation in a Split Hopkinson Pressure Bar Experiment. *Exp. Mech.* **2006**, *46*, 407–410. [CrossRef]
60. Xie, H.; Yang, L.; Zhang, Q.; Huang, C.; Chen, M.; Zhao, K. Research on energy dissipation and damage evolution of dynamic splitting failure of basalt fiber reinforced concrete. *Constr. Build. Mater.* **2022**, *330*, 127292. [CrossRef]

buildings

MDPI

Article

Bond Performance between Fiber-Wrapped Ribbed Basalt Fiber-Reinforced Polymer Bars and Seawater Sea-Sand Concrete

Min Lin [1], Chenyue Weng [2], Hesheng Xiao [2], Dong Zeng [2], Baifa Zhang [3,*], Xiaopan Chen [3], Shaohua He [3] and Lijuan Li [3,*]

[1] School of Surveying, Mapping and Remote Sensing Information, Guangdong Polytechnic of Industry and Commerce, Guangzhou 510550, China; 0001200037@gdgm.edu.cn

[2] Guangdong Bocheng Construction Co., Ltd., Guangzhou 510660, China; 17818588537@163.com (C.W.); hill_2023@163.com (H.X.); baishikefa@163.com (D.Z.)

[3] School of Civil and Transportation Engineering, Guangdong University of Technology, Guangzhou 510006, China; 2112009109@mail2.gdut.edu.cn (X.C.); 1112209006@mail2.gdut.edu.cn (S.H.)

* Correspondence: zhangbaifa@gdut.edu.cn (B.Z.); lilj@gdut.edu.cn (L.L.)

Abstract: The high corrosion resistance of fiber-reinforced polymers (FRPs) and related concrete structures means that they are suitable for application in the marine environment. Therefore, the replacement of steel bars with fiber-reinforced polymer (FRP) bars enhances corrosion resistance in seawater sea-sand concrete (SSC) structures. Geometric parameters significantly influence the performance of the bond between ribbed FRP bars and SSC, thereby affecting the mechanical properties of the concrete structures. In this study, the performance of the bond between ribbed (i.e., with fiber wrapping) basalt-fiber-reinforced polymer (BFRP) bars and SSC was investigated through pull-out tests that considered rib geometry and SSC strength. The results demonstrated that an increase in rib and dent widths reduced the bond stiffness, while an increase in rib height and SSC strength gradually increased the bond stiffness and strength. Additionally, the bond stiffness and bond strength were relatively low because the surface fiber bundles buffered the mechanical interlocking force between the BFRP ribs and the concrete, resulting in plastic bond failure during the loading process. Furthermore, the adhesion of the fiber bundles to the surface of the BFRP bars also influenced bond performance, with higher adhesion leading to greater bond stiffness and strength.

Keywords: fiber wrapping; ribbed BFRP bar; rib geometry; seawater sea-sand concrete; bond stress–slip curve

Citation: Lin, M.; Weng, C.; Xiao, H.; Zeng, D.; Zhang, B.; Chen, X.; He, S.; Li, L. Bond Performance between Fiber-Wrapped Ribbed Basalt Fiber-Reinforced Polymer Bars and Seawater Sea-Sand Concrete. *Buildings* **2024**, *14*, 38. https://doi.org/10.3390/buildings14010038

Academic Editor: Flavio Stochino

Received: 18 November 2023
Revised: 8 December 2023
Accepted: 10 December 2023
Published: 22 December 2023

1. Introduction

The construction material industry consumes large amounts of freshwater, river sand, and other natural resources [1,2]. Approximately 20 billion tonnes of concrete is produced annually, and the production process requires the use of approximately 15 billion tonnes of aggregates and 1.8 billion tonnes of water [3,4]. The extensive consumption of freshwater has exacerbated its scarcity, and the extensive exploitation of river sand has caused damage to riverbeds [5,6]. To alleviate the environmental damage caused by concrete preparation, seawater and sea sand have been used to replace freshwater and river sand [7,8]. Several studies have reported the feasibility of using seawater and sea sand to prepare concrete with workability and mechanical properties similar to those of traditional concrete (TC). Wang et al. [9] found that seawater could enhance the hydration of ordinary Portland cement (OPC), resulting in a denser microstructure, higher crystal content, and reduced autogenous shrinkage. Pan et al. [10] compared the compressive behavior and microstructures of seawater sea-sand concrete (SSC), sea-sand concrete (SC), and TC. They found that the utilization of seawater and sea sand increased the early strength but reduced the late compressive strength of concrete, with seawater and sea sand inducing the formation of the ettringite phase. Fang et al. [11] investigated the axial compressive performance of square SSC columns reinforced with glass-fiber-reinforced polymer (FRP) interlocking multi-spirals and proposed a design-oriented load–strain model for predicting monotonic and cyclic axial responses of this structure. Other studies have found that the chloride and sulfate content in seawater and sea sand can accelerate the corrosion of steel bars in reinforced concrete structures, significantly reducing the bearing capacity and durability of concrete structures [12–14]. To address this issue, FRP bars are used to replace steel bars in SSC-based structures owing to their light weight, low thermal conductivity, and strong corrosion resistance [15].

FRP bars are produced by mixing fibers and polymeric materials through the pultrusion process, and have been extensively studied as a promising alternative for the construction of marine structures [7,16,17]. The strength of the bond between the bars and the concrete plays a significant role in the ultimate and serviceability limit states (e.g., anchorage length and crack width) of concrete structures [18]. The differing composition and physical properties of FRP and steel are reflected in their distinct bond behavior with concrete. Typically, bond strength is influenced by factors such as bond length, cover thickness, concrete strength, transverse compression, bar diameter, and surface-treatment methods [19–21]. Among these factors, the surface treatments (involving deformation, sand-coating, or combinations of both) of FRP bars are some of the most critical [22].

Deformed FRP bars exhibit better adhesion to concrete than sand-coated FRP bars owing to the excellent mechanical interlocking force of the former [23,24]. Furthermore, as depicted in Figure 1, deformed FRP bars can be further categorized into three types according to their production processes: (1) fiber-wrapped bars, in which ribs are formed by helically wrapped fibers (remaining in the bar) on the smooth bar [25]; (2) indented bars, in which ribs are formed via a helix cutting method on the smooth bar [26]; and (3) ribbed bars, which are formed by squeezing strands onto the smooth bar [27]. Ribbed bars can be divided into two types depending on the presence or absence of fiber. The behavior of bar–concrete bonding varies for these two types of ribbed bars. Hao et al. [28] found that the bonding behavior of ribbed (without fiber) glass FRP-bar-reinforced concrete was affected by rib parameters. Wei et al. [29] compared the bond performances of concretes reinforced with different fiber-wrapped ribbed FRP bars and concluded that carbon FRP bars showed the greatest bond strength and durability, while glass FRP bars exhibited the smallest bond strength, as the fibers were easily cut off and attached to the concrete interface.

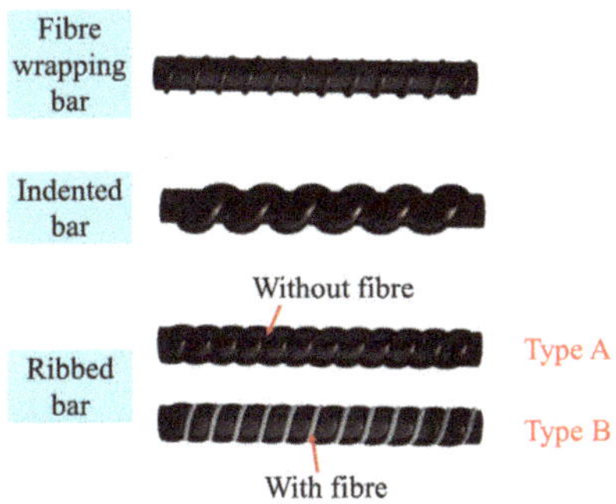

Figure 1. Three types of deformed FRP bar.

In addition to the treatment types, rib geometry, including dent width (DW), rib width (RW), rib height (RH), and rib spacing (RS), also has a significant effect on the behavior of the bond between FRP bars and concrete. Solyom and Balazs [26] reported that the bond strength of concrete reinforced with Type A ribbed glass FRP bars increased by 100% as the ratio of DW/RS increased from 0.17 to 0.30. Through finite element analysis, Sasmal et al. [30] found that a decrease in RS was beneficial in improving bond strength when RS was greater than 80% of the bar diameter. Shan et al. [31] reported that the bond strength of

Type A ribbed basalt FRP (BFRP) bars with concrete was significantly enhanced by a factor of 11 as the RH increased from 0 to 0.76 mm, and they proposed a novel bar-geometry-based model. Furthermore, bond failure between deformed FRP bars and concrete is also affected by the properties of the concrete and the geometry of the ribs on the FRP bars [32]. However, although existing research has confirmed that rib geometry has a significant effect on the bond performance between FRP bars and concrete, its influencing mechanism is still unclear, and the relevant theoretical models are not yet perfect [20,31,33]. For these reasons, in existing standards, the effects of rib geometry are not considered in calculating the development length. For example, Japan standard JSCE [34] and Canadian standard CSA S806 [35] only generally consider the effects of bar surface treatments in calculating the development length, without specific provisions for rib geometry. In the American standard [36], even the effects of bar surface treatment are not taken into account when calculating the development length. To determine the development, for FRP-bar-reinforced concrete structures, it is necessary to clarify the effects of rib geometry on the bond between the FRP bars and the concrete.

Fiber-wrapped ribbed bars have received less attention than ribbed bars without fiber. In our previous studies [37], we found that the damage to the bar ribs could be reduced by fiber floccules formed by strands damaged under friction. We have also observed that fiber-wrapped ribbed BFRP bars can undergo full deformation, unlike ribbed bars that do not have fiber wrapping, with the former showing higher bond strength than the latter. These results indicate that the presence of fiber in the rib favors the improvement of bond strength between the bars and the concrete. Therefore, clarifying the effects of rib geometry on the bond performance between the FRP bars and the concrete is necessary, especially for fiber-wrapped ribbed bars.

Among all the FRP bars, BFRP bars were used in this study because they are cost-effective and result in low pollution during manufacture [38,39]. In this study, we focused on the combination of FRP bars and concrete prepared with seawater sea sand, and especially on the effect of rib geometry (RW, DW, and RH of the ribbed bars) and concrete strength on the bond behavior (failure modes of the bond interface, bond stress–slip curves, and bond indexes) of the bonds between fiber-wrapped ribbed BFRP bars and SSC. These parameters are of great importance in promoting the application of SSC structures in the marine environment, and were studied in detail.

2. Experimental Programme

2.1. Seawater Sea-Sand Concrete (SSC)

OPC (P.O. 42.5R) was used as the binder. Its chemical composition, determined by X-ray fluorescence spectrometer, was as follows: SiO_2 (11.5 wt%), Al_2O_3 (2.5 wt%), CaO (76.2 wt%), Fe_2O_3 (3.9 wt%), K_2O (1.2 wt%), TiO_2 (0.6 wt%), SO_3 (3.7 wt%), and others (0.4 wt%). Both seawater and sea sand were collected at a shore in Zhuhai city, Guangdong (N22°13′; E113°33′) for use as mixed water and fine aggregates. The seawater mainly contained ions such as Na^+ (2500 mg/L), Cl^- (4126 mg/L), Mg^{2+} (981 mg/L), and $SO4^{2-}$ (379 mg/L). Crushed stones were used as coarse aggregates. The particle size distributions of fine and coarse aggregates are presented in Figure 2.

Following specification JGJ 55-2011 [40], we prepared SSC with three strength grades by adjusting the mix proportion of the concrete, as shown in Table 1. According to American standards ASTM C39/C39M [41] and ASTM C469/C469M [42], the compressive tests were carried out at the concrete age of 28 d. The compressive strengths were approximately 43, 58, and 68 MPa, and the elastic moduli were 26, 29, and 31 GPa, respectively.

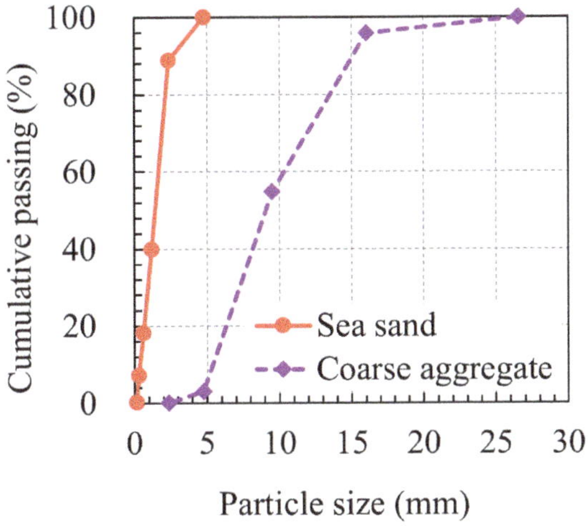

Figure 2. Particle size distributions of aggregates.

Table 1. Mix proportions of SSC.

Strength Grade	Cement (kg/m³)	Seawater (kg/m³)	Sea Sand (kg/m³)	Coarse Aggregates (kg/m³)
C30	400.83	230.00	721.82	1047.35
C45	540.07	230.00	573.57	1056.36
C60	686.50	230.00	438.82	1044.68

2.2. Ribbed BFRP Bars

Bar diameter, RW, DW, and RH are typically used to characterize rib geometry (Figure 3). In this study, fiber-wrapped ribbed BFRP bars with a nominal diameter of 10 mm were selected, and the effects of rib geometry on the mechanical properties of BFRP-reinforced SSC were investigated. The detailed rib geometry (with an error of <5%) and mechanical properties (tensile strength and elastic modulus) of the ribbed BFRP bars are listed in Table 2. The samples are denoted as $W_XD_YH_Z$ or $W_XD_YH_Z$-C_G, where X, Y, and Z represent the values of RW, DW, and RH, respectively. -C_G indicates different SSC strengths, and G represents the concrete grade. Notably, those sample IDs without -C_G indicate that the concrete strength was 45 MPa.

Table 2. Detailed rib geometry and mechanical properties of ribbed BFRP bars.

No.	Sample ID	Diameter (mm)	Rib Width (mm)	Dent Width (mm)	Rib Height (mm)	Equivalent Diameter (mm)	Tensile Strength (Mpa)	Elastic Modulus (Gpa)
1	$W_{1.5}D_{2.5}H_{0.6}$	10.67	1.60	2.68	0.61	10.98	1069.52	43.29
2	$W_{3.5}D_{2.5}H_{0.6}$	10.70	3.63	2.52	0.61	11.18	1089.95	47.20
3	$W_{5.5}D_{2.5}H_{0.6}$	10.61	5.69	2.67	0.63	11.18	1218.84	44.40
4	$W_{7.0}D_{2.5}H_{0.6}$	10.15	7.14	2.78	0.68	11.06	1158.43	44.08
5	$W_{7.5}D_{2.0}H_{0.7}$	10.75	7.65	2.06	0.67	11.45	975.86	43.34
6	$W_{7.5}D_{3.0}H_{0.7}$	10.35	7.90	2.97	0.73	10.80	1305.98	41.59
7	$W_{7.5}D_{3.3}H_{0.7}$	10.55	8.05	3.31	0.71	11.22	1012.06	44.78
8	$W_{7.5}D_{3.0}H_{0.4}$	10.18	7.52	2.83	0.45	10.62	1006.58	42.76
9	$W_{7.5}D_{3.0}H_{1.0}$	10.45	8.09	2.75	1.01	11.46	997.86	46.08
10	$W_{7.5}D_{3.0}H_{1.7}$	10.66	7.51	3.39	1.70	12.22	967.64	35.14
11	$W_{7.5}D_{3.0}H_{0.7}$-C_{30}	10.35	7.90	2.97	0.73	10.80	1305.98	41.59
12	$W_{7.5}D_{3.0}H_{0.7}$-C_{60}	10.35	7.90	2.97	0.73	10.80	1305.98	41.59

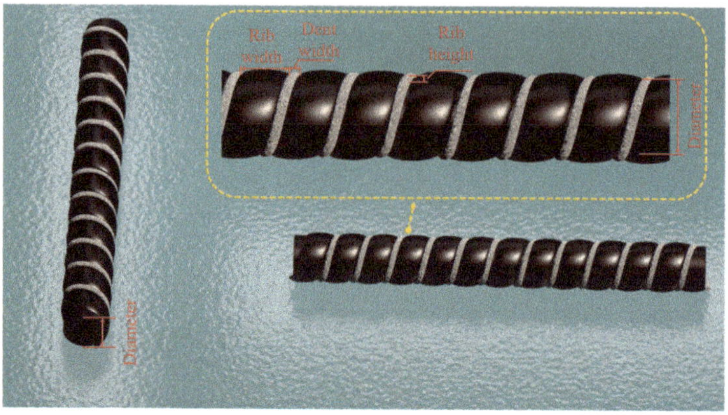

Figure 3. Geometric details of fiber-wrapped ribbed BFRP bars.

2.3. Specimen Preparation

The process used to prepare specimens is shown in Figure 4, and the detailed preparation process can be found in our previous study [7]. The preparation of SSC was divided into two steps: (I) the sea sand, coarse aggregates, and 70% of seawater were poured into the mixer and mixed for 30 s, and (II) the cement and the remaining seawater were poured in the mixer and mixed for 60 s. The size of the SSC cube was 200 × 200 × 200 mm^3, and the BFRP bar was 550 mm in length. Three replicates were prepared and tested for each sample. These pull-out specimens were cured for 28 days in a room with a temperature of 20 ± 3 °C and a humidity of 95%.

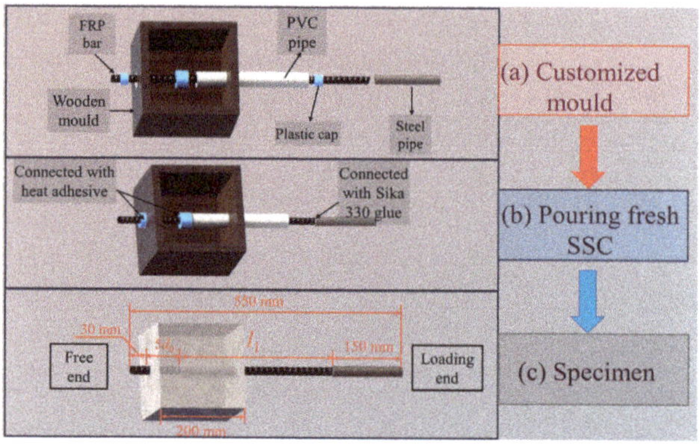

Figure 4. Specimen preparation process. (**a**) customized mould; (**b**) pouring fresh SSC; (**c**) Specimens.

2.4. Test Setup

A Matest 330 testing machine was used to test the pull-out behavior of SSC reinforced with fiber-wrapped ribbed BFRP. The specimens were covered with gypsum at the bottom to level the contact surface and reduce friction. According to American standard ACI 440.3R-12 [43], a displacement rate of 1.2 mm/min was applied during the loading, and the test was stopped when the displacement distance exceeded 20 mm. It is assumed that the bond stress is evenly distributed along the bonded length, so bond stress (τ) was calculated by Equation (1). The slip between the BFRP bar and SSC at the loading end (s) was calculated by Equation (2).

$$\tau = \frac{F}{\pi d_0 l_b} \tag{1}$$

$$s = \frac{s'EA - l_1F}{F + EA} \tag{2}$$

where A, E, and d_0 represents the cross-sectional area, elastic modulus, and equivalent diameter of the BFRP bar, respectively; F is the loading force; l_b is the bond length, which equals $5d_0$; s' is the displacement at the position of the steel pipe in Figure 4; and l_1 is the length in Figure 4.

3. Experimental Results and Discussions

3.1. Failure Modes of the Bond Interface between Bars and SSC

All specimens were split to observe the failure modes of the bond interface (Figure 5). All of these specimens experienced pull-out failure, and the rib geometry and SSC strength had little effect on the bond interface failure mode of the specimens. The appearances of different fiber-wrapped ribbed BFRP bars after the pull-out test were similar: crushed concrete was embedded on the surface of the fibers and ribs, the fibers were destroyed, and flocculent fibers were attached to the concrete surface.

As the RW and DW increased, the presence of damaged white fiber became more apparent, both on the bar's surface and in the concrete, because a wider rib or dent allowed more room for fiber slippage. Increasing the concrete strength also resulted in more noticeable peeled fiber, owing to the increased bonding strength between the concrete and bars [44]. With increasing RH, the amount of flocculent fiber in the concrete decreased, and the concrete was more severely damaged.

Despite these observations, the specimens could be categorized into two types according to the damage to the interface.

The first type exhibited brittle-fracture failure of the ribs, including the peeling of ribs from BFRP bars and the shearing of the concrete ribs. This was often accompanied by the breakage or severe wear of entangled fibers on the bar's surface (as in samples $W_{7.5}D_{3.3}H_{0.7}$, $W_{7.5}D_{3.0}H_{0.7}$, $W_{7.5}D_{3.0}H_{1.0}$, $W_{7.5}D_{3.0}H_{1.7}$, and $W_{7.5}D_{3.0}H_{1.7}$-C_{60}). The second type had ribs with a relatively intact shape (as seen in samples $W_{1.5}D_{2.5}H_{0.6}$, $W_{7.5}D_{3.0}H_{0.4}$, and $W_{7.5}D_{3.0}H_{1.7}$-C_{30}), in which only deformed and worn ribs occurred, owing to the compressive interaction between the BFRP bars, the wrapped fiber bundles, and the concrete.

The primary forces between fiber-wrapped BFRP bars and concrete are chemical adhesion, friction, and mechanical interlocking force [24]. During the initial loading stage, the bond stress was mainly provided by the chemical adhesion [33]. However, the chemical adhesive force disappeared once a slight slip occurred, owing to the separation of bars and concrete [45]. At this stage, relative slip between BFRP bars and concrete occurred on the inclined plane where their ribs came into contact (Figure 6). Simultaneously, the concrete was both sheared and compressed by the component forces along the pulling and vertical directions of the ribs [46]. Compared with the threaded surface of BFRP bars under similar conditions, the fiber wrapping around the dents reduced the gap between the BFRP bars and the concrete, significantly cushioning the strong mechanical interlocking effect in the bonding section, thus favoring the integrity of both the BFRP and the concrete ribs [33].

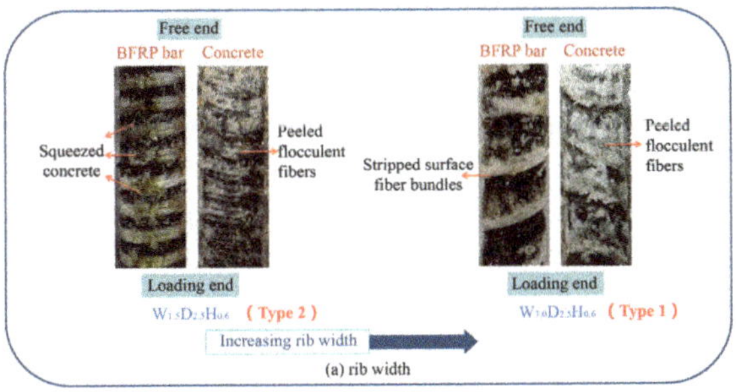

Figure 5. *Cont.*

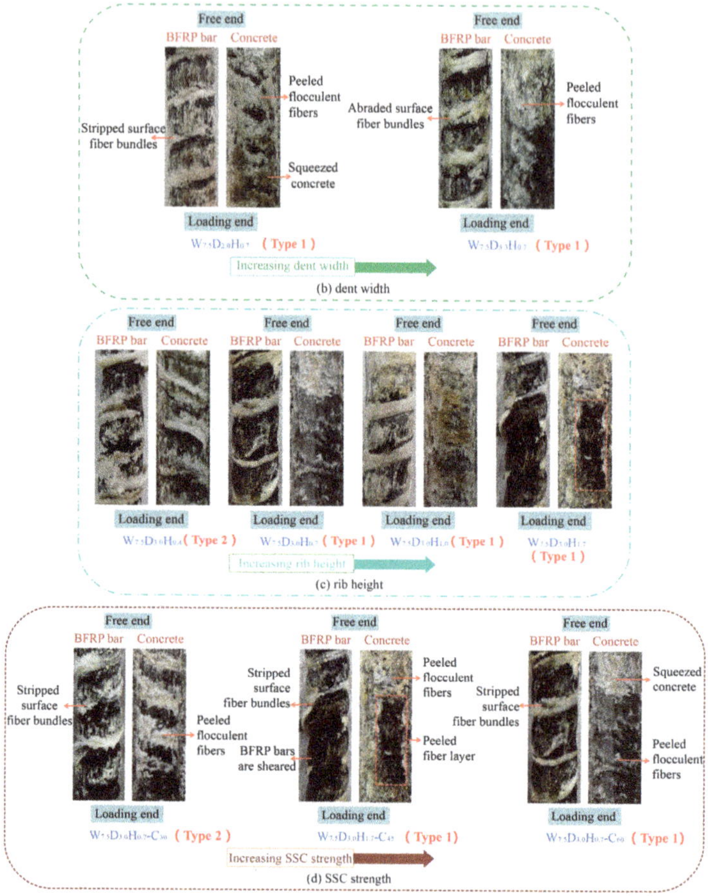

Figure 5. Failure modes of the bond interface between fiber-wrapped BFRP bars and SSC: (**a**) RW, (**b**) DW, (**c**) RH, and (**d**) SSC strength.

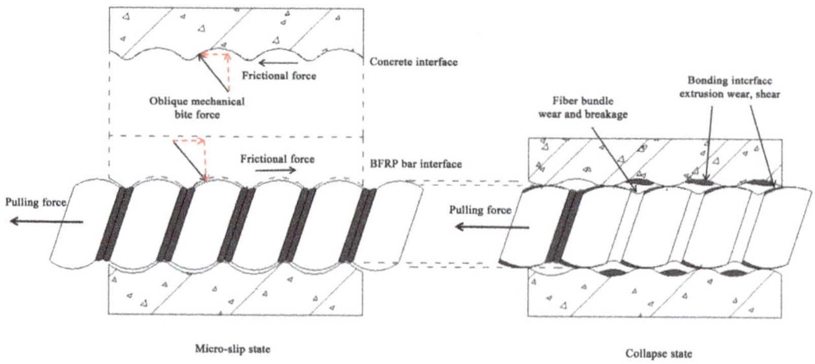

Figure 6. Bond mechanism between a fiber-wrapped BFRP bar and SSC.

As the pulling force increased further, the fibers and ribs of the BFRP and the concrete began to suffer damage owing to their relative slip. The deformed ribs of the BFRP bars and concrete interlocked, increasing the bond stress. However, with the accumulation of stress between the BFRP bar and the concrete, brittle-fracture failure of the ribs might occur when this stress surpasses the shear strength of either the BFRP bar's or the concrete's ribs. Moreover, the surface fibers were severely worn and broken, accompanied by a rapid degradation of the mechanical interlocking force until the bond ultimately failed. In cases in which the stress between the BFRP bar and the concrete did not exceed the shear strength of the ribs, the bar and concrete showed deformation and wear owing to mutual extrusion [8].

3.2. *Bond Stress–Slip Curves*

Figure 7 displays the bond stress–slip curve of BFRP-bar-reinforced SSC at varying rib geometries and concrete strengths. The curve can be categorized into three types:

Case 1, the bond stress–slip curve of $W_{7.5}D_{3.0}H_{0.5}$ can be divided into two parts—the ascending and descending parts (Figure 7a,m). The ascending part of the curve exhibited elastic characteristics, with slip increasing as stress increased. However, as slip continued to increase, bond damage occurred, and the curve entered a nonlinear upward phase. During this stage, the bonding stiffness of the BFRP-bar-reinforced concrete gradually decreased until it reached zero, while the bonding stress reached its maximum value. Afterward, the stress began to decline, and the slip continued to increase. In this phase, the bonding stress decreased rapidly until it reached zero.

Case 2, the bond stress–slip curve of $W_{7.0}D_{2.5}H_{0.6}$, $W_{7.5}D_{3.0}H_{0.7}$, $W_{7.5}D_{3.3}H_{0.7}$, $W_{7.5}D_{3.0}H_{0.4}$ $W_{7.5}D_{3.0}H_{1.7}$, $W_{7.5}D_{3.0}H_{0.7}$-C_{30}, and $W_{7.5}D_{3.0}H_{0.7}$-C_{60} can be divided into ascending, descending, and residual parts (Figure 7o). In the ascending part, the bonding stress increased rapidly owing to the high bond stiffness. After the bond stress reached its maximum value, it decreased significantly but did not reach zero, because the rib of the free end of the BFRP bar entered the bonding section, providing a new mechanical interlocking force before the bond stress dropped to zero. Subsequently, the bond stress–slip curve entered the residual part. Owing to the protection from the surface fiber bundles, mechanical occlusion between the bars and concrete remained the main effect in the residual stage. However, the bond stiffness and bond stress in the residual part were smaller than those in the ascending part owing to the accumulation of bond damage, resulting in a reduction in mechanical interlocking force [29].

Case 3, the bond stress–slip curve of $W_{1.5}D_{2.5}H_{0.6}$, $W_{3.5}D_{2.5}H_{0.6}$, $W_{5.5}D_{2.5}H_{0.6}$, and $W_{7.5}D_{2.0}H_{0.7}$ was also divided into three parts—ascending, descending, and residual parts (Figure 7n). In the ascending stage, the bond stress–slip curve rose linearly. As slip increased, bond damage occurred, and the curve entered a nonlinear rising stage. The bonding stiffness gradually decreased until it reached zero, and the bonding stress reached its peak. Subsequently, the curve entered the second stage (descending stage). In this stage, the bond stress decreased with increasing slip owing to the compression deformation and wear between BFRP ribs, surface fiber bundles, and concrete ribs, corresponding to plastic characteristics. Similar to Case 2, the stress–slip curve entered a residual stage after a rib of the BFRP bar at the free end entered the bonding part.

Changes in the rib geometry altered the shape of the bond stress–slip curve for these samples. For example, when the RW increased from 5.5 to 7.0 mm, the type of bond stress–slip curve transformed from Case 3 to Case 2. The effects of rib geometry and SSC strength on bond stiffness, bond strength, slip, and residual strength and slip were further investigated.

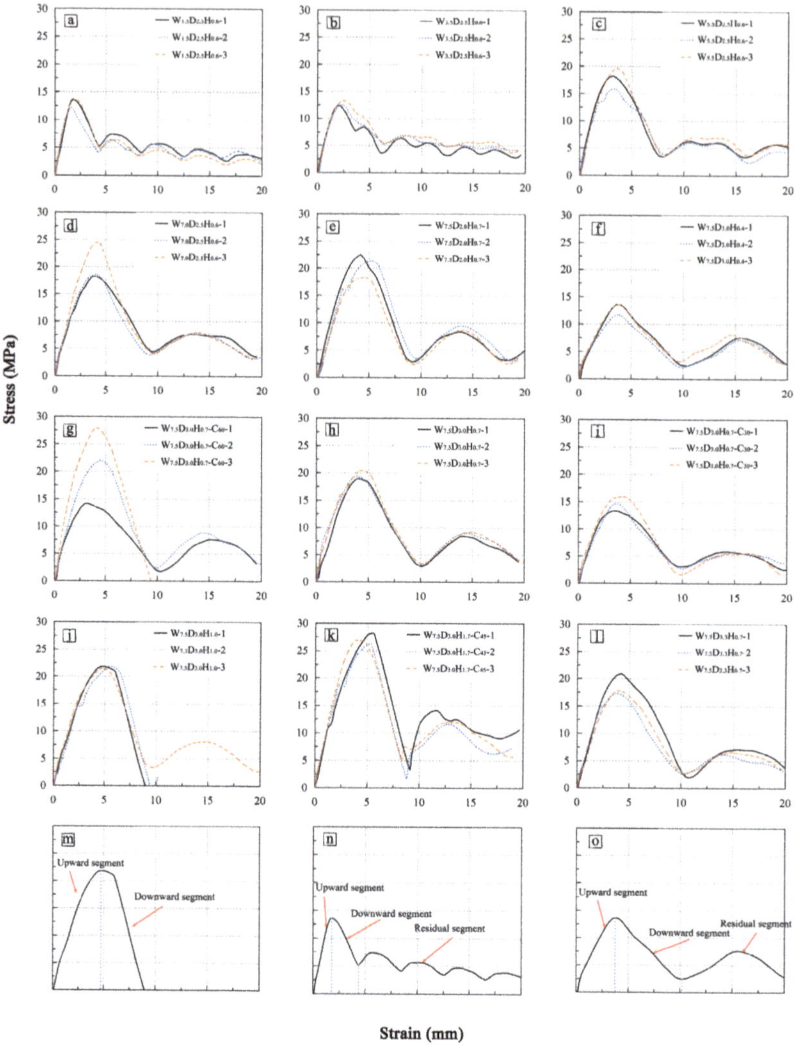

Figure 7. Bond stress–slip curves of all samples. (**a**) $W_{1.5}D_{2.5}H_{0.6}$; (**b**) $W_{3.5}D_{2.5}H_{0.6}$; (**c**) $W_{5.5}D_{2.5}H_{0.6}$; (**d**) $W_{7.0}D_{2.5}H_{0.6}$; (**e**) $W_{7.5}D_{2.0}H_{0.7}$; (**f**) $W_{7.5}D_{3.0}H_{0.4}$; (**g**) $W_{7.5}D_{3.0}H_{0.7}$-C_{60}; (**h**) $W_{7.5}D_{3.0}H_{0.7}$; (**i**) $W_{7.5}D_{3.0}H_{0.7}$-C_{30}; (**j**) $W_{7.5}D_{3.0}H_{1.0}$; (**k**) $W_{7.5}D_{3.0}H_{1.7}$-C_{45}; (**l**) $W_{7.5}D_{3.3}H_{0.7}$; and typical model: (**m**) Case 1; (**n**) Case 2; (**o**) Case 3.

3.3. Bond-Index Analysis

3.3.1. Bond Stiffness

In the bond stress–slip curve, the secant value at a slip of 0.5 mm was defined as the bond stiffness. Figure 8 illustrates the influence of rib geometry and SSC strength on bond stiffness.

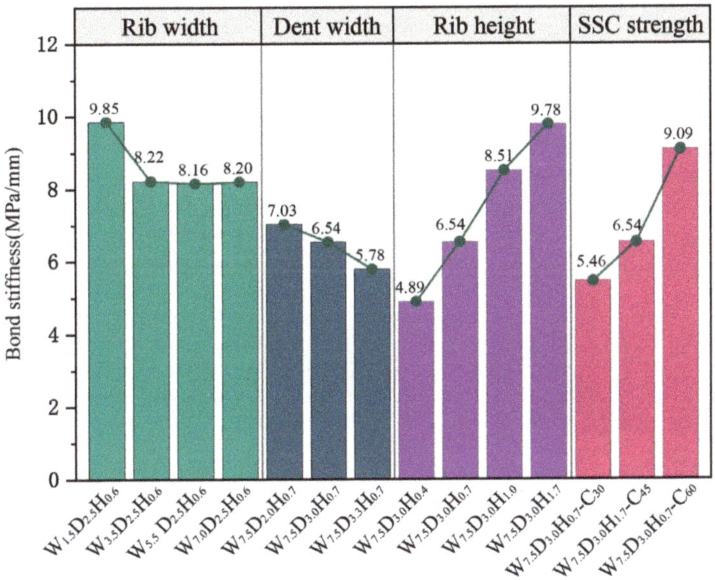

Figure 8. Bond stiffness of all samples.

Bond stiffness was highest (9.84 MPa/mm) when the RW was 1.5 mm. However, increasing the RW reduced the bond stiffness, and the bond stiffness values (approximately 8.20 MPa/mm) under RW values of 3.5, 5.5, and 7.0 mm were similar. This was attributable to the following: (1) Bond stiffness was affected by the contact angle of the inclined plane between the BFRP rib and the concrete, which is inversely proportional to the RW. When the RW was 1.5 mm, the inclination angle was at its maximum. (2) The smaller width of the BFRP ribs resulted in more concrete ribs within a certain bonding length. Consequently, the number of biting surfaces between the BFRP ribs and the concrete also increased. Therefore, 1.5 mm-width ribs exhibited greater mechanical interlocking force and bonding stiffness than other ribs owing to the combined effect of the above factors.

In addition, the increase in the DW also reduced the bond stiffness. However, the changes in DWs were caused by different processing technologies for fiber-wrapped BFRP bars. For example, only a thick fiber bundle was wrapped around the surface of the BFRP bar with a DW of 2.0 mm, whereas two thin fiber bundles of 3.0 and 3.3 mm were used in other cases. Therefore, in addition to the DW, the occurrence of different fiber thicknesses also influenced the bond stiffness. According to the bond interface damage (Figure 5b), the decrease in bond stiffness was likely due to the presence of gaps between the fiber and dent when the DW increased; the presence of gaps reduced the frictional force. Therefore, the higher the adhesion of the fiber bundles on the surface of BFRP bars in the dents, the greater the bonding stiffness and the better the bonding performance.

As shown in Figure 8, the bond stiffness increased from 4.89 to 9.78 MPa/mm as the RH increased from 0.61 to 1.7 mm. This was also attributable to the angle (which is proportionate to the RH) of the inclined plane between the BFRP rib and the concrete. A larger angle leads to a greater mechanical interlocking force and, consequently, greater bond stiffness. As the RH increased from 0.7 to 1.0 mm, the failure mode transformed into brittle-fracture failure owing to the shearing of the BFRP bars.

Figure 8 also shows that the bond stiffness increased with the increase in concrete strength. Higher concrete strength reduced the deformation capacity and thus increased the mechanical interlocking force [47]. When the concrete strength reached C60, the failure mode of the sample transformed into brittle-fracture failure.

3.3.2. Bond Strength and Bond Slip

The peak bond stress on the bond stress–slip curve was defined as the bond strength ($\tau_{b,x}$), and the slip corresponding to this bond strength was the bond slip ($s_{b,x}$).

Figure 9 illustrates the influence of rib geometry and SSC strength on the bond strength. The increased RW favored the bond strength, which increased to 20.34 MPa as the RW increased to 7.0 mm. Although the number of ribs and the slope decreased as the RW increased, the presence of fiber enhanced the bond strength. This was also supported by the similar bond stiffness but increased bond slip.

Figure 9 also shows that the bond strength decreased with the increase in DW. As the DW increased from 2.0 to 3.3 mm, the bond strength decreased by 13.89% because the number of bar ribs per unit bond length reduced with the increase in DW.

The increase in RH and SSC strength increased the bond strength, which was attributable to the increase in mechanical interlocking force between the BFRP bar and the SSC. Regarding the RH, the increased slope caused by the increasing RH resulted in greater bond strength [48]. For the SSC, where bond failure mainly occurred, increasing SCC strength enhanced the bond strength.

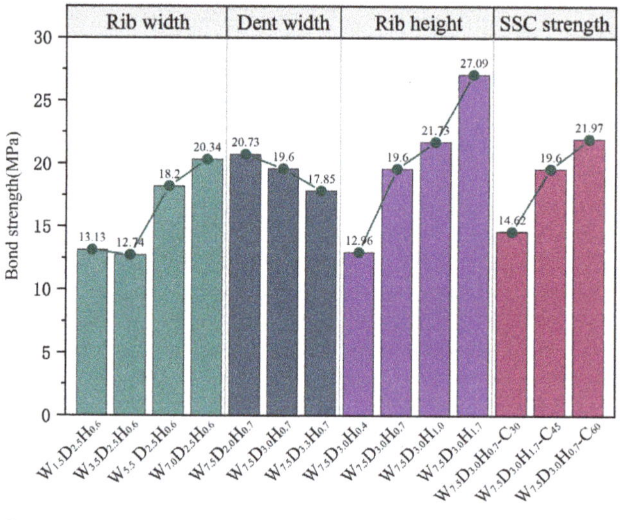

Figure 9. Bond strength of all samples.

Figure 10 illustrates the influence of rib geometry and SSC strength on the bond slip. The bond slip increased with increases in the RW and the RH. For example, as the RW increased from 1.5 to 7.0 mm, the bond slip increased from only 1.75 to 3.94 mm. As the RH increased from 0.4 to 1.7 mm, the bond slip increased by 37.1%. In addition, the increase in DW decreased the bond slip. However, the bond slip appeared to be not significantly affected by the changes in concrete strength. As the concrete strength increased from C30 to C60, the bond slip remained at ~3.90 mm.

Similar to the case of Type A BFRP-bar-reinforced SSC, bond strength reached its maximum value once the contact occurred at the highest point of the BFRP rib and the SSC, where the BFRP bar was pulled to approximately $\frac{1}{2}$ of the actual RS (rib width (W) + dent width (D)). However, for a fiber-wrapped BFRP bar, the thickness of the fiber bundle and the snugness of the winding should be considered, not just the W and D values.

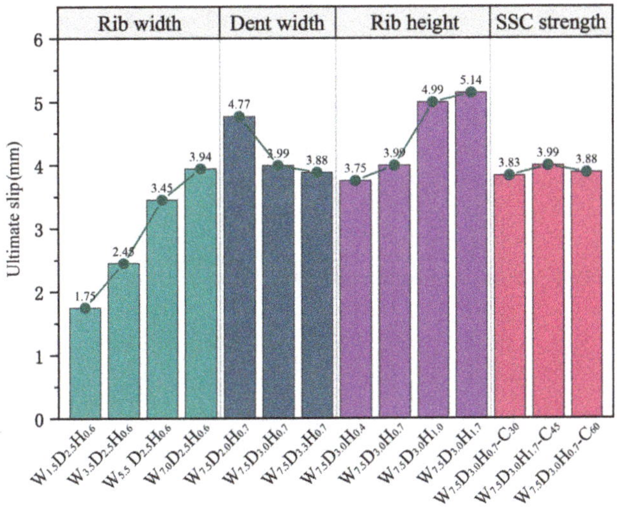

Figure 10. Bond slip of all samples.

3.3.3. Residual Bond Strength and Bond Slip

Figure 11 shows the influence of rib geometry and SSC strength on the residual bond strength. As the RW increased from 1.5 to 5.5 mm, the residual bond strength decreased. However, as the RW further increased to 7.0 mm, the residual bonding strength increased. Additionally, when the DW increased from 2.0 to 3.3 mm, the residual bond strength initially increased and then decreased, but to a small extent. In addition, the residual bond strength increased with the increase in RH. As for the increasing concrete strength, the residual bond strength first increased and then decreased. For the SSC reinforced with fiber-wrapped BFRP bars, the changes in bond strength in response to the changes in the rib geometry and SSC strength were complex because the factors influencing the reduction of the wedge cone were complex. These factors included the height of the remaining BFRP bar, the height of the SSC ribs, and the two slip interfaces.

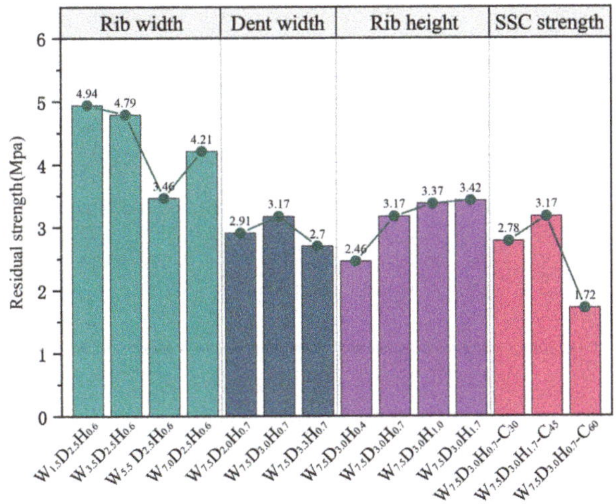

Figure 11. Residual bond strength of all samples.

Figure 12 shows that the residual bond slip increased with the increase in the RW and the DW but slightly decreased with the increase in RH. The concrete strength appeared to have little effect on the residual bond slip, with the residual bond slip being ~10.0 mm. For example, the residual bond slips increased by 119.77% and 12.39% as the RW increased from 1.5 to 7.0 mm and as the DW increased from 2.0 to 3.3 mm, respectively. According to the pull-out mechanism, the residual slip marked both the end of the former slip and the beginning of the next slip process, which was approximately one concrete rib distance, i.e., (W + D). Therefore, the changes in RW and DW had a more significant effect on the bond residual slip than the RH and SSC strength. For fiber-wrapped BFRP bars, the thickness of the fiber bundle and the snugness of the winding should be considered, not just the W and D values.

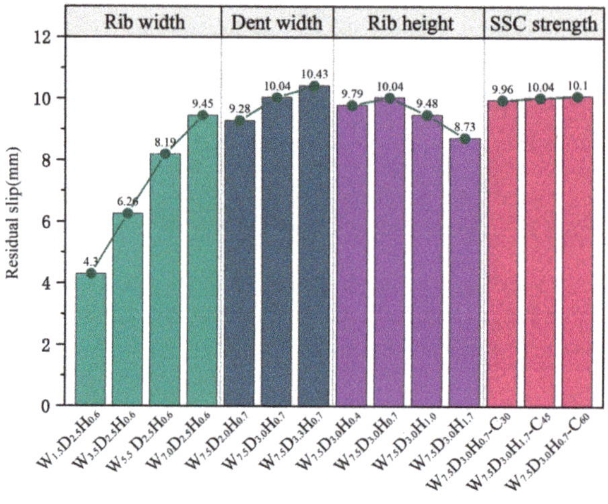

Figure 12. Residual bond slip of all samples.

Therefore, the increased dent width generally decreased the bond stiffness and residual strength but increased the bond strength and the ultimate and residual slips due to the greater width of the rib. The increased dent width also decreased bond stiffness, but increased the residual strength and slip. Meanwhile, the bond strength was also decreased. In fact, the dent width affected the technology process of the fiber-wrapped BFRP bar, causing the different fiber thicknesses and complicated interaction between the bars and the SSC. In contrast, the increase in rib height increased most of the bond indexes, including bond stiffness, bond strength, residual strength, and ultimate slip owing to the increasing interlocking strength. Consequently, the residual slip was also decreased. In addition, the increased concrete strength increased the bond stiffness and strength significantly, but had little effect on the bond and residual slip. This is because the interaction between bars and SSC was improved by increasing the strength of the SSC.

4. Conclusions

This study focused on the performance of the bond between fiber-wrapped BFRP bars and seawater sea-sand concrete (SSC) under different rib geometries and concrete strengths. Through the pull-out test, their bond behavior was investigated.

Geometric parameters had a significant effect on the bond index of BFRP bars and SSC. The increase in rib width decreased the bond stiffness from 9.85 to 8.20 MPa/mm but increased the bond strength from 13.13 to 20.34 MPa. The ultimate slip and residual slip were also increased owing to changes in the number of ribs. The increase in the dent width (DW) reduced the bond stiffness, bond strength, and bond slip. This change was related not only the DW but also to the adhesion of the fiber bundles on the surface of the BFRP bars, which was related to the processing technology. In fact, the fiber also significantly affected the bond performance of BFRP-bar-reinforced SSC. The fiber wrapped on the surface of rib could, to a certain extent, protect against the destruction of the bar. In particular, the processing technology and wrapping type resulting from different DWs significantly affected the failure mode of the SSC. The increase in both rib height (RH) and SSC strength contributed to the improvement of bond stiffness and bond strength owing to the enhancement of the mechanical interlocking force. The increase in RH also increased the bond slip and residual strength. However, the changes in bond slip, residual slip, and residual strength due to changes in SSC strength were complicated because of the low water content during hydration.

According to the bond interface damage, the specimens could be divided into two categories. One was characterized by the brittle-fracture failure of the ribs, which included the detachment of BFRP bar ribs and the shearing of the concrete ribs, accompanied by the collapse or severe wear of the surface-wrapped fiber bundles. The other category was categorized by deformation and wear caused by the compression between the ribs of BFRP bars, the surface-wrapped fiber bundles, and the concrete ribs. By altering the geometric parameters of the BFRP bars and the SSC strength, the bond performance, including bond interface damage, and the bond index could be controlled.

These findings show the relationship between the geometric parameters of fiber-wrapping rib and its bond strength in seawater sea-sand concrete, and provide an understanding of bond behavior of fiber-wrapped ribbed BFRP bars and seawater sea-sand concrete.

Author Contributions: Conceptualization, M.L., C.W. and L.L.; data curation, B.Z., D.Z. and L.L.; formal analysis, X.C.; funding acquisition, L.L. and M.L.; investigation, H.X.; methodology, S.H. and B.Z.; project administration, L.L. and M.L.; resources, L.L.; software, C.W.; supervision, S.H. and L.L.; writing—original draft, M.L. and X.C.; writing—review and editing, H.X., D.Z. and L.L. All authors have read and agreed to the published version of the manuscript.

Funding: This research was funded by the Guangzhou Science and Technology Plan Project (Grant No. 202201011636), the 2022 Higher Education Science Research Plan Project (Grant No. 22GDZY0105), the 2021 Academic Ideological and Political Education Project of Guangdong Provincial Department of Education (Grant No. 2021GXSZ143), and the National Natural Science Foundation of China (Grant No. 12032009).

Data Availability Statement: The original data are available upon request. The data are not publicly available due to the project privacy.

Acknowledgments: The authors gratefully thanks all technical personnel from the Structural Laboratory of Guangdong University of Technology for their assistance during the experiment.

Conflicts of Interest: Authors Chenyue Weng, Hesheng Xiao and Dong Zeng were employed by the company Guangdong Bocheng Construction Co., Ltd. The remaining authors declare that the research was conducted in the absence of any commercial or financial relationships that could be construed as a potential conflict of interest.

References

1. Gautam, L.; Bansal, S.; Sharma, K.V.; Kalla, P. Bone-china ceramic powder and granite industrial by-product waste in self-compacting concrete: A durability assessment with statistical validation. *Structures* **2023**, *54*, 837–856. [CrossRef]
2. Gautam, L.; Kalla, P.; Jain, J.K.; Choudhary, R.; Jain, A. Robustness of self-compacting concrete incorporating bone china ceramic waste powder along with granite cutting waste for sustainable development. *J. Clean. Prod.* **2022**, *367*, 132969. [CrossRef]
3. Shirani, S.; Cuesta, A.; Morales-Cantero, A.; Santacruz, I.; Diaz, A.; Trtik, P.; Holler, M.; Rack, A.; Lukic, B.; Brun, E.; et al. 4D nanoimaging of early age cement hydration. *Nat. Commun.* **2023**, *14*, 2652. [CrossRef]
4. Habert, G.; Miller, S.A.; John, V.M.; Provis, J.L.; Favier, A.; Horvath, A.; Scrivener, K.L. Environmental impacts and decarbonization strategies in the cement and concrete industries. *Nat. Rev. Earth Environ.* **2020**, *1*, 559–573. [CrossRef]
5. Salehi, M. Global water shortage and potable water safety; Today's concern and tomorrow's crisis. *Environ. Int.* **2022**, *158*, 106936. [CrossRef]
6. Rentier, E.; Cammeraat, L. The environmental impacts of river sand mining. *Sci. Total. Environ.* **2022**, *838*, 155877. [CrossRef]
7. Xiong, Z.; Mai, G.; Qiao, S.; He, S.; Zhang, B.; Wang, H.; Zhou, K.; Li, L. Fatigue bond behaviour between basalt fibre-reinforced polymer bars and seawater sea-sand concrete. *Ocean Coast. Manag.* **2022**, *218*, 106038. [CrossRef]

8. Zhang, B.; Zhu, H.; Cao, R.; Ding, J.; Chen, X. Feasibility of using geopolymers to investigate the bond behavior of FRP bars in seawater sea-sand concrete. *Constr. Build. Mater.* **2021**, *282*, 122636. [CrossRef]
9. Wang, J.; Liu, E.; Li, L. Multiscale investigations on hydration mechanisms in seawater OPC paste. *Constr. Build. Mater.* **2018**, *191*, 891–903. [CrossRef]
10. Pan, D.; Yaseen, S.A.; Chen, K.; Niu, D.; Leung, C.K.Y.; Li, Z. Study of the influence of seawater and sea sand on the mechanical and microstructural properties of concrete. *J. Build. Eng.* **2021**, *42*, 103006. [CrossRef]
11. Fang, S.; Li, L.; Lin, L.; Wang, H.; Fang, Z.; Li, Z.; Xiong, Z.; Liu, F. FRP interlocking multi-spiral reinforced square concrete columns: A promising compression application for marine engineering. *Eng. Struct.* **2021**, *244*, 112733. [CrossRef]
12. Pan, D.; Niu, D.; Li, Z. Corrosion products of low-alloy steel bars and their induction of cracking in seawater sea-sand concrete cover. *Constr. Build. Mater.* **2023**, *389*, 131800. [CrossRef]
13. Dong, Z.; Han, T.; Ji, J.; Zhu, H.; Wu, G. Durability of discrete BFRP needle-reinforced seawater sea-sand concrete-filled GFRP tubular columns in the ocean environment. *Constr. Build. Mater.* **2023**, *365*, 130017. [CrossRef]
14. Xu, J.; Wu, Z.; Cao, Q.; Yu, R.C. Eccentric compression behavior of seawater and sea sand concrete columns reinforced with GFRP and stainless steel bars. *Eng. Struct.* **2023**, *291*, 116486. [CrossRef]
15. Bakis, C.E.; Bank, L.C.; Brown, V.L.; Cosenza, E.; Davalos, J.F.; Lesko, J.J.; Machida, A.; Rizkalla, S.H.; Triantafillou, T.C. Fiber-Reinforced Polymer Composites for Construction—State-of-the-Art Review. *J. Compos. Constr.* **2002**, *6*, 73–87. [CrossRef]
16. Ahmed, A.; Guo, S.; Zhang, Z.; Shi, C.; Zhu, D. A review on durability of fiber reinforced polymer (FRP) bars reinforced seawater sea sand concrete. *Constr. Build. Mater.* **2020**, *256*, 119484. [CrossRef]
17. Bai, Y.-L.; Yan, Z.-W.; Ozbakkaloglu, T.; Gao, W.-Y.; Zeng, J.-J. Mechanical behavior of large-rupture-strain (LRS) polyethylene naphthalene fiber bundles at different strain rates and temperatures. *Constr. Build. Mater.* **2021**, *297*, 123786. [CrossRef]
18. Yan, Z.-W.; Bai, Y.-L.; Ozbakkaloglu, T.; Gao, W.-Y.; Zeng, J.-J. Axial impact behavior of Large-Rupture-Strain (LRS) fiber reinforced polymer (FRP)-confined concrete cylinders. *Compos. Struct.* **2021**, *276*, 114563. [CrossRef]
19. Godat, A.; Aldaweela, S.; Aljaberi, H.; Al Tamimi, N.; Alghafri, E. Bond strength of FRP bars in recycled-aggregate concrete. *Constr. Build. Mater.* **2021**, *267*, 120919. [CrossRef]
20. Liu, S.; Bai, C.; Zhang, J.; Zhao, J.; Hu, Q. Experimental and theoretical study on bonding performance of FRP bars-Recycled aggregate concrete. *Constr. Build. Mater.* **2022**, *361*, 129614. [CrossRef]
21. Zhou, W.; Feng, P.; Lin, H.; Zhou, P. Bond behavior between GFRP bars and coral aggregate concrete. *Compos. Struct.* **2023**, *306*, 116567. [CrossRef]
22. Mai, G.; Li, L.; Lin, J.; Wei, W.; He, S.; Zhong, R.; Xiong, Z. Bond durability between BFRP bars and recycled aggregate seawater sea-sand concrete in freezing-thawing environment. *J. Build. Eng.* **2023**, *70*, 106422. [CrossRef]
23. Solyom, S.; Balázs, G.L. Bond of FRP bars with different surface characteristics. *Constr. Build. Mater.* **2020**, *264*, 119839. [CrossRef]
24. Fahmy, M.F.; Ahmed, S.A.; Wu, Z. Bar surface treatment effect on the bond-slip behavior and mechanism of basalt FRP bars embedded in concrete. *Constr. Build. Mater.* **2021**, *289*, 122844. [CrossRef]
25. Basaran, B.; Donmez, E.T. Effects of fibre wrapping degree and ratio on the tensile properties of carbon FRP-steel hybrid reinforcements. *J. Build. Eng.* **2023**, *76*, 107189. [CrossRef]
26. Chen, J.; Fang, Z.; Chen, X.; Jiang, R. Experimental study on lap behavior of CFRP indented bars in UHPC. *Constr. Build. Mater.* **2022**, *344*, 127959. [CrossRef]
27. Majain, N.; Rahman, A.B.A.; Adnan, A.; Mohamed, R.N. Pullout behaviour of ribbed bars in self-compacting concrete with steel fibers. *Mater. Today Proc.* **2021**, *39*, 1034–1040. [CrossRef]
28. Hao, Q.; Wang, Y.; He, Z.; Ou, J. Bond strength of glass fiber reinforced polymer ribbed rebars in normal strength concrete. *Constr. Build. Mater.* **2009**, *23*, 865–871. [CrossRef]
29. Wei, W.; Liu, F.; Xiong, Z.; Lu, Z.; Li, L. Bond performance between fibre-reinforced polymer bars and concrete under pull-out tests. *Constr. Build. Mater.* **2019**, *227*, 116803. [CrossRef]
30. Sasmal, S.; Khatri, C.P.; Ramanjaneyulu, K.; Srinivas, V. Numerical evaluation of bond–slip relations for near-surface mounted carbon fiber bars embedded in concrete. *Constr. Build. Mater.* **2013**, *40*, 1097–1109. [CrossRef]
31. Shan, Z.; Liang, K.; Chen, L. Bond behavior of helically wound FRP bars with different surface characteristics in fiber-reinforced concrete. *J. Build. Eng.* **2023**, *65*, 105504. [CrossRef]
32. El-Nemr, A.; Ahmed, E.A.; Barris, C.; Joyklad, P.; Hussain, Q.; Benmokrane, B. Bond performance of fiber reinforced polymer bars in normal- and high-strength concrete. *Constr. Build. Mater.* **2023**, *393*, 131957. [CrossRef]
33. Liang, K.; Chen, L.; Shan, Z.; Su, R. Experimental and theoretical study on bond behavior of helically wound FRP bars with different rib geometry embedded in ultra-high-performance concrete. *Eng. Struct.* **2023**, *281*, 115769. [CrossRef]
34. JSCE. *Recommendation Fordesign and Construction of Concrete Structures Using Continuous Fiber Reinforcing Materials*; Japan Society of Civil Engineers: Tokyo, Japan, 1997.
35. *CSA S806*; Design and Construction of Building Components with Fibre Reinforced Polymers. Canadian Standards Association: Rexdale, ON, USA, 2002.
36. *ACI 440*; 1R-Guide for the Design and Construction of Structural Concrete Reinforced with FRP Bars. American Concrete Institute: Farmington Hills, MI, USA, 2015.

37. Xiong, Z.; Wei, W.; He, S.; Liu, F.; Luo, H.; Li, L. Dynamic bond behaviour of fibre-wrapped basalt fibre-reinforced polymer bars embedded in sea sand and recycled aggregate concrete under high-strain rate pull-out tests. *Constr. Build. Mater.* **2021**, *276*, 122195. [CrossRef]
38. Lu, Z.; Su, L.; Lai, J.; Xie, J.; Yuan, B. Bond durability of BFRP bars embedded in concrete with fly ash in aggressive environments. *Compos. Struct.* **2021**, *271*, 114121. [CrossRef]
39. Yan, Z.-W.; Bai, Y.-L.; Ozbakkaloglu, T.; Gao, W.-Y.; Zeng, J.-J. Rate-dependent compressive behavior of concrete confined with Large-Rupture-Strain (LRS) FRP. *Compos. Struct.* **2021**, *272*, 114199. [CrossRef]
40. *JGJ 55-2011*; Specification for Mix Proportion Design of Ordinary Concrete. Ministry of Housing and Urban Rural Development of the People's Republic of China: Beijing, China, 2011.
41. *ASTM C39/C39M*; Standard Test Method for Compressive Strength of Cylinder Specimens. ASTM International: West Conshohocken, PA, USA, 2018.
42. *ASTM C469/C469M*; Standard Test Method for Static Modulus of Elasticity and Poisson's Ratio of Concrete in Compression. ASTM International: West Conshohocken, PA, USA, 2014.
43. *ACI 440.3R-12*; Guide Test Methods for Fiber-Reinforced Polymer (FRP) Composites for Reinforcing or Strengthening Concrete and Masonry Structures. American Concrete Institute: Farmington Hills, MI, USA, 2012.
44. Xiong, Z.; Zeng, Y.; Li, L.; Kwan, A.; He, S. Experimental study on the effects of glass fibres and expansive agent on the bond behaviour of glass/basalt FRP bars in seawater sea-sand concrete. *Constr. Build. Mater.* **2021**, *274*, 122100. [CrossRef]
45. Chen, L.; Liang, K.; Shan, Z. Experimental and theoretical studies on bond behavior between concrete and FRP bars with different surface conditions. *Compos. Struct.* **2023**, *309*, 116721. [CrossRef]
46. Otoom, O.F.; Lokuge, W.; Karunasena, W.; Manalo, A.C.; Ozbakkaloglu, T.; Ehsani, M.R. Flexural behaviour of circular reinforced concrete columns strengthened by glass fibre reinforced polymer wrapping system. *Structures* **2022**, *38*, 1326–1348. [CrossRef]
47. Al-Hamrani, A.; Alnahhal, W. Bond durability of sand coated and ribbed basalt FRP bars embedded in high-strength concrete. *Constr. Build. Mater.* **2023**, *406*, 133385. [CrossRef]
48. Ke, L.; Liang, L.; Feng, Z.; Li, C.; Zhou, J.; Li, Y. Bond performance of CFRP bars embedded in UHPFRC incorporating orientation and content of steel fibers. *J. Build. Eng.* **2023**, *73*, 106827. [CrossRef]

MDPI AG
Grosspeteranlage 5
4052 Basel
Switzerland
Tel.: +41 61 683 77 34

Buildings Editorial Office
E-mail: buildings@mdpi.com
www.mdpi.com/journal/buildings

www.ingramcontent.com/pod-product-compliance
Lightning Source LLC
Chambersburg PA
CBHW040823250926
51850CB00007BA/1242
9783725820184